*Soumyajyoti Biswas, Purusattam Ray,
and Bikas K. Chakrabarti*

**Statistical Physics of Fracture,
Breakdown, and Earthquake**

Related Titles

Klages, R., Just, W., Jarzynski, C. (eds.)

Nonequilibrium Statistical Physics of Small Systems

Fluctuation Relations and Beyond

2013
ISBN: 978-3-527-41094-1, also available in digital formats

Hertzberg, R. W., Vinci, R. P., Hertzberg, J. L.

Deformation and Fracture Mechanics of Engineering Materials

2012
ISBN: 978-0-470-52780-1, also available in digital formats

Freiman, S., Mecholsky, J. J.

The Fracture of Brittle Materials

Testing and Analysis

2012
Print ISBN: 978-0-470-15586-8, also available in digital formats

Recho, N.

Fracture Mechanics and Crack Growth

2012
ISBN: 978-1-84821-306-7, also available in digital formats

Sahimi, M.

Flow and Transport in Porous Media and Fractured Rock

From Classical Methods to Modern Approaches

2011
ISBN: 978-3-527-40485-8, also available in digital formats

Guyer, R. A., Johnson, P. A.

Nonlinear Mesoscopic Elasticity

The Complex Behaviour of Granular Media including Rocks and Soil

2009
ISBN: 978-3-527-40703-3, also available in digital formats

Reichl, L. E.

A Modern Course in Statistical Physics

2009
ISBN: 978-3-527-40782-8, also available in digital formats

Tjong, S. C.

Carbon Nanotube Reinforced Composites

Metal and Ceramic Matrices

2009
ISBN: 978-3-527-40892-4, also available in digital formats

Klages, R., Radons, G., Sokolov, I. M. (eds.)

Anomalous Transport
Foundations and Applications

2008
ISBN: 978-3-527-40722-4, also available in digital formats

Kagan, Y.

Earthquakes
Models, Statistics, Testable Forecasts

2014
Print ISBN: 978-1-118-63792-0, also available in electronic formats

Goehring, L., Nakahara, A., Dutta, T., Kitsunezaki, S., Tarafdar, S.

Desiccation Cracks and their Patterns
Formation and Modelling in Science and Nature

2015
Print ISBN: 978-3-527-41213-6; also available in electronic formats

Soumyajyoti Biswas, Purusattam Ray,
and Bikas K. Chakrabarti

Statistical Physics of Fracture, Breakdown, and Earthquake

Effects of Disorder and Heterogeneity

Verlag GmbH & Co. KGaA

The Authors

Dr. Soumyajyoti Biswas
Saha Institute of Nuclear Physics
1/AF Bidhan Nagar
Kolkata 700064
India

Prof. Purusattam Ray
CIT Campus
Institute of Mathematical Sciences
Chennai 600113
India

Prof. Bikas K. Chakrabarti
Saha Institute of Nuclear Physics
Applied Math. and Comp. Science
1, Bidhannagar
Kolkata 700064
India

All books published by **Wiley-VCH** are carefully produced. Nevertheless, authors, editors, and publisher do not warrant the information contained in these books, including this book, to be free of errors. Readers are advised to keep in mind that statements, data, illustrations, procedural details or other items may inadvertently be inaccurate.

Library of Congress Card No.: applied for

British Library Cataloguing-in-Publication Data
A catalogue record for this book is available from the British Library.

Bibliographic information published by the Deutsche Nationalbibliothek
The Deutsche Nationalbibliothek lists this publication in the Deutsche Nationalbibliografie; detailed bibliographic data are available on the Internet at <http://dnb.d-nb.de>.

© 2015 Wiley-VCH Verlag GmbH & Co. KGaA, Boschstr. 12, 69469 Weinheim, Germany

All rights reserved (including those of translation into other languages). No part of this book may be reproduced in any form – by photoprinting, microfilm, or any other means – nor transmitted or translated into a machine language without written permission from the publishers. Registered names, trademarks, etc. used in this book, even when not specifically marked as such, are not to be considered unprotected by law.

Print ISBN: 978-3-527-41219-8
ePDF ISBN: 978-3-527-67267-7
ePub ISBN: 978-3-527-67265-3
Mobi ISBN: 978-3-527-67266-0
oBook ISBN: 978-3-527-67264-6

Cover Design Schulz Grafik-Design, Fußgönheim, Germany
Typesetting Laserwords Private Ltd., Chennai, India
Printing and Binding Markono Print Media Pte Ltd, Singapore

Printed on acid-free paper

Contents

Series Editors' Preface *XIII*
Preface *XV*
Notations *XVII*

1	**Introduction** *1*	
2	**Mechanical and Fracture Properties of Solids** *7*	
2.1	Mechanical Response in Materials *8*	
2.1.1	Elastic and Plastic Regions *8*	
2.1.2	Linear Elastic Region *9*	
2.1.3	Nonlinear Plastic Region *10*	
2.2	Ductile, Quasi-brittle, and Brittle Materials *11*	
2.3	Ductile and Brittle Fracture *11*	
2.3.1	Macroscopic Features of Ductile and Brittle Fractures *12*	
2.3.2	Microscopic Features of Ductile and Brittle Fractures *14*	
3	**Crystal Defects and Disorder in Lattice Models** *17*	
3.1	Point Defects *17*	
3.2	Line Defects *18*	
3.3	Planar Defects *20*	
3.4	Lattice Defects: Percolation Theory *22*	
3.5	Summary *25*	
4	**Nucleation and Extreme Statistics in Brittle Fracture** *27*	
4.1	Stress Concentration Around Defect *27*	
4.1.1	Griffith's Theory of Crack Nucleation in Brittle Fracture *30*	
4.2	Strength of Brittle Solids: Extreme Statistics *32*	
4.2.1	Weibull and Gumbel Statistics *32*	
4.3	Extreme Statistics in Fiber Bundle Models of Brittle Fracture *34*	
4.3.1	Fiber Bundle Model *34*	
4.3.1.1	Strength of the Local Load Sharing Fiber Bundles *35*	
4.3.1.2	Crossover from Extreme to Self-averaging Statistics in the Model *35*	

4.4	Extreme Statistics in Percolating Lattice Model of Brittle Fracture *37*	
4.5	Molecular Dynamics Simulation of Brittle Fracture *39*	
4.5.1	Comparisons with Griffith's Theory *39*	
4.5.2	Simulation of Highly Disordered Solids *41*	
4.6	Summary *42*	
5	**Roughness of Fracture Surfaces** *45*	
5.1	Roughness Properties in Fracture *45*	
5.1.1	Self-affine Scaling of Fractured Surfaces *46*	
5.1.2	Out-of-plane Fracture Roughness *47*	
5.1.3	Distribution of Roughness: Mono- and Multi-affinity *49*	
5.1.3.1	Nonuniversal Cases *50*	
5.1.3.2	Anisotropic Scaling *54*	
5.1.4	In-plane Roughness of Fracture Surfaces *56*	
5.1.4.1	Waiting Time Distributions in Crack Propagation *60*	
5.1.5	Effect of Probe Size *62*	
5.1.6	Effect of Spatial Correlation and Anisotropy *65*	
5.2	Molecular Dynamics Simulation of Fractured Surface *66*	
5.3	Summary *68*	
6	**Avalanche Dynamics in Fracture** *69*	
6.1	Probing Failure with Acoustic Emissions *70*	
6.2	Dynamics of Fiber Bundle Model *74*	
6.2.1	Dynamics Around Critical Load *77*	
6.2.2	Dynamics at Critical Load *81*	
6.2.3	Avalanche Statistics of Energy Emission *81*	
6.2.4	Precursors of Global Failure in the Model *82*	
6.2.5	Burst Distribution: Crossover Behavior *84*	
6.2.6	Abrupt Rupture and Tricritical Point *85*	
6.2.7	Disorder in Elastic Modulus *87*	
6.3	Interpolations of Global and Local Load Sharing Fiber Bundle Models *88*	
6.3.1	Power-law Load Sharing *89*	
6.3.2	Mixed-mode Load Sharing *90*	
6.3.3	Heterogeneous Load Sharing *92*	
6.3.3.1	Dependence on Loading Process *93*	
6.3.3.2	Results in One Dimension *94*	
6.3.3.3	Results in Two Dimensions *96*	
6.3.3.4	Comparison with Mixed Load Sharing Model *101*	
6.4	Random Threshold Spring Model *101*	
6.5	Summary *107*	
7	**Subcritical Failure of Heterogeneous Materials** *111*	
7.1	Time of Failure Due to Creep *111*	

7.1.1	Fluctuating Load	*112*
7.1.2	Failure Due to Fatigue in Fiber Bundles	*119*
7.1.3	Creep Rupture Propagation in Rheological Fiber Bundles	*122*
7.1.3.1	Modification for Local Load Sharing Scheme	*126*
7.2	Dynamics of Strain Rate	*129*
7.3	Summary	*134*
8	**Dynamics of Fracture Front**	*135*
8.1	Driven Fluctuating Line	*135*
8.1.1	Variation of Universality Class	*140*
8.1.2	Depinning with Constant Volume	*142*
8.1.3	Infinite-range Elastic Force with Local Fluctuations	*144*
8.2	Fracture Front Propagation in Fiber Bundle Models	*146*
8.2.1	Interfacial Crack Growth in Fiber Bundle Model	*146*
8.2.2	Crack Front Propagation in Fiber Bundle Models	*149*
8.2.3	Self-organized Dynamics in Fiber Bundle Model	*151*
8.3	Hydraulic Fracture	*161*
8.4	Summary	*163*
9	**Dislocation Dynamics and Ductile Fracture**	*165*
9.1	Nonlinearity in Materials	*165*
9.2	Deformation by Slip	*165*
9.2.1	Critical Stress to Create Slip in Perfect Lattice	*166*
9.3	Slip by Dislocation Motion	*167*
9.4	Plastic Strain due to Dislocation Motion	*169*
9.5	When Does a Dislocation Move?	*170*
9.5.1	Dislocation Width	*170*
9.5.2	Dependence on Grain Boundaries in Crystals	*171*
9.5.3	Role of Temperature	*171*
9.5.4	Effect of Applied Stress	*172*
9.6	Ductile–Brittle Transition	*172*
9.6.1	Role of Confining Pressure	*172*
9.6.2	Role of Temperature	*173*
9.7	Theoretical Work on Ductile–Brittle Transition	*174*
10	**Electrical Breakdown Analogy of Fracture**	*177*
10.1	Disordered Fuse Network	*178*
10.1.1	Dilute Limit ($p \to 1$)	*179*
10.1.2	Critical Behavior ($p \to p_c$)	*180*
10.1.3	Influence of the Sample Size	*181*
10.1.4	Distribution of the Failure Current	*182*
10.1.4.1	Dilute Limit ($p \to 1$)	*182*
10.1.4.2	At Critical Region ($p \to p_c$)	*183*
10.1.5	Continuum Model	*183*
10.1.6	Electromigration	*184*

10.2	Numerical Simulations of Random Fuse Network	*185*
10.2.1	Disorders in Failure Thresholds	*187*
10.2.2	Avalanche Size Distribution	*188*
10.2.3	Roughness of Fracture Surfaces in RFM	*191*
10.2.4	Effect of High Disorder	*193*
10.2.5	Size Effect	*196*
10.3	Dielectric Breakdown Problem	*197*
10.3.1	Dilute Limit ($p \to 1$)	*198*
10.3.2	Close to Critical Point ($p \to p_c$)	*199*
10.3.3	Influence of Sample Size	*199*
10.3.4	Distribution of Breakdown Field	*200*
10.3.5	Continuum Model	*200*
10.3.6	Shortest Path	*201*
10.3.7	Numerical Simulations in Dielectric Breakdown	*201*
10.3.7.1	Stochastic Models	*201*
10.3.7.2	Deterministic Models	*202*
10.4	Summary	*205*
11	**Earthquake as Failure Dynamics**	*207*
11.1	Earthquake Statistics: Empirical Laws	*207*
11.1.1	Universal Scaling Laws	*209*
11.2	Spring-block Models of Earthquakes	*214*
11.2.1	Computer Simulation of the Burridge–Knopoff Model	*215*
11.2.2	Train Model of Earthquake	*219*
11.2.3	Mapping of Train Model to Sandpile	*221*
11.2.3.1	Mapping to Sandpile Model	*222*
11.2.4	Two-fractal Overlap Models	*223*
11.2.4.1	Model Description	*224*
11.2.4.2	GR and Omori Laws	*225*
11.3	Cellular Automata Models of Earthquakes	*227*
11.3.1	Bak Tang Wiesenfeld (BTW) Model	*228*
11.3.2	Zhang Model	*232*
11.3.3	Manna Model	*234*
11.3.4	Common Failure Precursor for BTW and Manna Models and FBM	*237*
11.3.4.1	Precursor in BTW Model	*238*
11.3.4.2	Precursor in Manna Model	*240*
11.3.4.3	Precursor in Fiber Bundle Model	*240*
11.3.5	Olami–Feder–Christensen (OFC) Model	*240*
11.3.5.1	Moving Boundary	*242*
11.4	Equivalence of Interface and Train Models	*246*
11.4.1	Model	*248*
11.4.2	Avalanche Statistics in Modified Train Model	*250*
11.4.3	Equivalence with Interface Depinning	*253*

11.4.4	Interface Propagation and Fluctuation in Bulk	*255*
11.5	Summary	*261*
12	**Overview and Outlook**	*265*
A	**Percolation**	*269*
A.1	Critical Exponent: General Examples	*269*
A.1.1	Scaling Behavior	*270*
A.2	Percolation Transition	*270*
A.2.1	Critical Exponents of Percolation Transition	*272*
A.2.2	Scaling Theory of Percolation Transition	*273*
A.3	Renormalization Group (RG) Scheme	*274*
A.3.1	RG for Site Percolation in One Dimension	*276*
A.3.2	RG for Site Percolation in Two-dimensional Triangular Lattice	*278*
A.3.3	RG for Bond Percolation in Two-dimensional Square Lattice	*279*
B	**Real-space RG for Rigidity Percolation**	*281*
C	**Fiber Bundle Model**	*285*
C.1	Universality Class of the Model	*285*
C.1.1	Linearly Increasing Density of Fiber Strength	*285*
C.1.2	Linearly Decreasing Density of Fiber Strength	*286*
C.1.3	Nonlinear Stress–Strain Relationship	*288*
C.2	Brittle to Quasi-brittle Transition and Tricritical Point	*290*
C.2.1	Abrupt Failure and Tricritical Point	*292*
D	**Quantum Breakdown**	*293*
E	**Fractals**	*295*
F	**Two-fractal Overlap Model**	*297*
F.1	Renormalization Group Study: Continuum Limit	*297*
F.2	Discrete Limit	*299*
F.2.1	Gutenberg-Richter Law	*299*
F.2.2	Omori Law	*300*
G	**Microscopic Theories of Friction**	*303*
G.1	Frenkel-Kontorova Model	*303*
G.2	Two-chain Model	*304*
G.2.1	Effect of Fractal Disorder	*305*
	References	*309*
	Index	*323*

Book Series:
Statistical Physics of Fracture and Breakdown

Editors: Bikas K. Chakrabarti and Purusattam Ray

Why does a bridge collapse, an aircraft or a ship break apart? When does a dielectric insulation fail or a circuit fuse, even in microelectronic systems? How does an earthquake occur? Are there precursors to these failures? These remain important questions, even more so as our civilization depends increasingly on structures and services where such failure can be catastrophic. How can we predict and prevent such failures? Can we analyze the precursory signals sufficiently in advance to take appropriate measures, such as the timely evacuation of structures or localities, or the shutdown of facilities such as nuclear power plants?

Whilst these questions have long been the subject of research, the study of fracture and breakdown processes has now gone beyond simply designing safe and reliable machines, vehicles and structures. From the fracture of a wood block or the tearing of a sheet of paper in the laboratory, the breakdown of an electrical network on an engineering scale, to an earthquake on a geological scale, one finds common threads and universal features in failure processes. The ideas and observations of material scientists, engineers, technologists, geologists, chemists and physicists have all played a pivotal role in the development of modern fracture science.

Over the last three decades, considerable progress has been made in modeling and analyzing failure and fracture processes. The physics of nonlinear, dynamic, many-bodied and non-equilibrium statistical, mechanical systems, the exact solutions of fibre bundle models, solutions of earthquake models, numerical studies of random resistor and random spring networks, and laboratory-scale innovative experimental verifications have all opened up broad vistas of the processes underlying fracture. These have provided a unifying picture of failure over a wide range of length, energy and time scales.

This series of books introduces readers – in particular, graduate students and researchers in mechanical and electrical engineering, earth sciences, material science, and statistical physics – to these exciting recent developments in our understanding of the dynamics of fracture, breakdown and earthquakes.

Preface

The last few decades have witnessed an enormous interest in the study of varieties of physical processes, all of which have in common that their dynamics is activated when a system variable exceeds a threshold. Examples of such processes are sandpile models, excitable media, granular flow, Barkhausen noise, and to say the least fracture and breakdown phenomena. These systems, representing widely different physical phenomena, have distinctive physical processes, exhibit several features such as burst-like relaxation, and avalanches with power-law statistics, which are common to most of them.

Fracture and failure in materials involve dynamics over various length and timescales. The ideas of stress enhancement and fracture nucleation around defects, weakest point in a material, extreme and non-self-averaging statistics related to the strength of a solid are generally introduced when the problem is seen at a microscopic or mesoscopic scale. At this scale, the ideas of percolation, fractals, and random fields from statistical physics have been applied to the study of material failure. Similarly, ideas such as growth processes, aggregation phenomena, and dispersion in disordered systems have been applied to understand the universal behavior in the dynamics of the growth and development of fracture across different length scales ranging from sub-micron levels, for example electromigration failure, through daily examples of bricks or wood pieces to the geological scale of earthquakes. Models of failure and its dynamics based on these ideas, successful comparisons of the results with observations, and universal behaviors have led to major and exciting developments in this field in the last two or three decades.

This textbook is intended to introduce the studies of failure properties of materials applying the ideas of statistical physics. Following the first three introductory chapters, where basic definitions of fracture and ideas of defects are discussed, we move on to describe the various features of fracture. We discuss the roughness of fractured Front; the intermittent dynamics before failure, failure induced by fatigue, dislocation dynamics, and ductile fracture; and the attempts in modeling them. We discuss the discrete element models such as fiber bundle model and random fuse model to study the behavior of materials near the fracture point. We discuss the breakdown of dielectric materials, which often bears close resemblance with mechanical failures in several fronts. Finally, we go over to the failure phenomena in the geological scale, that is, the dynamics of earthquakes. We also discuss the models here and show their close relationship with failure in much smaller scales both in terms of the empirical laws and modeling approaches. We hope to have covered the width of the topics to a substantial extent and hopefully it will be useful for both researchers and students in physics, engineering, geology, and materials science.

We gratefully acknowledge collaborations over the years with M. Acharyya, J. H. Bakke, K. K. Bardhan, L. G. Benguigui, P. Bhattacharyya, D. Chowdhury, A. Delaplace, M. K. Dey, A. Hansen, T. Hatano, N. Kato, H. Kawamura, K. J. Mly, S. S. Manna, J. Mathiesen, S. Pradhan, S. Roy, D. Samanta, J. Schmittbuhl, H. E. Stanley, D. Stauffer, R. B. Stinchcombe, L. Vanel, A. Vespignani, S. Zapperi. We also wish to thank E. Bouchaud, S. Zapperi for discussions. We also acknowledge R. Biswas and S. Saha from Chennai Mathematical Institute for the figure in the cover page. The INDNOR project (No. 217413/E20) is also acknowledged for partial financial support. BKC acknowledges JC Bose Fellowship, DST, Govt. of India.

March 2015   *Soumyajyoti Biswas, Purusattam Ray, Bikas K. Chakrabarti*

Notations

Following are the notations to be followed throughout the book

- β avalanche size distribution exponent
- β' order parameter exponent
- ν correlation length divergence exponent
- ξ correlation length
- τ relaxation time (duration)
- τ_f failure time
- ζ roughness exponent
- ϵ strain
- σ stress
- σ_f failure/critical stress
- A area
- d dimension
- d_f fractal dimension
- D distribution function
- P distribution function
- Y elasticity modulus
- v velocity
- V volume.
- L size
- t time
- i current
- E energy
- s avalanche size
- U fraction of surviving fibers
- u displacements
- W interface width
- x position
- z relaxation time exponent
- z' dynamical exponent
- w width
- H hamiltonian

1
Introduction

Fracture is the result of driving a solid beyond its mechanical limit. It is immensely important to know the limit or how materials behave as they approach the limit and the factors that influence them. The failure properties of materials are very distinct from the other properties, such as elasticity, in the sense that their predictions are not always straightforward. For example, typically, fracture strength of a solid has a very wide distribution, and a larger object has lower failure strength than a smaller one of same composition. In brittle materials, fracture is catastrophic, that is, the solid fails without a precursor. It is this intriguing nature of failure phenomena that has led scientists to think about this problem over the centuries. It was Leonardo da Vinci (see Figure 1.1) who apparently first noticed that a longer wire has lower strength. Galileo also recognized the importance of this problem and commented about the limitation of sizes of an object for improvement in its strength (see Figure 1.2).

The understanding of fracture of materials has progressed enormously since those days. However, it is still far from being complete. Present-day understanding of fracture in homogeneous materials is based primarily on linear elastic fracture mechanics which deals with the stress concentration around notches and cracks in a model of linear elastic material. It starts with Griffith's theory (Griffith, 1921) of energy balance. The basic idea here is that when a solid gets strained, and if the elastic energy stored is sufficient to create new surfaces, then a crack becomes unstable and a fracture takes place. The theory was made more accurate by introducing a small plastic zone in front of the crack tip by Irwin and Dugdale (see e.g., Anderson, 1995). This picture, however, cannot handle fracture with plastic deformation and dissipation as it happens in ductile fracture, besides it cannot handle the effect of disorder. Disorder plays a vital role in the behavior of solids, especially before fracture. The strength of a material is determined by the weakest part of it, which leads to the extreme value statistics in failure properties.

After summarizing the basic characterizations of fracture, namely brittle and ductile fracture, the linearity of the stress–strain relationship in the elastic region and subsequent departure to nonlinearity in the plastic region, we go over to the properties of defects in solids in Chapter 3. The lattice defects are quantified in the form of the percolation theory, which gives the limit of high disorder in solids. These characterizations help us understand the nature of extreme statistics led by

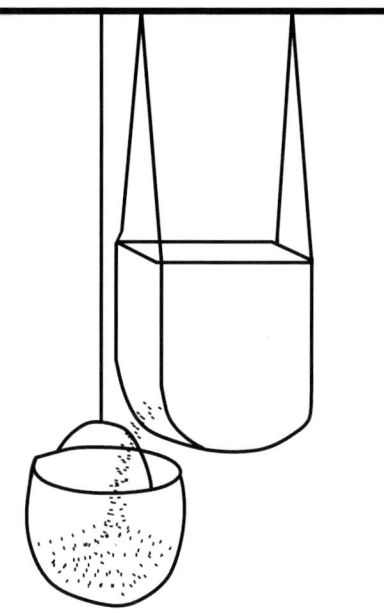

Figure 1.1 Leonardo di ser Piero da Vinci (1452–1519): da Vinci was a diversely talented person and a leader of the Italian Renaissance movement. He displayed his talent in many areas of arts and science. Best known as a painter (for his famous Mona Lisa, The Last Supper, Virgin of the rocks to name a few), he was also a great engineering designer. However, apart from his well-known inventions and sketches, comparatively less known is his contribution to fracture mechanics. In his experiment titled "Testing the strengths of iron wires of various lengths," he suspended a basket by an iron wire and slowly added sand to it from a pot hanging adjacent to the basket. The failure point of the wire was noted for its different lengths. In his own words (translated by Parsons, 1939): "The object of this test is to find the load an iron wire can carry. Attach an iron wire 2 braccia long to something which will firmly support it, then attach a basket or similar container to the wire and feed into the basket some fine sand through a small hole placed at the end of the hopper. A spring is fixed so that it will close the hole as soon as the wire breaks. The basket is not upset while falling, since it falls through a very short distance. The weight of sand and the location of the fracture of the wire are to be recorded. The test is repeated several times to check the results. Then a wire of 1/2 the previous length is tested and the additional weight it carries is recorded; then a wire of 1/4 length is tested and so forth, noting the ultimate strength and the location of the fracture." As we will see in Section 4.2, because of the extreme nature of the breaking statistics, the strength of solids decrease with their volume typically as $1/\ln V$.

the stress nucleation around defects, which is the topic of discussion in Chapter 4. In addition to the continuum approach, we introduce a discrete element model, called the fiber bundle model, which is a simple one depicting many essential features of failure statistics, including the stress nucleation, and extreme statistics as discussed there.

While disorder in solids governs the failure strength, it also steers the path of the crack. A defect can deflect a propagating crack. Since it is the impression of this

Figure 1.2 Galileo Galilei (1564–1642): Galileo was an Italian physicist and astronomer who is called the "Father of Modern Science" to honor his many contributions to our present-day understanding of science. Particularly, he produced telescopic evidence of phases of Venus, the four largest satellite of Jupiter, sun spots, and also confirmed the earlier ideas of Copernicus and Kepler that the earth and other planets move around the sun. Because of his conflicting views with the church, he was put under house arrest for the last part of his life. There he wrote his famous book "Two new sciences," where he described his works on the two sciences "kinematics" and "strength of matter." There he had observed (see discussions in Section 4.2) the size effects of fracture and described how the natural sizes are limited by their own strengths. In his own words: "From what has already been demonstrated, you can plainly see the impossibility of increasing the size of structures to vast dimensions either in art or in nature; likewise the impossibility of building ships, palaces, or temples of enormous size in such a way that their oars, yards, beams, iron-bolts, and, in short, all their other parts will hold together; nor can nature produce trees of extraordinary size because the branches would break down under their own weight; so also it would be impossible to build up the bony structures of men, horses, or other animals so as to hold together and perform their normal functions if these animals were to be increased enormously in height; for this increase in height can be accomplished only by employing a material which is harder and stronger than usual, or by enlarging the size of the bones, thus changing their shape until the form and appearance of the animals suggest a monstrosity." [From: http://ebooks.adelaide.edu.au/g/galileo/dialogues/chapter2.html]

crack front that creates the roughness of the fracture surfaces, in a way presence of disorder is responsible for the roughness. It is our everyday observation that fractured surfaces are not smooth but are rough. However, it is not until the pioneering work of Mandelbrot *et al.* (1984) that a universal feature was found in the roughness in fracture surfaces. It was found that the fracture surfaces of various materials were self-affine, meaning they looked similar, no matter to what part of it—small or large—one focuses. Roughness can be quantified by a number called the "roughness exponent". A surprising observation was that the value was almost same for various materials. This idea of scale invariance and universality led to substantial activities in this field using the tools of statistical physics and critical phenomena. Many subsequent studies revealed facts both supporting and opposing this universality, also noting a crossover behavior in the exponent value, signifying that the fractured surfaces are not self-affine in all scales after all! Furthermore, an anisotropic feature has also been observed in the roughness properties, distinguishing the direction of crack propagation from the direction perpendicular to it. The experimental observations and theoretical modeling of roughness of fracture surfaces are discussed in Chapter 5.

Another familiar experience with fracture is the accompanying noise. One can experience that in day-to-day activities such as tearing a paper or eating potato chips to failure in geological scale, that is, earthquakes, where the precursor can be lifesaving. The so-called "crackling noise" or emission of acoustic noise is a common fact of fracture, where a portion of energy is released in the form of sound. The intriguing feature, however, once again is the scale-invariant response of the solids in terms of size distributions of acoustic emissions (bursts). When force is applied on a material, some portions (probably weaker) will fail but not the entire solid, since the solid is disordered. When further strained, some weak parts will break again and increase the stress on the remaining part initiating a chain reaction, called an "avalanche." Since a proportional fraction will be emitted as sound, it can be detected to measure the size of the avalanche. One avalanche may not be sufficient to break the entire solid as the remaining stronger parts may not break. But understandably that part will be highly stressed and a small increase of force may cause an avalanche of much larger size than one usually expects with a small perturbation. Those who are familiar with self-organized criticality, the process may indeed sound like one—there is external forcing and dissipation in terms of acoustic emissions of any size. Under general circumstances, the scale-free distribution of avalanche sizes is a common manifestation of disordered solids. In Chapter 6 we discuss these issues related to the dynamics of fracture.

All structures around us carry finite load for a long time. Even though they may support the load initially, there is no guarantee that they will not fail to do so later. Indeed, imagining fracture as an energy barrier problem, it is easy to see that under the influence of any finite noise, supplied from the environment in the form of, say, temperature, the solid may overcome the barrier and break after a long time. It is therefore vital that one understands the precursors to the so-called "creep rupture". Similarly, a solid being exposed to cyclic loading may suffer permanent deformation in its structure (fatigue) and can eventually fail below its critical limit. The properties of these subcritical failures are the topics of discussions in Chapter 7. Subcritical failures also show avalanche dynamics and are often quite similar (with same exponent values in case of scale-free size distribution) with those under continuous loading. The chapter also contains relevant discussions of avalanche properties of subcritical failures. In addition to the experimental observations, attempts to model these phenomena theoretically by fiber bundle model are also discussed.

As mentioned earlier, the roughness properties of a fractured surface are due to the trace left behind by the fluctuating fracture front. When a crack is opened by pulling a solid (say, rectangular) from one side (much like opening a book), which is called the mode-I fracture, the fracture front propagates through the disordered solid facing obstacles of different magnitudes. The dynamics of this fracture front is also intermittent and gives the avalanche properties when averaged over time, as well as the roughness properties at one instant of time. The crack opened in this way can be made to remain confined approximately in one plane. This helps in modeling the crack front as an elastic line, which is driven through a disordered medium. Nevertheless, the elastic string model predicts a roughness exponent which does not match with the experiments in smaller length scales. There have been considerable efforts in explaining that theoretically, including the interfacial self-organized crack front propagation model from fiber bundles, the local, long-range dynamics, as well as by considering the loading plate as a semi-infinite elastic plane. Chapter 8 deals with the studies of fracture front propagation.

The dynamics of fracture mentioned so far is mainly the brittle fracture dynamics. However, for many solids the linear stress–strain region is not immediately followed by fracture point. The linearity may be lost for the ductile material beyond a certain stress. In this nonlinear regime, a large strain develops in the system, in response to a much smaller stress. This is due to the appearance and motion of the dislocations. We discuss the dynamics of the ductile fracture starting from the conditions of motion of the dislocation in Chapter 9.

The mechanical fracture is a tensor problem, which can be reduced to an analogous scalar field problem by considering the corresponding electrical breakdown problem. Indeed the two problems are similar in terms of stress concentration around defects. In fact, a mapping between the two problems exists. Like the fiber bundle model for mechanical fracture, the random fuse model has been studied extensively for electrical breakdown. The statistics of avalanche sizes as well as the roughness properties of the fractured surfaces are also studied in terms of this discrete element model. Although the existence of the avalanche dynamics has been claimed to be a finite size effect except for the limit of strong disorder. These issues are discussed in Chapter 10.

Finally, fractures in the largest scale (geological) that we experience are the earthquakes. Study of earthquakes has grown as an independent field of research over the years. It mainly follows the empirical observations about the size and duration distributions of earthquakes, which people have keenly studied over the last past years because of its catastrophic influence. The size distribution of energy emissions and the rate of aftershocks are known to follow scale-free distributions, and the observations go by the same of Gutenberg–Richter law and Omori law, respectively. This is also what is generally seen in much smaller scale of the laboratory in fracturing of rocks. The scale invariance has prompted researchers to think in the line of self-organized criticality for earthquake modeling. The simplest is the spring-block type model, where a train of blocks, connected by linear springs, is slowly pulled over a rough surface. The steady state dynamics of the model shows the intermittent nature that matches with many observations in earthquakes that are "critical." One wonders if the universality between earthquake dynamics and laboratory scale fracture propagations is also reflected in the respective models. We take up the earthquake dynamics and its different modeling approaches of fracture and their equivalence in Chapter 11.

2
Mechanical and Fracture Properties of Solids

The response of a material when subjected to an external force constitutes the mechanical properties of the material. When a force is applied on a material, it causes changes in the shape and size of it. Removal of the applied force may lead to the restoration of the initial shape and size or sometimes the material remains in the deformed state. As a reaction to the applied forces, internal restoring forces arise in the system which cause the restoration of the system to its original state. When the system is in equilibrium with the applied forces, the restoring forces must be equal and opposite to balance the applied forces. A measure of the restoring forces is given by the stress. Quantitatively, it is determined by the average force applied per unit area on the system. Strain is a measure of the deformation that occurs due to the application of the stress. It is the ratio of the change in dimensions to the initial dimensions. If we keep increasing the external forces, the material will reach a limiting point where the restoring forces developed in it cannot balance the applied forces any longer and the material fails and fracture occurs. It is seen from our day-to-day experience that some materials show appreciable deformation before fracture and some materials break at once without any deformation or with very small deformation.

It is important to be familiar with some of the standard tests carried out to characterize these properties and to understand the significance of the information obtained from them. The capacity of a material to withstand a static load can be determined by testing the material under tension or compression. Information about its resistance to permanent deformation is obtained from hardness tests. Information about toughness of a material under shock loading conditions is obtained form impact tests. Fatigue tests measure a material's lifetime under a cyclic load. Creep and stress rupture tests are conducted to evaluate the behavior of a material at high temperatures and subjected to a load for a long time.

In this chapter, we first discuss the typical response of a material under an applied force (tensile test) and then discuss about the characteristics of fracture in materials.

Statistical Physics of Fracture, Breakdown, and Earthquake: Effects of Disorder and Heterogeneity, First Edition.
Soumyajyoti Biswas, Purusattam Ray, and Bikas K. Chakrabarti.
© 2015 Wiley-VCH Verlag GmbH & Co. KGaA. Published 2015 by Wiley-VCH Verlag GmbH & Co. KGaA.

2.1
Mechanical Response in Materials

A material responds to an applied force by deforming itself or by developing a strain and a restoring force or stress that allows the system to come to mechanical equilibrium, when there is no further change in the shape or size of the system and the net force on the system becomes zero. The relationship between stress and strain in materials was given by British physicist Robert Hooke during the seventeenth century in Hooke's law. The law suggests that in the limit of small stress and strain, the stress is linearly proportional to the strain. The deformation in this region is called elastic deformation. The stress–strain characteristic curve tells us how strain in a system varies with stress. A general stress–strain curve that one finds in a large class of materials can be schematically represented as in Figure 2.1.

As force is applied, initially a straight line (OA) is obtained in stress–strain curve. The end of this linear region is called the proportional limit, up to which Hooke's law is valid. The largest stress (point C on the curve) is called the ultimate stress. In a force-controlled experiment, the specimen breaks at the ultimate stress. In a displacement-controlled experiment, a decrease in stress (region CE) is observed. The stress at breaking point E is called fracture or rupture stress.

2.1.1
Elastic and Plastic Regions

If a specimen is loaded up to the proportional limit and then unloaded, its response retraces the original stress–strain curve OA, and the system regains its original state represented by point O when the load is totally removed. This region, where the stress grows linearly with strain obeying Hooke's law, is the elastic region. If we start unloading the system after the system reaches point C, where the system shows appreciable nonlinearity in the stress–strain behavior, the system will follow not the original path OA but a different path CF on lowering or removal of stress. At point F, the stress is zero, but the strain is nonzero. C thus lies in

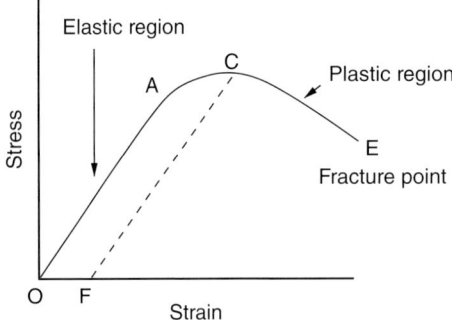

Figure 2.1 Stress–strain curve showing the variation of stress with strain.

the plastic region of the stress–strain curve, where the material is deformed permanently even when the stress is removed. The system shows hysteresis, and the permanent strain at point F is called the plastic strain. The nonlinear region in the stress–strain curve, straining up to which the material shows hysteresis and permanent deformation, is called the plastic region. In ductile materials one gets the full stress–strain behavior as given by Figure 2.1. The brittle materials only show up to linear region (OA) and break suddenly after that point.

2.1.2
Linear Elastic Region

In the linear elastic region of the stress–strain curve, Hooke's law is valid.

In this region, the stress s is defined as the applied force per unit length (or area or volume), F/L_0, where F is the applied force and L_0 is the initial length of sample in the experiment. The strain e is defined as the change in length referred to the original length;

$$e = \frac{\Delta L}{L_0} = \frac{1}{L_0} \int_{L_0}^{L} dL$$

Hooke's law suggests that in the linear elastic region, $s \propto e$ or $s = Ye$, where the proportionality constant Y is called the Young's modulus. In a material, the Young's modulus is generally related to other material parameters. For example, if v is the velocity of sound in the material and ρ is the density of the material, then $v = \sqrt{Y/\rho}$.

Several other elastic proportionality constants are in common use. These differ in the types of stress and strain which they relate: For example, the shear modulus (G) is the ratio of shear stress to shear strain, similarly the bulk modulus (K) is the ratio of hydrostatic tensile or compressive stress to fractional volume expansion or contraction. While a tensile force along a direction, say x-direction, produces an extension along that direction, it also produces a contraction in the transverse directions (y- and z-directions). The transverse strain s_t has been observed to be a constant fraction of the strain s_l in the longitudinal direction. This fraction is called Poisson's ratio $v = -s_t/s_l$.

In a homogeneous and isotropic material, where the elastic properties remain the same from point to point and along different directions, there are only two independent elastic constants that describe the elastic behavior of the material. In this case, the four elastic constants Y, G, K, and v are related by the following relations:

$$K = \frac{Y}{3(1-2v)}$$
$$G = \frac{Y}{2(1+v)}$$
$$v = \frac{Y}{2G} - 1$$

Macroscopic elastic strain results from a change in the interatomic spacing, and the elastic moduli can be related to the parameters that define the interatomic potential of a material.

2.1.3
Nonlinear Plastic Region

In the region CE of the stress–strain curve in Figure 2.1 Hooke's law is no longer valid, and nonlinearity appears. This region is called plastic region. Many aspects of plastic deformation make the mathematical formulation of theory of plasticity difficult:

1) Plastic deformation is not a reversible process like elastic deformation.
2) The elastic deformation is dependent only upon the initial and final states of stress and strain, while plastic strain depends on loading path by which the final state is achieved.
3) In plastic deformation, there is no easily measured constant relating stress to strain as with Young's modulus for elastic case.

In this region, of the material is strained up to point C and the load is released, the total strain will immediately decrease to a value F by an amount of stress/Young's modulus. This decrease in strain from C to F is the recoverable elastic strain. Since the true stress–strain curve gives the stress required to make the metal flow plastically, it is often called a flow curve. The most common mathematical expression used to fit the curve is a power expression of the form

$$s = K e^n$$

where K is a material-dependent constant and n is called the strain-hardening coefficient.

Since the strain in the plastic region is very large and it no longer follows linearity with stress, the conventional definition of strain as described earlier is not applicable. Quantities such as true stress and true strain are often used in this region. In this definition of strain, the change in length is compared with the instantaneous length of the sample, rather than its original length. The true strain is given by

$$\epsilon = \frac{L_1 - L_0}{L_0} + \frac{L_2 - L_1}{L_1} + \frac{L_3 - L_2}{L_2} + \cdots$$
$$= \int_{L_0}^{L} \frac{dL}{L} = \ln \frac{L}{L_0}$$

Conventional strain is related to the true strain by

$$e = \frac{\Delta L}{L_0} = \frac{L - L_0}{L_0} = \frac{L}{L_0} - 1$$
$$e + 1 = \frac{L}{L_0}$$
$$\epsilon = \ln(1 + e)$$

True stress is similarly defined as the ratio of load to the cross-sectional area over which it acts. The conventional stress is the ratio of load to the original area. In considering elastic behavior, it is not necessary to make this distinction. We denote true stress by the symbol σ, while conventional stress by s. True stress is given by $\sigma = \frac{P}{A} = \frac{P}{A_0}\frac{A_0}{A} = s\frac{A_0}{A}$. By the consistency-of-volume relationship, $\frac{A_0}{A} = \frac{L}{L_0} = e + 1$. This will finally give us the relationship between true and conventional stresses as: $\sigma = s(e + 1)$.

2.2
Ductile, Quasi-brittle, and Brittle Materials

The behavior to an external stress and the stress–strain curve that we discussed so far is the most general case that one encounters in experimental samples. The stress–strain relationship can be more complicated. However, even in general cases, materials do not always show appreciable plastic deformation. There are many materials that show the initial linear elastic response before fracture and then break abruptly. These are called brittle materials. Cracks through such materials propagate very quickly and cause fracture without any prior indication. Examples include ceramics, glasses, and some polymeric materials such as PMMA and polystyrene.

Materials that show some plastic deformation, but return to the initial configuration (zero strain) as the stress is withdrawn are called quasi-brittle materials.

Materials that show appreciable plastic deformation are called ductile materials. In practice, a material is said to be ductile when it shows a strain greater than 0.5%. Such materials give warning of fracture through flow and deformation in structure. Most metals, for example, copper, platinum, tungsten, and gold are ductile and show the features of ductility at normal temperature, pressure, and atmospheric condition (Figure 2.2).

2.3
Ductile and Brittle Fracture

The stress–strain curve ends at a point called the breaking point, where, the stress developed in the system cannot balance the external stress and the system breaks by fracturing into pieces. The nature of fracture differs with materials and is often affected by the nature of the applied stress, geometrical features of the sample, and conditions of temperature and strain rate. Broadly, fracture is characterized as either ductile or brittle as it is exhibited by the ductile or brittle material, respectively. Fracture processes in ductile and brittle materials show characteristic differences both macroscopically and microscopically (see Figure 2.3 for Mott's early works on brittle fracture).

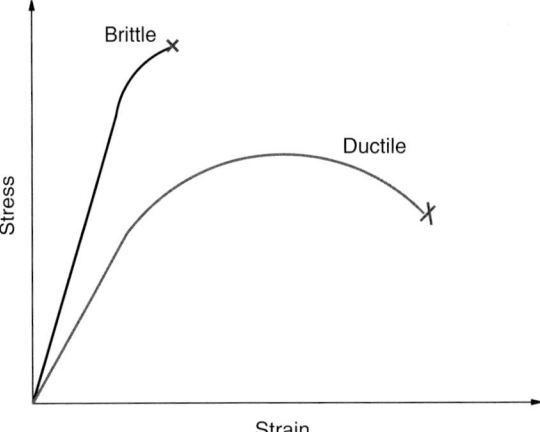

Figure 2.2 Stress–strain curve for ductile and brittle materials.

2.3.1
Macroscopic Features of Ductile and Brittle Fractures

Ductile fracture occurs after an appreciable plastic deformation. It is often associated with necking at the point of plastic instability. In general, ductile fracture in engineering samples in tension has three distinct stages: (i) the sample begins necking, and cavities form in the necked region, (ii) cavities coalesce to form a crack at the center of the sample and the crack grows perpendicular to the applied stress, and (iii) the crack spreads to the surface of the sample in a direction 45° to the tensile axis to form a cone-like part of the fracture. The result of this series of processes is the "cup-and-cone" fracture (see Figure 2.4). Ductile fractures can be of several types. Single crystals of hexagonal close-packed metals may slip on successive basal planes until the crystal separates by shear. Polycrystalline specimens of ductile metals, such as gold or lead, can be drawn down to a point before the rupture.

On the other hand, in brittle materials, appreciable plastic deformation does not occur. The crack develops and propagates very fast and nearly perpendicular to the direction of the applied stress. The crack often propagates by cleavage—breaking of atomic bonds along specific crystallographic planes. Figure 2.5 (a) shows the transgranular brittle fracture where the crack propagates through grains. Similarly, Figure 2.5 (b) shows the intergranular fracture where the propagation of cracks is through the grains.

In brittle fractures, the materials fracture by separating normal to the tensile stress. Externally, there is no evidence of deformation, although with X-ray diffraction analysis it is possible to detect even a thin layer of deformed metal at the fracture surface. In general, the fracture surfaces in brittle fractures are much smoother than those in ductile fractures.

16

BRITTLE FRACTURE IN MILD-STEEL PLATES—II.

In October, 1945, as a consequence of failures of deck and shell plating in certain strips, a Conference on Brittle Fracture in Mild Steel Plates was held at Cambridge University. On page 532 of the previous volume of ENGINEERING, we began to reprint a report of the proceedings at the Conference, further instalments appearing on pages 556, 581 and 605. We commence below the second part of our report, by reprinting the paper contributed by Professor N. F. Mott, F.R.S., on "Fracture of Metals: Some Theoretical Considerations."

FRACTURE OF METALS: THEORETICAL CONSIDERATIONS.

Professor Mott, in his introduction, said that much information existed about the conditions under which solids fracture, but it had not been possible to formulate a theory to provide a satisfactory synthesis of the experimental data. This was true particularly of metals and of ductile materials in general; for brittle material, such as glass, there was the theory put forward by A. A. Griffith in 1921, which accounted for some of the more important facts. The purpose of his paper, Professor Mott explained, was to outline the present state of the theory for brittle substances, and to discuss its application to ductile materials such as metals.

Figure 2.3 Sir Nevill Francis Mott (1905–1996) won the Nobel Prize in Physics for his works on electronic structure of magnetic and disordered systems along with Phillip Anderson and J. H. Van Vleck in 1977. In addition to his excellent contribution to this field, he has also done significant research on fracture of solids and fragmentations (the figure shows the first page of his article on fragmentation). For these, he won also the A. A. Griffith Medal and Prize in 1973. The above excerpt is from Mott (1948).

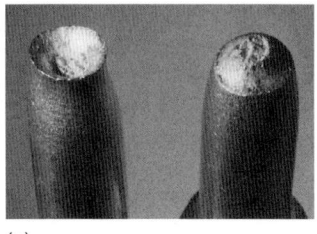
(a)

(b)

Figure 2.4 (a) Cup-and-cone ductile fracture surface and (b) sharp brittle fracture surface.

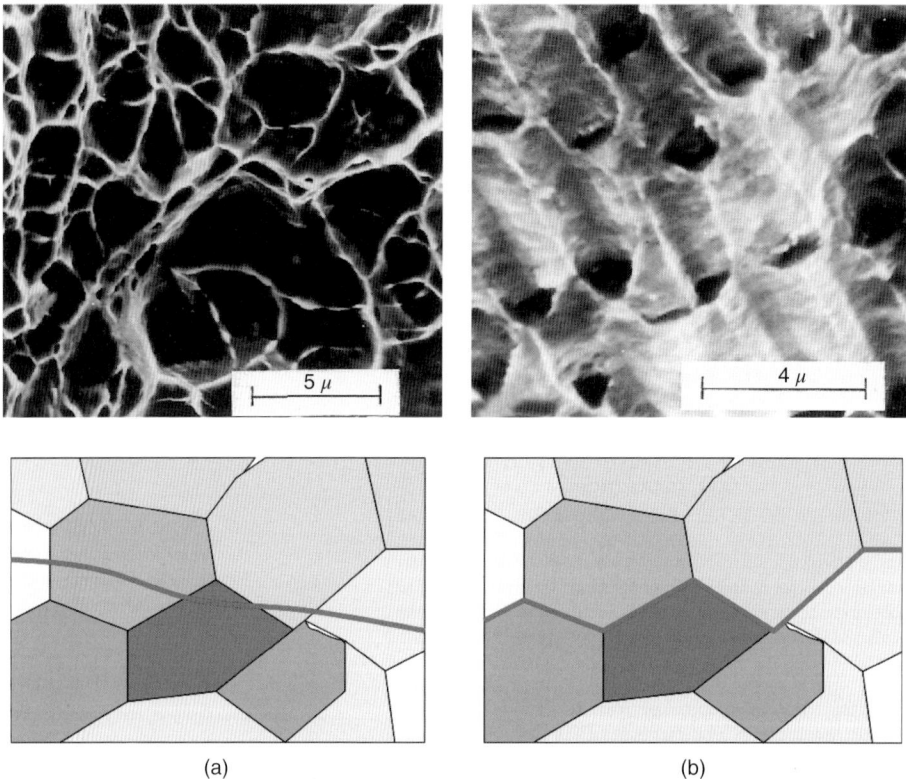

Figure 2.5 Crack propagation in ductile (a) and brittle (b) materials.

2.3.2
Microscopic Features of Ductile and Brittle Fractures

Since brittle materials show no deformation before fracture, the nature of their crack tip differs from that of the ductile ones. The crack tip in brittle fracture is very sharp, whereas in ductile fracture the crack tip has a blunt structure because of the accumulation of defects such as dislocations (Figure 2.6).

The stages in the development of a ductile cup-and-cone fracture are illustrated in Figure 2.7. Under applied stress, the shape of the ductile sample changes, and at the microscopic level, defects such as dislocations (will be described later) start

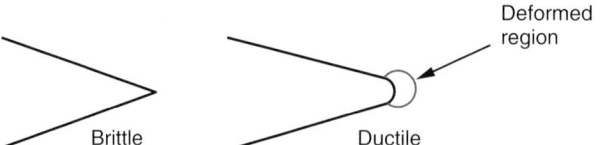

Figure 2.6 Difference in crack tip for ductile and brittle materials.

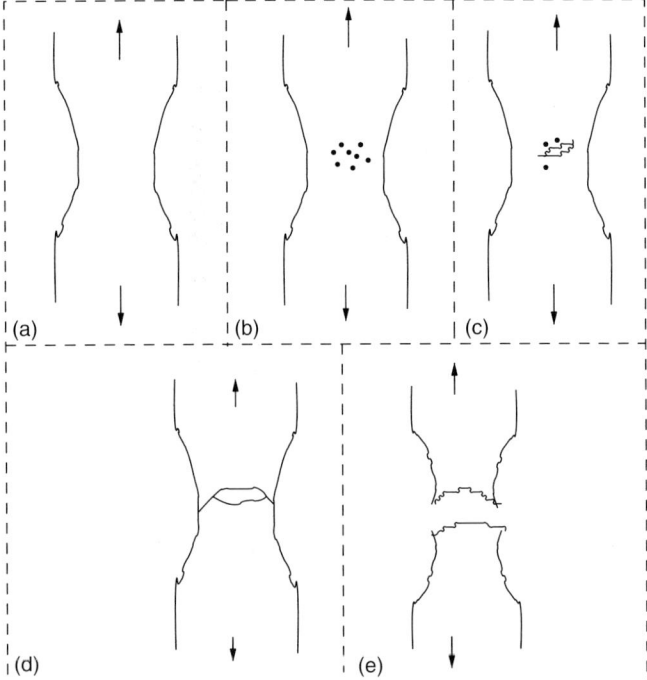

Figure 2.7 Different stages of ductile fracture.

moving and piling up at the grain boundaries causing the "necking" in the material. Necking begins at the point of plastic instability. With increasing applied stress, the piled-up dislocations give birth to small cavities or microvoids. Under continued straining, these cavities grow and coalesce into a central crack (Figure 2.7 c). The crack propagates along localized shear planes at roughly 45° to the tensile axis to form the cone part of the fracture (Figure 2.7 d).

Crack growth in ductile fracture takes place essentially by the process of void coalescence. Fractographic studies at high resolution show that the spherical dimples correspond to microvoids that initiate crack formation. Coalescence occurs by elongation of the voids and elongation of the bridges of materials between the voids. This leads to the formation of a fracture surface consisting of elongated dimples, as if it had formed from numerous holes which were separated by thin walls until it fractures.

If the defects are pinned, and cannot move with stress, the nature of fracture becomes very different from ductile fracture. This is what happens in brittle fractures. Experimentally it is observed that where the stress needed to fracture bulk glass is around 100 MPa (15 000 psi), the theoretical stress needed for breaking atomic bonds is approximately 10 000 MPa (1 500 000 psi). This conflicting scenario was resolved during the times of World War I by English aeronautical engineer, A. A. Griffith, who suggested that brittle fractures originate from the large

Table 2.1 Comparison of fracture properties of brittle and ductile materials.

Properties	Ductile materials	Brittle materials
Deformation	Appreciable deformation	No deformation
Crack propagation	Very slow, needs continuous stress	Very fast
Type of materials	Metals	Glass, Ceramics
Indication before fracture	Permanent elongation	No indication, breaks suddenly
Strain energy	High	Low
Fracture surface	Rough	Smoother

stress concentration at the tips of a sharp microcrack preexisting inside the material. At a particular stress depending on the crack size, the microcrack becomes unstable and grows spontaneously, leading to the fracture of the entire sample. The brittle cracks are generally sharp and the fracture surface is smoother than that of ductile materials.

In the Table 2.1, we present the distinguishing features of brittle and ductile fractures.

3
Crystal Defects and Disorder in Lattice Models

Defects in materials play a crucial role in the behavior of the materials under stress and on the nucleation and propagation of fracture. In engineering and materials science, materials are broadly classified into three types: brittle, quasi-brittle, and ductile. The stress–strain response and fracture characteristics are very different for these three types of materials, and the guiding factors behind these different behaviors are the defects and their kinetics in these materials.

A perfect solid may be visualized by orderly array of lattice points. Experimental and naturally occurring materials are hardly perfect. Defects in crystals, generally, arise from the deviation of the lattice structure from the ordered one. If the deviation is localized to the vicinity of only a few atoms, it is called a point defect. However, if the defect extends through macroscopic regions of the crystal, it is the lattice imperfection. Lattice imperfections can be of the form of line defects and surface or planar defects. The edge and screw dislocations are the common line defects and are line discontinuities in the otherwise regular crystal structure. Dislocation is one of the important crystal defects responsible for plastic deformations in materials. Planar defects arise from the clustering of line defects into a plane, and are discontinuities of the crystal structure across a plane. Low-angle tilt boundaries and grain boundaries are surface defects. Microcracks are another example of planar defects which play a crucial role in the nucleation and propagation of fracture in brittle materials. In this chapter, we discuss the aforementioned three kinds of defects in more detail.

3.1
Point Defects

A point defect is the change in a single atom within the otherwise normal crystal array. The three major types of point defects are vacancies, interstitials, and impurities (see Figure 3.1). These defects often are built-in with the original crystal growth, or activated by heat. The defects are often generated by radiation or electric current.

A vacancy is the absence of an atom from a lattice site normally occupied in the lattice. In otherwise pure metals, few vacancies are always present due to thermal

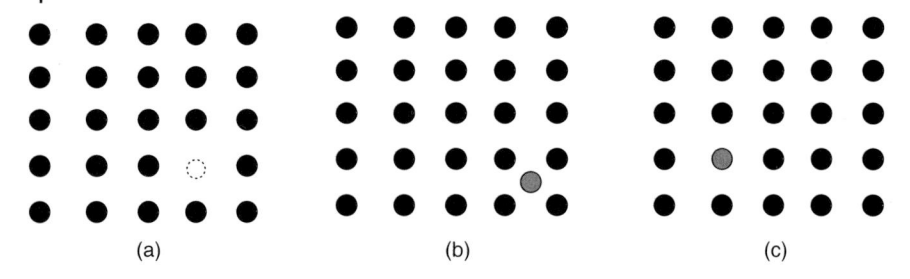

Figure 3.1 Point defects. (a) Vacancy; (b) Interstitial; (c) Impurity atom.

fluctuations. It can be estimated as $n \sim Ne^{-E_v/k_BT}$, where N is the total number of atoms in the system, E_v is the energy required to displace an atom from the interior of the crystal to its surface, T is the temperature in the absolute scale, and k_B is the Boltzmann constant. Vacancies can be created by rapidly quenching the system from a high temperature to a temperature close to its melting point, by extensive plastic deformation, or by bombarding with high-energy particles as in radiation damage. More number of vacancies cluster together to form voids.

An interstitial is an atom present on a nonlattice site. There needs to be enough room for an interstitial atom to fit in the otherwise regular lattice. This type of defects occur in open covalent structures, or metallic structures with large atoms. A large amount of energy is required for an atom to go to the interstitial position, and it seldom happens by thermal activation. Radiation damage can cause this kind of defect.

If an atom that does not normally occupy a site in the lattice substitutes an atom that normally sits there, the defect is called an impurity. The replacement atom can come from within the crystal (e.g., a chlorine atom on a sodium site in an NaCl crystal) or from the addition of impurities. The concentration of point defects in a crystal is typically between 0.1% and 1% of the atomic sites. However, extremely pure crystals can now be grown using modern techniques. The concentrations and dynamics of point defects in a solid are very important in controlling color and deformation of the solid.

3.2
Line Defects

Dislocation is a line discontinuity in the regular crystal structure. It is the defect responsible for the slip phenomena, by which most metals deform plastically. Dislocations may exist in crystals as a result of growth faults; they can also be produced by dislocation sources, which under stress disgorge dislocations successively. One way to think about a dislocation is to consider it as the region of localized lattice disturbance that separates the slipped and unslipped regions of crystals. The two basic types of dislocations are the edge dislocation and screw dislocation. In dislocations, the part of the crystal above the slip plane has been

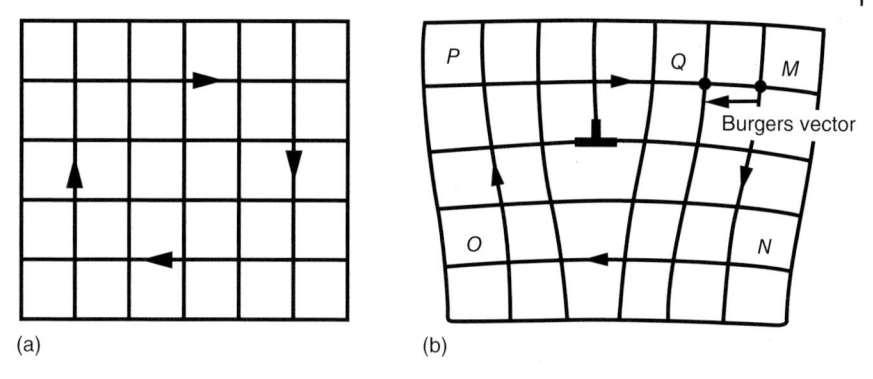

Figure 3.2 Geometry and Burgers vector in edge dislocation.

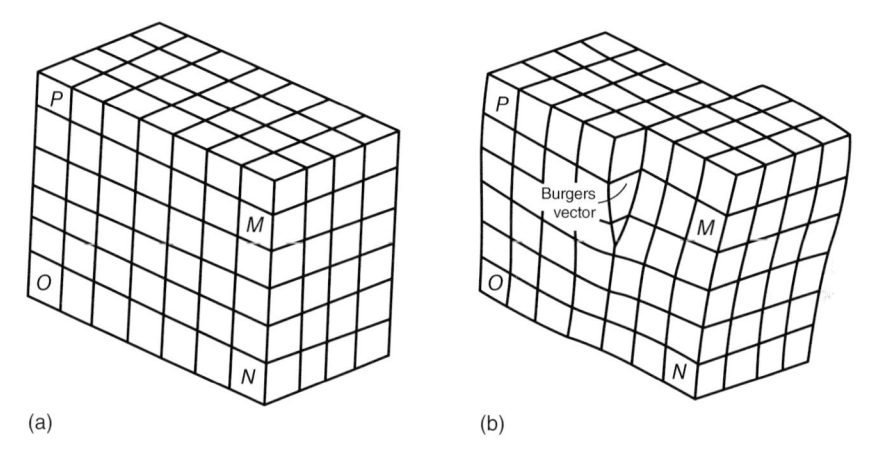

Figure 3.3 Geometry and Burgers vector in screw dislocation.

displaced in the direction of slip, with respect to the part of the crystal below the slip plane by a certain amount. All points of crystal that were originally coincident across the slip plane have been displaced relative to each other by the same amount. The amount of displacement is called Burgers vector $\vec{b}$ of the dislocation (see Figures 3.2 and 3.3).

A defining characteristic of an edge dislocation is that its Burgers vector is always perpendicular to the dislocation line (see Figure 3.2). Dislocation is a topological defect, and the strength of the defect is measured by traversing along a closed contour on the lattice around the defect. For example, in Figure 3.2a, a closed contour is shown on a perfect lattice starting from a lattice point and by traversing four lattice steps along the horizontal direction and three lattice steps along the vertical direction. On a perfect lattice, because of the lattice symmetry, the contour returns to the starting point making it a closed contour. But, while traversing round the edge dislocation, as shown in Figure 3.2b, the contour ends at Q and falls short of the starting point M. The distance between M and Q gives the magnitude of the Burgers vector. Similarly, the direction from M to Q gives the direction of the vector. The only major change in the crystal structure occurs along the dislocation line (perpendicular to the page). Perpendicular to the page, the dislocation line may step up or down. These steps are known as jogs. In edge dislocation, the Burgers vector is always perpendicular to the dislocation line.

In screw dislocation, Burgers vector is parallel to the dislocation line (see Figure 3.3b). Here also, the Burgers vector is defined in the same way as it was defined in edge dislocation. In a screw dislocation, atomic planes no longer exist separately from each other. A single surface, like a screw thread, is formed which "spirals" from one side of the crystal to the other. Combinations of edge and screw dislocations are often formed; for example, edge dislocations can be formed by branching off a screw dislocation. (It is in fact a helical structure because it winds up in three dimensions, not like a spiral which is flat.)

As a dislocation moves, slip occurs in the area over which it moves. In the absence of any obstacle, a dislocation can move easily on application of even a small force; this explains why real crystals deform much more readily than what is expected for a crystal with a perfect lattice. As mentioned earlier, dislocations are primarily responsible for nonlinear plastic behavior of the metals. We discuss the motions of the dislocations and slip later.

Dislocations are not thermodynamically stable as their presence increases the free energy of a crystal. However, it is, in general, very difficult to eliminate these defects completely from crystal lattices as it requires rearrangement of lattice structure over a large region requiring more energy. The crystals remain in metastable configurations with the dislocations. The energy of a dislocation can be estimated by considering a cylinder of length l with a screw dislocation of Burgers vector $\vec{b}$ along its axis. Consider a thin annular section of radius r and thickness dr. The shear strain of the section is $\gamma = b/2\pi r$, where $b = |\vec{b}|$. The elastic strain energy per unit volume is then $dE/dV = G\gamma^2/2 = Gb^2/8\pi^2 r^2$, where G is the elastic shear modulus. The volume of the annular ring is $dV = 2\pi r l dr$. This gives $dE = lG\, b^2 dr/4\pi r$. On integrating from some lower limit r_0 to some upper limit R, the strain energy due to the presence of a dislocation becomes $E = \frac{lG\, b^2}{4\pi} \ln(\frac{R}{r_0})$. We should not choose $r_0 = 0$, as at the dislocation core, Hooke's law is not valid. We see that $E \simeq lG\, b^2$. We can define a line tension $\mathcal{T} = \frac{\partial E}{\partial l} \simeq Gb^2$ of the dislocation line. This shows that the most stable dislocations are those with minimum Burgers vector.

3.3
Planar Defects

A discontinuity across a plane in a perfect crystal structure is called planar defect. For example, a grain boundary is a planar defect that separates regions of different crystalline orientations (i.e., grains) in a polycrystalline solid (see Figure 3.4). The atoms in the grain boundary will not be in a perfect crystalline arrangement. Grain boundaries arise when there is uneven growth when a solid crystallizes. In usual crystals, grain sizes vary from 1 μm to 1 mm. If the mismatch in the crystallographic orientation is low (typically less than 10 degrees), the grain

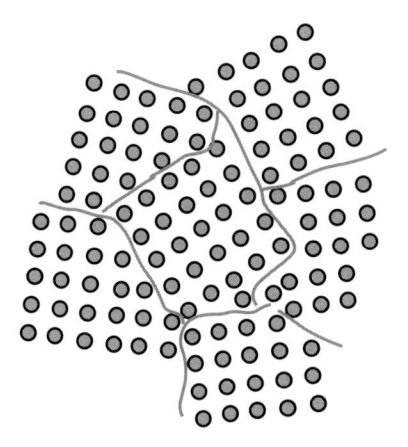

Figure 3.4 Schematic diagram of grain boundaries and grains with different crystallographic orientations.

boundary is semicoherent, and is called a low-angle boundary. Low-angle grain boundaries are delineated by dislocation walls and are often represented as an array of edge dislocations (tilt boundary) or screw dislocations (twist boundary).

The presence of grain boundaries in polycrystalline materials strongly affects their stress–strain behavior. The grain boundaries are usually stronger than the material in the grain themselves. That is why the fracture in most metals at low temperatures is transcrystalline, the fracture goes through the interior of the grains rather than along the grain boundaries. Dislocation motion is severely constrained in the presence of grain boundaries. If the grain boundary structure is to be preserved, each grain must deform in a manner compatible with the deformations of all its neighboring grains. On this account, the macroscopic behavior of a polycrystalline metal can, in general, be only qualitatively described from the behavior of a single crystal of that metal.

A microcrack occurs where internal bonds are broken to create new surfaces (see Figure 3.5). Microcracks occur on the surface of a solid rather than in the bulk. They can also occur at grain boundaries and other regions of disorder in the solid. The region across which the bonds are broken is known as the separation plane. Microcracks are important in determining how, and where, a solid may fracture. In amorphous materials and polymers, since there is no perfect crystal structure, microcracks and voids can form easily.

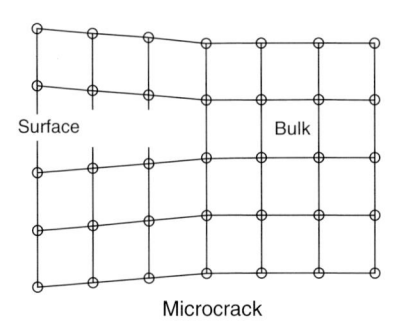

Figure 3.5 Schematic diagram of a microcrack formed due to breaking of atomic bonds.

3.4
Lattice Defects: Percolation Theory

Percolation is a simple model of lattice defects that can be applied to various physical phenomena, and can be studied analytically and numerically to a large extent (Stauffer and Aharony, 1994). We have discussed disorders induced by vacancies. The percolation theory formalizes the statistics of such defects in terms of universal scaling functions in the limit of high disorders.

Consider a nonconducting plate, on which some conducting dye is sprayed uniformly. If one applies a potential difference across any two opposite ends of the plate, with an ammeter in series, there will be no current initially when no dye is spread. But, if the entire area of the plate is covered by the dye, then obviously the plate conducts. Is it necessary to cover the entire plate to get a nonvanishing current? The answer is, no. Current starts to flow when there is a marginally connected path of the overlapping clusters of the dye grains across the plate (see Figure 3.6). This phenomenon is called percolation, and the point at which conduction first takes place is called the percolation threshold (see Appendix A for details).

In order to make the discussion more quantitative and precise, let us consider the lattice percolation model. There are two versions of this model: site percolation and bond percolation. In site percolation, each site of a large lattice is randomly occupied with probability p. Clusters are defined as graph of neighboring lattice sites. In bond percolation, each bond of a lattice is occupied randomly, with a probability p. A cluster is defined as a graph of overlapping bonds, sharing a common site. Most of the physical properties of such random systems depend on the geometric properties of these random clusters, and in particular, on the existence of an infinite connected cluster, which spans the system (see Figure 3.7). Percolation theory deals with the statistics of the clusters formed (the structural constraints are, however, often determined by rigidity percolation threshold, see Appendix B).

Let us define some quantities of interest in percolation theory. Let $n_s(p)$ denote the number of clusters (per lattice site) of size s. A detailed knowledge of $n_s(p)$ would give a lot of information about percolation statistics, as most of the quantities of interest can be extracted from various moments of the cluster size distribution n_s.

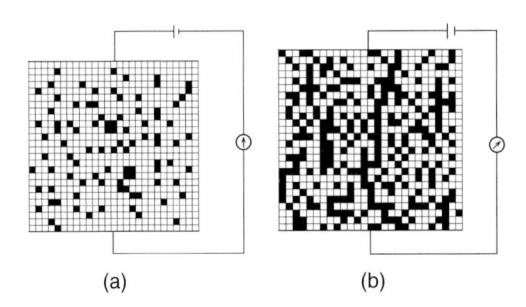

Figure 3.6 Sample below and above the percolation threshold.

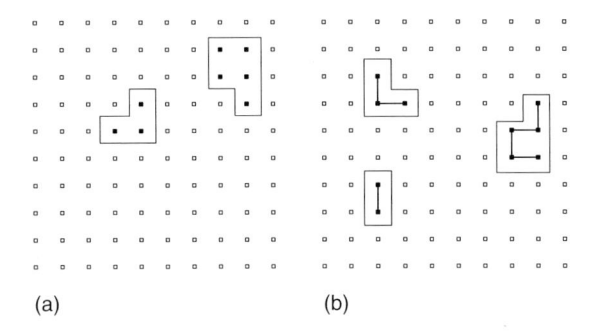

Figure 3.7 (a) Clusters in site percolation. (b) Clusters in bond percolation.

The probability that a given site (bond) is occupied and is a part of an s-size cluster is $sn_s(p)$. Let, $P(p)$ denote the probability that any occupied site (bond) belongs to the infinite (lattice spanning) cluster. Then, we have the obvious relation:

$$\sum_s sn_s + P = 1, \qquad (3.1)$$

where the summation extends over all finite clusters. Clearly, at $p = 1, P(p) = 1$ and $P(p) = 0$ for $p < p_c$ as the infinite cluster does not exist for $p < p_c$. $P(p)$ can therefore be taken as the order parameter of the percolation phase transition. Another quantity of interest is the mean size of the finite clusters, denoted by $S(p)$, which is related to $n_s(p)$ through the relation

$$S(p) = \frac{\sum_s s^2 n_s(p)}{\sum_s sn_s(p)}, \qquad (3.2)$$

where the summation is again over all finite clusters. One can also define a pair connectedness (or two-point correlation) function $C(p, r)$ as the probability that two occupied sites (bonds) at a distance r are members of the same cluster. The sum over the pair connectedness over all distances gives the mean cluster size: $S(p) = \sum_r C(p, r)$.

Percolation is a phenomenon that shows critical properties can have a critical point such as magnet and superconductors, and can be studied with the machinery of statistical physics. Critical phenomena refers to the behavior of a system tuned to be near a critical point (usually second-order phase transition point) across which, the behavior of the system is qualitatively different. It is generally characterized by divergence of a correlation length, in the sense that near critical point, the system becomes almost fully correlated, and hence the correlation length scale determines the responses of the system near criticality. In the absence of any other length scale, than the diverging correlation length, the physical quantities show power-law (scale-free)-type behavior near critical point. Fracture and failure of a material can also be considered as a phase transition. In addition to the static properties under applied load (relaxation time, correlations in roughness, etc.), it is also important to study the dynamic transitions, in terms of fracture front propagation and its universalities. In addition, a structural transition in terms of substitutional disorder, namely the percolation transition, plays an important role in understanding disorder properties. We discuss the basic features of scaling in percolation transition.

In general, if x is the tuning parameter of the system, changing which system can be led to the critical point, the different quantities near critical point behaves as

$$O(x) \sim |x - x_c|^{\beta'}, \tag{3.3}$$

$$\xi(x) \sim |x - x_c|^{-\nu}, \tag{3.4}$$

where x_c is the critical point. The different quantities are associated with different observables of the system: $O(x)$ denotes the order parameter, which is zero in the disordered state and nonzero in the ordered state; a phase transition is said to have occurred when the order parameter becomes nonzero. $\xi(x)$ is the correlation length of the system, that is, a disturbance in the system typically affects up to a distance ξ. The functional form suggests that this quantity diverges as the critical point is approached. The exponents β', ν, and so on, are called critical exponents.

Most of the quantities defined in the context of percolation have power-law variations near p_c. For example, the variation of the total number of clusters per site $G(p) = \sum_s n_s(p)$ (sum extends over all finite clusters), the decay of the order parameter $P(p)$, and the divergence of the mean cluster size: $S(p)$, as $p \to p_c$ can be expressed by power-law variations of these quantities with the concentration interval $p - p_c$ as

$$G(p) \equiv \sum_s n_s(p) \sim p - p_c^{2-\alpha'}. \tag{3.5}$$

The correlation function behaves as

$$C(p, r) \sim \frac{\exp(-r/\xi(p))}{r^{d-2+\eta}} \tag{3.6}$$

where the correlation length

$$\xi(p) \sim p - p_c^{-\nu} \tag{3.7}$$

diverges at $p = p_c$. These powers $\alpha', \beta', \gamma, \eta$, and ν are called the critical exponents. These exponents are observed to be universal in the sense that although p_c depends on the details of the model under study, these exponents depend only on the lattice dimensionality.

Scaling theory assumes that the cluster distribution function $n_s(p)$ is a homogeneous function near $p = p_c$. Thus $n_s(p)$ is basically a function of the single-scaled variable $s/S_\xi(p)$, where S_ξ denotes the typical cluster size,

$$n_s(p) \sim s^{-\beta} f\left(\frac{s}{S_\xi(p)}\right), \tag{3.8}$$

with $S_\xi(p) \sim p - p_c^{-1/y}$. Here, β and y are two independent exponents, and the scaling theory intends to relate all the aforementioned exponents to these exponents through the scaling relations.

One can show (see Appendix A) the following scaling relations between the exponents.

$$\alpha' + 2\beta' + \gamma = 2. \tag{3.9}$$

$$\gamma = \nu(2 - \eta). \tag{3.10}$$

And the hyperscaling relation,

$$d\nu = (\beta - 1)/y = 2 - \alpha'. \tag{3.11}$$

These scaling relations are satisfied by the numerically estimated values and in special cases the exact values, of the critical exponents.

3.5
Summary

The role of defects is vital in the study of failure properties of fracture. Strength of a solid decreases substantially because of the presence of defects in it. Around the defects, stress concentration takes place, which ultimately leads to failure of the solids (to be discussed in Chapter 4).

Here we have dealt with characterization of defects. The point defect, line defect and plane defect are discussed in sections 3.1, 3.2 and 3.3 respectively. In particular, we have also discussed the statistics of point defects within the formalism of percolation theory in section 3.4.

4
Nucleation and Extreme Statistics in Brittle Fracture

Intuitively one may think that the strength of a material is determined only by the parameters of that material, such as intermolecular bond structure and bond strength and crystal or material structure. The actual picture of failure differs significantly from it. It is found that the typical strength of a material is orders of magnitude less than what can be estimated from the bond strength. It is now realized that the prime reason for this low failure strength is the presence of defects such as microcracks and dislocations. The reason is that any applied stress gets hugely concentrated at the sharp ends of the defects such as voids and microcracks. This makes the stress field within the material highly nonuniform. When the stress field grows to a level such that the released elastic energy is sufficient to provide the surface energy of the newly created crack surfaces, the defect becomes unstable and fracture nucleates. Hence, to estimate the strength of a material, one needs to understand the precise forms of distortion of the stress field in a material due to the defects and also the criterion for which a given defect will start propagating as a crack and subsequently cause the failure of that material. In this chapter, we discuss the role of disorders in modifying the stress field and hence initiating crack propagation satisfying Griffith's criterion.

4.1
Stress Concentration Around Defect

As mentioned earlier, a uniformly applied stress will get distorted within a material containing defects. For a material with voids, this is easy to see. In the absence of voids (pure material), the stress lines within the material will be uniform. If voids are present, those parts within the material cannot support any stress, therefore the stress lines will be expelled from the voids, hence creating a higher concentration at its boundaries.

However, as can be guessed, the precise form of stress concentration will depend on the geometry of the defect and the mode of application of stress. One of the very important geometry in which stress concentration can be calculated is an elliptical void. Clearly, in the limit of the ratio of minor axis to major axis being very small, an elliptical defect resembles a crack within the material.

Statistical Physics of Fracture, Breakdown, and Earthquake: Effects of Disorder and Heterogeneity, First Edition.
Soumyajyoti Biswas, Purusattam Ray, and Bikas K. Chakrabarti.
© 2015 Wiley-VCH Verlag GmbH & Co. KGaA. Published 2015 by Wiley-VCH Verlag GmbH & Co. KGaA.

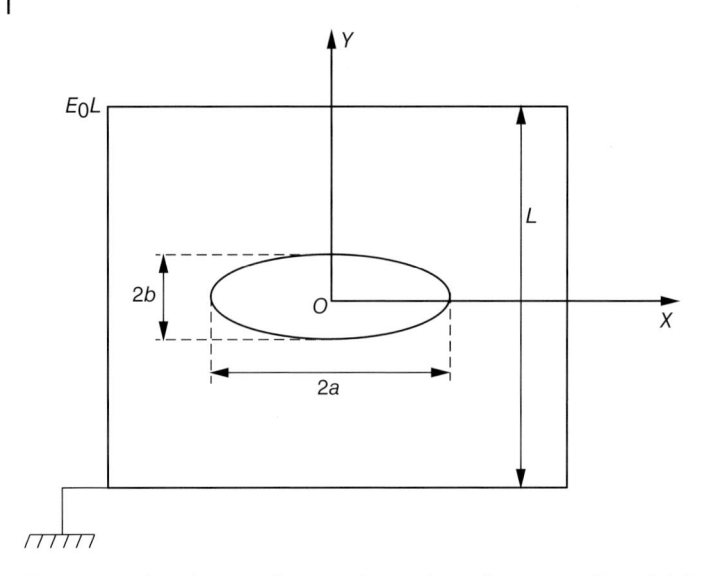

4 Nucleation and Extreme Statistics in Brittle Fracture

Figure 4.1 The schematic diagram of a conductor having an elliptical defect.

It was Inglis (1913) who first calculated the precise form of stress concentration around an elliptical defect. Before we discuss that, let us consider an easier but equivalent form of stress concentration, that is, the voltage concentration around an elliptic void placed within a conducting material.

Let the lengths of the semi-major axis and semi-minor axis of the elliptical defect be a and b, respectively (see Figure 4.1), which are much smaller than the linear size L of the conductor. Now suppose a potential difference is applied along Y-axis. In order to solve the voltage distribution within the conductor, one has to solve the Laplace's equation

$$\nabla^2 V = \frac{\partial^2 V}{\partial^2 x} + \frac{\partial^2 V}{\partial^2 y} = 0, \tag{4.1}$$

with the boundary condition

$$\left[\frac{\partial V}{\partial b}\right]_{\phi=\phi_0} = 0, \tag{4.2}$$

along the boundary ($\phi(x, y) = \phi_0$) of the void, since no current flows across the boundary. The equation of the elliptical boundary in the Cartesian coordinates is $x^2/a^2 + y^2/b^2 = 1$. But, to simplify the following algebra, let us work in an elliptical coordinate system, where the transformation equations read

$$x = c \cosh(\phi) \cos(\theta), \quad y = c \sinh(\phi) \sin(\theta) \tag{4.3}$$

with

$$a = c \cosh(\phi_0), \quad b = c \sinh(\phi_0) \quad \text{and,} \quad c = \sqrt{a^2 - b^2}. \tag{4.4}$$

Clearly, the value of the radius of curvature will be the smallest at $\theta = 0$ and π. The transformation of Laplacian to elliptic coordinate keeps it invariant (except for a constant multiplier $2/c^2(\cosh(2\phi) - \cos(2\theta)))$, therefore the Laplace's equation remains the same, and a simple separation of variable will lead to its solution, which, in the general form, reads

$$V(\phi, \theta) = A_1 x + A_2 y + B_1 e^{-\phi}[\sin(\theta) + B_2 \cos(\theta)] + C_1 e^{-\phi}[\sin(\theta) + C_2 \cos(\theta)] \quad (4.5)$$

where A_is, B_is, and C_is are constants. In order to satisfy the boundary conditions, A_1, B_1, B_2, and C_2 must vanish and $A_2 = E_0$, where E_0 is the field applied along Y-axis. Therefore

$$V = E_0 y + C_1 e^{-\phi} \sin(\theta)$$
$$= -cE_0 \sinh(\phi) \sin(\theta) + C_1 e^{-\phi} \sin(\theta) \quad (4.6)$$

where

$$C_1 = -cE_0 e^{\phi_0} \cosh(\phi_0). \quad (4.7)$$

If S is the conductance of the conductor, then the current density in the Y-direction can be calculated by

$$i_y = S \frac{\partial V}{\partial y}$$
$$= -\left(\frac{S}{c}\right) \frac{\cosh(\phi) \sin(\theta)[\partial V/\partial \phi] + \sinh(\phi) \cos(\theta)[\partial V/\partial \theta]}{\cosh^2(\phi) - \cos^2(\theta)}$$
$$= E_0 S - \left(\frac{C_1 S}{c}\right) e^{-\phi} \frac{\cosh(\phi) \sin^2(\theta) - \sinh(\phi) \cos^2(\theta)}{\cosh^2(\phi) - \cos^2(\theta)}. \quad (4.8)$$

The current concentration at the point of maximum radius of curvature of the defect ($\phi = \phi_0$ and $\theta = 0$) is then

$$i_{tip} = SE_0[1 + \coth(\phi_0)]$$
$$= SE_0\left[1 + \frac{a}{b}\right]$$
$$= i\left[1 + \sqrt{\frac{a}{\rho}}\right] \quad (4.9)$$

where i is the current density within the conductor, far away from the defect and $\rho = b^2/a$ is the maximum radius of curvature (at the tip) of the ellipse. Therefore, in the limit $\rho \to 0$, which is the limit of a sharp crack, the current density at the sharp edges becomes very high thereby inducing further growth of the defect that finally leads to failure.

If one applies a mechanical stress (σ_0) in place of the electrical potential difference in the same geometry, an equivalent picture regarding the mechanical stress concentration around the defect can be obtained. The stress at the tip of the defect is given by

$$\sigma_{tip} = \sigma_0(1 + 2\frac{a}{b})$$
$$= \sigma_0\left[1 + 2\sqrt{\frac{a}{\rho}}\right] \quad (4.10)$$

As mentioned before, one can note that for $\rho \to 0$, the stress concentration can be very high, which may lead to mechanical failure of the material.

4.1.1 Griffith's Theory of Crack Nucleation in Brittle Fracture

The stress field distortions due to the presence of defects is an important step toward understanding the failures in brittle materials. However, stress field nucleation is not sufficient to predict when a defect will start to grow and thereby induce failure of the material. In order to get that information, one needs to consider a detailed energy balance equation, an approach first considered by Griffith (1921). He considered cracks as a system in equilibrium, which would minimize its free energy (see Figure 4.2). When a crack propagates, there are two competing terms in its energy $E = E_e + E_s$. The first term, E_e is the strain energy of the material, which is released when the crack propagates and the second term, E_s is the surface

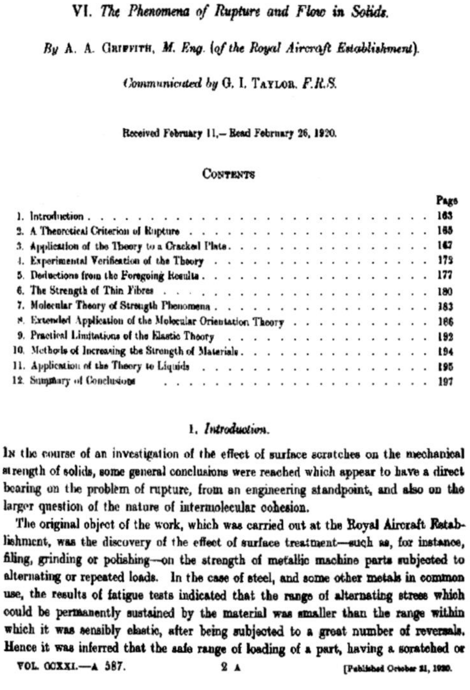

VI. *The Phenomena of Rupture and Flow in Solids.*

By A. A. GRIFFITH, *M. Eng.* (*of the Royal Aircraft Establishment*).

Communicated by G. I. TAYLOR, *F.R.S.*

Received February 11,—Read February 26, 1920.

CONTENTS

1. *Introduction.*

IN the course of an investigation of the effect of surface scratches on the mechanical strength of solids, some general conclusions were reached which appear to have a direct bearing on the problem of rupture, from an engineering standpoint, and also on the larger question of the nature of intermolecular cohesion.

The original object of the work, which was carried out at the Royal Aircraft Establishment, was the discovery of the effect of surface treatment—such as, for instance, filing, grinding or polishing—on the strength of metallic machine parts subjected to alternating or repeated loads. In the case of steel, and some other metals in common use, the results of fatigue tests indicated that the range of alternating stress which could be permanently sustained by the material was smaller than the range within which it was sensibly elastic, after being subjected to a great number of reversals. Hence it was inferred that the safe range of loading of a part, having a scratched or

VOL. CCXXI.—A 587. 2 A [Published October 21, 1920.

Figure 4.2 Alan Arnold Griffith (1893–1963) was a mechanical engineer with a doctorate degree from Liverpool university in 1915. Among his many contributions such as developing the basis of jet engine, he is mostly remembered for his work on strength of brittle materials. He formulated the idea of fracture propagation from an energy balance principle, known as Griffith's criterion. It states that fracture propagates through a medium when the cost to form the surface area is compensated by the strain energy released. The work was published in Griffith (1921), the first page of which is shown in the figure with their kind permission.

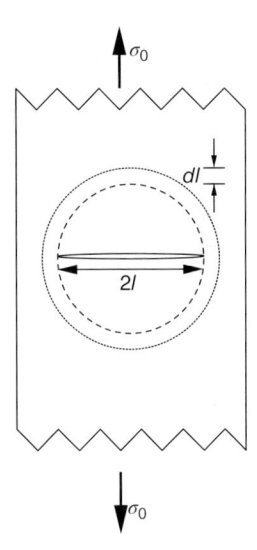

Figure 4.3 A portion of the plate of thickness δw having a crack of length l under a load σ_0 inducing mode-I fracture.

energy gained due to the work done against the molecular attractive forces in creating the extra surface area. Therefore, the crack would propagate when the released strain energy is higher than the gained surface energy. Along with the knowledge of stress concentration, Griffith's criterion gives a clear picture about the failure of a material when stressed.

Let us consider an example where their terms can be estimated and one can then predict other consequences regarding the failure properties. Consider a thin linear crack of length $2l$ in a infinite elastic medium with thickness δw (see Figure 4.3). A tensile stress σ_0 is applied on the material, which is perpendicular to the length of the crack. As discussed earlier, the presence of a preexisting crack would distort the stress field around it, since the crack itself cannot support any stress (will exclude the stress lines). Without going into the details of the stress field distortion as before, let us just assume that the stress field (and for that matter its energy density $\sigma_0^2/2Y$; Y being Young's modulus) will get distorted in a circular cross section around the crack, keeping the crack in its diameter. The actual geometry will be different, but in any case the distortion will be up to the distance of the order of the crack length. The other details will only change some prefactors, in which we are not currently interested. When the length of the crack is increased by $2dl$, we assume that half of the strain energy stored in the cylindrical volume, $2\pi l\delta w dl$ is released; again the factor half may change due to detailed considerations of geometry, but that would not change the order of magnitude estimates. In addition, we assume linear elastic behavior up to failure point and no plastic deformation. Because of the advancement of the crack, the total new surface area created will be $4\delta w dl$. The Griffith's criterion for the propagation of the crack is, therefore,

$$\frac{1}{2}\frac{\sigma_0^2}{2Y} \geq \Gamma(4\delta w dl) \qquad (4.11)$$

where, Γ is the surface energy density of the material. This implies

$$\sigma_f = \frac{K}{\sqrt{2l}}; \quad K = \left(\frac{4}{\sqrt{\pi}}\right)\sqrt{Y\Gamma} \quad (4.12)$$

being the critical stress at which the crack starts propagating. K is called the stress intensity factor. The point of fracture propagation can also be thought of as the point at which the stress intensity factor exceeds (due to nucleation at sharp edges) the elastic limit of the material.

In general, for any dimension d (≥ 2), if a crack of length l exists in the medium and a stress is applied uniformly in the direction perpendicular to the crack, then the elastic energy will be $E_e \approx \left(\frac{\sigma_0^2}{2Y}\right) l^d$ and surface energy $E_s \approx \Gamma l^{d-1}$. Hence, the Griffith's criterion reads

$$\left(\frac{\sigma_f^2}{2Y}\right) l^d = \Gamma l^{d-1}, \quad (4.13)$$

where Y is the appropriate elastic constant. Hence

$$\sigma_f \sim \frac{K}{\sqrt{l}}; \quad K \sim \sqrt{Y\Gamma}, \quad (4.14)$$

is generally valid for any dimension higher or equal to two.

4.2
Strength of Brittle Solids: Extreme Statistics

Unlike the linear response of material properties such as elastic modulus, the fracture properties of materials are non-self-averaging, because fracture is determined by the weakest point of the material (weakest link of a chain concept). As a consequence, there is a large sample to sample variation of the fracture strength as is generally seen in brittle materials. In a non-self-averaging distribution, the most probable value is different from the mean. Here, we discuss the distribution functions that describe the failure probability of a disordered sample of linear dimension L. As we shall see, such distributions can be cast into a few generic forms. In this chapter, we discuss those distribution functions and the effect of system sizes on the failure properties of disordered materials.

4.2.1
Weibull and Gumbel Statistics

Assume that the solid has N microcracks, each of which has independent failure probabilities $f_i(\sigma_0)$, $i = 1, 2, \ldots, N$ under an applied stress σ_0 (the stress released regions of these cracks are assumed not to overlap). Now, if the cumulative probability of failure of the entire sample under the stress σ_0 is denoted by $F(\sigma_0)$, then (Ray and Chakrabarti, 1985a,b; Chakrabarti and Benguigui, 1997)

$$1 - F(\sigma_0) = \prod_{i=1}^{N}(1 - f_i(\sigma_0)) \approx \exp\left[-\sum_{i=1}^{N} f_i(\sigma_0)\right] = \exp[-L^d \tilde{\rho}(\sigma_0)] \quad (4.15)$$

4.2 Strength of Brittle Solids: Extreme Statistics

where $\tilde{p}(\sigma_0)$ is the density of the cracks that fail at stress level σ_0. The survival probability, as written above, is the joint survival probability in the presence of all these cracks, that is, when none of these has failed.

In the literature of extreme statistics (Gumbel, 1958), one finds few distributions which characterize such statistics. Two of these distributions that find most applications are Weibull and Gumbell statistics. We readily get Weibull and Gumbell statistics from the above analysis, if we assume $\tilde{p}(\sigma_0) \sim \sigma_0^w$ and $\tilde{p}(\sigma_0) \sim \exp(-A/\sigma_0^2)$, respectively.

Therefore, in Weibull statistics, defects at every scale play a role in determining the fracture in the material. As the Weibull distribution is given by

$$F_W(\sigma_0) \sim 1 - \exp[-L^d \sigma_0^w],$$

one gets for the typical fracture strength $\tilde{\sigma}_0$ of the material having a system size dependence: $\sim L^{-\frac{d}{w}}$ (assuming that $F(\tilde{\sigma}_0)$ is finite). As has been mentioned before, lowering of the strength of a system with its size is a telltale signature of fracture in brittle materials.

In Gumbell statistics, one assumes that the defects of a certain characteristic dimension a only lead to the nucleation and development of fracture and hence $\tilde{p}(\sigma_0) \sim \exp(-Aa)$. Griffith's law (Eq. (4.14)) suggests that $a \sim 1/\sigma_0^2$. This gives the Gumbell distribution

$$F_G(\sigma_0) \sim 1 - \exp\left[-L^d \exp\left(-\frac{A}{\sigma_0^2}\right)\right].$$

Gumbell distribution, therefore, predicts the system size dependence of the typical fracture strength as: $\sim \frac{A}{\sqrt{\ln L^d}}$.

Generically, these distribution functions have the characteristics that the failure probability monotonically increases with σ_0 and for bigger system sizes, the rate of increase is faster (see Figure 4.4). This can be intuitively understood from the

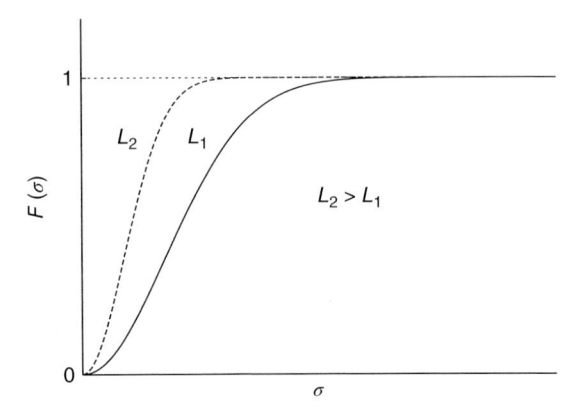

Figure 4.4 The schematic variation of failure probability $F(\sigma)$ for disordered solids with volumes L_1^d and L_2^d with $L_2 > L_1$.

fact that a larger sample has a higher probability of having a bigger (hence weaker) defect. The most probable failure stress then becomes a decreasing function of system size (Bazant and Planas, 1997; Alava et al., 2009b).

In addition to the simplified models to describe fracture (to be discussed later), there have been some early attempts to do molecular dynamics simulation, using Lennard-Jones interatomic interaction, of fracture propagation in a diluted medium (Chakrabarti et al., 1986). In a triangular lattice, it was seen that the typical failure stress vanishes and relaxation time diverges near the percolation threshold of a triangular lattice (to be discussed in Section 4.5).

4.3
Extreme Statistics in Fiber Bundle Models of Brittle Fracture

The emergence of extreme statistics in the failure properties of solids has also been attempted to be modeled by simple discrete element models. Fiber bundle model is one such generic model, which has been used to reproduce many basic properties of failure dynamics of solids. The model was introduced in Peirce (1926) as a model for testing the strength of cotton yarns in textile industry. Much later, it emerged as a very rich model for detecting fracture in disordered solids (see Pradhan et al. 2010 for a review). We discuss its properties mainly in Chapter 6 (see also Appendix C), but here we focus, after a brief introduction of the model, on the stress nucleation and extreme value statistics shown in the model.

4.3.1
Fiber Bundle Model

The basic model consists of N fibers hanging from a rigid ceiling and a plate hanging from the ends of the fiber. From this bottom plate, an external load (W) hangs. The fibers are considered as massless elastic objects having identical elastic constants in general. The springs are linear, elastic (Hookean), and brittle, that is, fails irreversibly beyond a certain stress (there are variations of the model that relax this). In order to model the disordered solid, the failure strengths of the fibers are generally taken randomly from a (normalized) distribution function. Initially, each fiber has to carry a load $\sigma = W/N$. The model has two extreme versions depending on the choice of elastic properties of the bottom plate. In one extreme, the bottom plate is considered fully rigid. The initial value of the load may cause some of the fibers to break. The load is now to be carried by the surviving fibers, that is, the load carried by the failed fibers are to be redistributed effectively. The process of redistribution crucially depends on the elastic properties of the bottom plate. When the bottom plate is fully rigid, there is no stress concentration around the failed fibers and the load carried by the failed fibers are now equally distributed among the surviving ones. This is known as global load sharing (GLS) or equal load sharing (ELS). The other extreme is the one where the bottom plate is absolutely soft and gets deformed in the places where fibers have failed. The load carried by those fibers is now redistributed among the nearest surviving fiber(s). This is known as local load sharing (LLS).

As can be easily seen, the first case does not give rise to a length scale. Indeed, it is a mean-field model and the notion of distance is absent. Nevertheless, on putting the fibers on a square lattice, random percolation process is observed. For a broad class of the distribution function of the failure threshold (Andersen et al., 1997), the "critical load," where the system fails, is a self-averaging quantity, that is, it converges to a well-defined value as the system size is increased. The situation, however, is markedly different for LLS, as we discuss here.

4.3.1.1 Strength of the Local Load Sharing Fiber Bundles

Unlike the global load sharing model, the local load sharing model is difficult to tackle analytically. One can of course estimate the strength σ_f as $1/l$, where the typical defect size l goes as $\ln\ L$ (following Lifhitz argument for one-dimensional system of size L; see below).

As indicated earlier, LLS gives rise to stress concentration around failed fibers, which actually leads to extreme statistics such that the strength $\sigma_f \to 0$ in the macroscopic size limit ($L \to \infty$). Essentially, for any finite load (σ), depending on the fiber strength distribution, the size of the defect cluster can be estimated using Lifshitz argument: Roughly speaking, if p is the probability (depends on threshold distribution and value of load) of failure of a fiber under the load per fiber value σ, then the probability of a "crack" of length l, that is, successive l fibers broken, is $p^l(1-p)^2$ (considering that the fibers are arranged in an one-dimensional array). In a system of L fibers, a crack of size l will surely occur when $Lp^l(1-p)^2 = 1$. On the other hand, this crack can be fatal if the stress concentration around it (i.e., two fibers on either sides) is such that the crack starts propagating and never stops. An overestimate of this criterion can be made by assuming the stress values on either sides to have reached the maximum value of failure threshold of the fibers (say, unity). Then, $\sigma_f + l\sigma_f/2 = 1$, assuming the load to be critical (the highest threshold value of any fiber to be normalized). These two equations together give $\sigma_f \sim 1/\ln\ L$ up to constant factors. This rough estimate agrees with rigorous calculations done for load transfer on one side (Gómez et al., 1993) with Weibull distribution of failure thresholds. Numerical estimates match with this logarithmic decay (see Figure 4.5) (Pradhan and Chakrabarti, 2003b). Thus the local load sharing version of the fiber bundle model shows extreme statistics, in the sense that its strength vanishes in the large system size limit.

4.3.1.2 Crossover from Extreme to Self-averaging Statistics in the Model

As we have discussed so far, the equal load sharing version of the fiber bundle model shows self-averaging behavior and the local load sharing version shows extreme statistics in the sense that the presence of one weak patch in the system causes the damage to grow, and leads to catastrophic failure as soon as any load is applied in the system (in the large system size limit).

There have been several attempts to determine the crossover from self-averaging to extreme statistics by varying different parameters of the model.

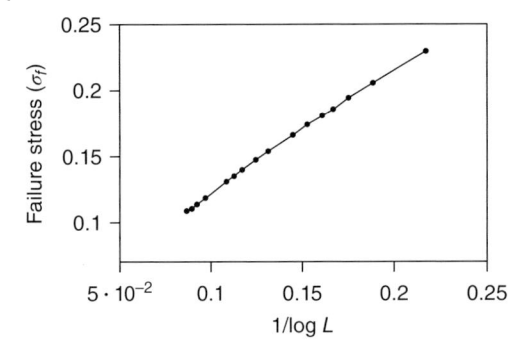

Figure 4.5 The variations of failure load of the local load sharing fiber bundle model with local load sharing scheme. It indicates a $1/\ln L$-type decay in the failure load. Figure courtesy S. Roy.

As has been discussed earlier, the emergence of extreme statistics comes from stress field deformation around defects and for the largest one, the deformation (stress concentration) can be sufficient to lead to a catastrophic failure. However, on one end, one has the global load sharing model, where the notion of length is absent. Hence, the question of stress field concentration does not arise. The failures of fibers are completely random, if one puts them in a hypothetical square lattice. The broken patch statistics is same as that of random percolation, with no spatial correlations induced by the successive failures of fibers, making the statistics self-averaging.

A competition arises when the load sharing has a local part. This can be done in many ways, the details of which will be discussed in Chapter 6. In general, a local part in the load sharing brings the notion of distance in the model, which in turn leads to stress concentration. For example, Hidalgo *et al.* (2002b) introduced a power-law load sharing scheme, in which the load of a failed fiber is to be shared with all surviving fibers in inverse proportion of their distances ($1/r^\gamma$). While $\gamma \to 0$ gives the global load sharing limit, $\gamma \to \infty$ is the local load sharing limit. It was numerically shown for a square lattice that the crossover from global to local (i.e., from self-averaging to extreme statistics) critical behavior takes place near $\gamma = 2$. Similar crossover was seen for the so-called mixed mode load sharing (Pradhan *et al.*, 2005a). A fraction of the load of the failed fiber was shared with the nearest surviving neighbor(s) and the rest was shared equally among all the other fibers. This also led to a crossover from global to local critical behavior for a finite value of the sharing fraction for both one and two (Biswas and Chakrabarti, 2013a) dimensions. Remembering that the load sharing mechanism is the manifestation of the elastic properties of the bottom plate (from which the load is hanging) of the system, Gjerden *et al.* (2013) considered the changes in the failure properties when the stiffness of the elastic bottom plate was varied. The elastic nature of the bottom plate in fiber bundle models of failure was first explicitly considered in Batrouni *et al.* (2002). The force transmission to the surviving fibers (i.e., the load sharing mechanism) comes from the Green's function (to be discussed in

Chapter 6). It was explicitly shown by considering statistics of distances of successive failures and cluster distributions that in the limit of hard plate, the model is global load sharing as expected. But, beyond a certain extent of softness, the failures are correlated and localization occurs, leading to stress-nucleated failure. It was also noted that the localization sets in only after a finite fraction of fibers failed and that the setting in of localization is a crossover behavior and not a transition one. On a similar note, the elastic properties were considered to be heterogeneous in the presence of random disorders in the model. In Biswas and Chakrabarti (2013a), it was argued that random "soft" patches in the bottom plate will cause the fibers there to share their load only locally when failed. It was shown that beyond a finite fraction of this local load sharing concentration, the critical behavior becomes local. In two dimensions, it was argued that the percolation threshold ($p_c \approx 0.59$) will be an upper bound for the crossover limit. This is because, beyond that point, a percolating cluster of local load sharing fibers will form (that will scale linearly with system size) and that will essentially behave as a local load sharing model, making the entire system follow its nature. Indeed, the crossover thresholds were smaller than the percolation point. It was also shown in the study that the partially failed system retains the memory of the loading mechanism. Hence, the step size of load increase will change the fraction of surviving fibers in the model, even when the fraction of local fibers is much smaller than the crossover point. This could help in monitoring mechanisms in determining the presence of impurities much before it reaches the limit of catastrophic failure.

Recently, a criterion for the effective range of stress redistribution in unit time (range of load transfer) has been proposed (Biswas et al., 2014) that distinguishes between the nucleation driven by failure from that driven by uncorrelated damage (or percolation). In a linear arrangement of fibers, it was proposed that if the range scales slower than $L^{2/3}$, the failure mode is nucleation driven. For faster increase, the mode is percolation-like (uncorrelated damage). Consequently, for power-law load transfer, $\gamma_c = 4/3$ for one dimension (calculating the average range and applying the scaling criterion).

4.4
Extreme Statistics in Percolating Lattice Model of Brittle Fracture

As discussed earlier, we have modeled the disordered solid as a random percolating system. The response functions such as elastic moduli of such a system can be obtained from the average of the cluster statistics. This description is valid for these quantities, since they are "self-averaging," in the sense that all connected paths contribute to that quantity, making it the average of all contributions. This situation, however, is not valid for breakdown statistics: The failure properties are solely dictated by the weakest part (often the largest defect) of the material and not by the average weakness. The smaller defects have no direct role to play. In the following, we discuss the origin of the extreme statistics for a disordered solid of linear dimension L from the point of view of a random percolation process.

In the presence of a disorder (say, site dilution), the linear elastic properties (say, Y) are analytic for low disorder concentration. But, near the percolation threshold, at and near which the system is only marginally connected, the elastic responses become singular. The leading order singularity can be described by a critical exponent, which is independent of the details of the material and depends mainly upon the dimensionality. Hence, $Y \sim (p_c - p)^{T_e}$, where p is the disorder concentration (vacancies). Here, we intend to relate the nonlinear irreversible problem of failure strength variation with the corresponding linear response exponents. If we model the disordered solid as a percolating system, the dimension of a preexisting crack (a) will be of the order of the correlation length ξ, which in turn is related to the disorder concentration via $\xi \sim (p_c - p)^{-\nu}$. Assuming that the surface energy scales as ξ^{d_B}, where d_B is the fractal dimension of the back-bone cluster, comparing the elastic and surface energies, one gets

$$\left(\frac{\sigma_c^2}{2Y}\right) \xi^d \sim \xi^{d_B}, \tag{4.16}$$

leading to

$$\sigma_f \sim (p_c - p)^{T_f}, \tag{4.17}$$

where

$$T_f = \frac{1}{2}[T_e + (d - d_B)\nu]. \tag{4.18}$$

Since in an actual case, the strain would be higher for singly connected bonds than for multiply connected ones, the average strain per bond calculated here is less than the actual strain for a singly connected bond (Ray and Chakrabarti, 1988). Thus, the fracture stress estimated in this way is always higher than the actual one, and the expression for T_f given above is a lower-bound. It is only for $d \geq 6$ that the expression is exact. An upper bound of T_f could also be derived by considering only the singly connected bonds in the superlattice network noting $d_B = 1/\nu$ here. The bounds of the exponent T_f are then given as

$$\frac{T_e + d\nu - 1}{2} \geq T_f \geq \frac{T_e + (d - d_B)\nu}{2}. \tag{4.19}$$

For $d \geq 6$, the two bounds coincide to give $T_f = 3$. For two dimensions, the bounds are $2.8 \geq T_f \geq 2.3$, which are in agreement with the exponents obtained (2.5 ± 0.4) in the experiments performed with randomly punched holes in a metal plate (Benguigui et al., 1987).

In addition to the aforementioned static properties, the dynamics response of diluted solids is important in understanding the relaxation processes in such solids. In Ghosh et al. (1989), stress relaxation dynamics was studied in randomly perforated aluminum and copper foils. The samples were anelastic to start with, having a single relaxation time. The strain growth in time was recorded in the linear region. A crossover was noted from a considerably slow-stretched exponential behavior (for $t < t_c$)

$$\epsilon(t) \sim \epsilon(\infty) - A \exp\,(-t/\tau)^a \quad \text{with} \quad a \approx 0.8. \tag{4.20}$$

to the simple exponential behavior $\epsilon(t) \sim \epsilon(\infty) - A\exp(-t/\tau)$ (for $t > t_c$). Both the crossover time t_c and the relaxation time τ tend to diverge as the percolation point is approached.

4.5 Molecular Dynamics Simulation of Brittle Fracture

An important aspect in modeling the fracture processes in solids is the molecular dynamics (MD) simulation, which, in some sense, is a bridge between continuum models and experiments. Here, one has the luxury of choosing and controlling the disorders, brittleness, and other properties of a sample. There have been extensive studies in MD simulation of brittle fracture (see Rountree et al., 2002 for a review) including the verifications of the Griffith's criterion mentioned above, which we discuss here. In addition, the roughness properties of fractured surfaces are compared with those obtained from experiments and different discrete models, which we will discuss in Chapter 5.

4.5.1 Comparisons with Griffith's Theory

The basic purpose of molecular dynamics simulation is to solve the Newton's equation of motion in a different form for a finite number of atoms in a lattice. The most crucial ingredient of MD simulation is the choice of the interatomic potential. The Lennard-Jones-type potential of the form

$$\phi(r_{ij}) = \phi_0 \left[\left(\frac{d}{r_{ij}}\right)^{12} - 2\left(\frac{d}{r_{ij}}\right)^{6} \right], \tag{4.21}$$

has been applied to fracture of amorphous solids extensively. In Falk (1999), it was shown that the brittle as well as the ductile behavior can be recovered from a simple Lennard-Jones-type potential (see also Bonamy and Bouchaud, 2011). The two-dimensional system consisted of 90 000 particles. In order to avoid local crystallization, eight different species of particles were considered, which differed in their radii

$$r_i = 1.1 r_{i-1}, \quad \sum_{i=1}^{8} \pi r_i^2 = 8\pi \left(\frac{d_0}{2}\right)^2. \tag{4.22}$$

Thus, the total volume is same as if all particles were of radius $d_0/2$. Two types of potentials were used to see the ductile and brittle behaviors. The ductile behavior was seen in the usual Lennard-Jones-type potential, whereas the brittle behavior was seen in a modified Lennard-Jones-type potential of the form

$$\phi_{ij}^{(2)}(r_{ij}) = \phi_{ij}^{LJ}[\lambda r_{ij} + (1-\lambda)(r_i + r_j)] \tag{4.23}$$

where r_{ij} is the interparticle distance and r_i is the radius of the ith particle. Note that $\lambda = 1$ is the LJ potential limit. When $\lambda = 0$, all particles interact with all others equivalently. For $\lambda \gg 1$, the interaction becomes nearest neighbor type. In general, when $\lambda > 0$, the potential is called compressed Lennard-Jones potential (CLJ). In CLJ simulation, the dissipation was comparatively lower than that in LJ (as one expects in the brittle fracture compared to ductile one). With CLJ potential, the crack starts to propagate for about 7% higher stress than that estimated by Griffith's theory for an ideal brittle crack. Even though the stress drops significantly once the crack propagates, it does not release the entire stress as the crack is arrested. The crack tip for CLJ remains atomically sharp throughout, reaching about 30% of the speed of shear wave. On the other hand, for LJ potential, the cracks start propagating after 48% increase beyond the estimated critical stress. The crack tip gets blunted and propagates slower than the CLJ case. Thus the CLJ potential shows much more brittle behavior than the LJ potential.

For ceramics, a combination of two-body and three-body interaction terms is considered in the potential (Vashishta et al., 1990) (see also Rountree et al., 2002; Bonamy and Bouchaud, 2011 for reviews). The system size of the simulations goes up to 130 million particles. Fractures in notched samples show that they propagate through coalescence of damage cavities in front of them (Rountree et al., 2002, 2007). The same mode of fracture is also seen in amorphous materials, such as $Si_3 N_4$.

In Karimi et al. (2006), the MD simulations with embedded atom methods were done for Ni slab with 160 000 atoms in (001)[100] crack system, where (001) is the crack-free Surface, and the crack propagates along the direction [100]. Even though it is a ductile material in general, one is able to study brittle crack propagation in Ni for this particular choice of crack system and periodic boundary along y[010]-axis that prevents emission of dislocations in the system. In the simulations without defects, the crack propagation starts with about 20% higher stress than is predicted in Griffith's theory. This is attributed to the local potential wells created due to discrete lattice and interatomic potential. A much smaller departure (4%) from the Griffith's theory was obtained in Gumbsch et al. (1997) with a different interatomic potential.

When a stress higher than the critical value is applied in breaking a material, the excess energy is dissipated in various ways. The crack velocity might increase, and branching, dislocation, and so on can take place (Cramer et al., 2000; Gao, 1996). In MD simulation with modified embedded atom method for silicon (Swadener et al., 2002), the Griffith's criterion was shown to be valid (no dissipation) for low speed. However, for higher speeds, dissipations in the form of dislocation were seen (although no branching was seen). In 3C-SiC single-crystal simulation (Kikuchi et al., 2005), significant orientation dependence was found, that is, (110) fracture is cleavage, but (111) and (001) show branching of cracks. The energy release rate also agrees with theoretical estimates.

In the simulations of brittle fracture, a common property is that the crack gets arrested before the complete rupture of the material, if the boundaries are not continuously pulled. In experiments, however the crack keeps moving under a

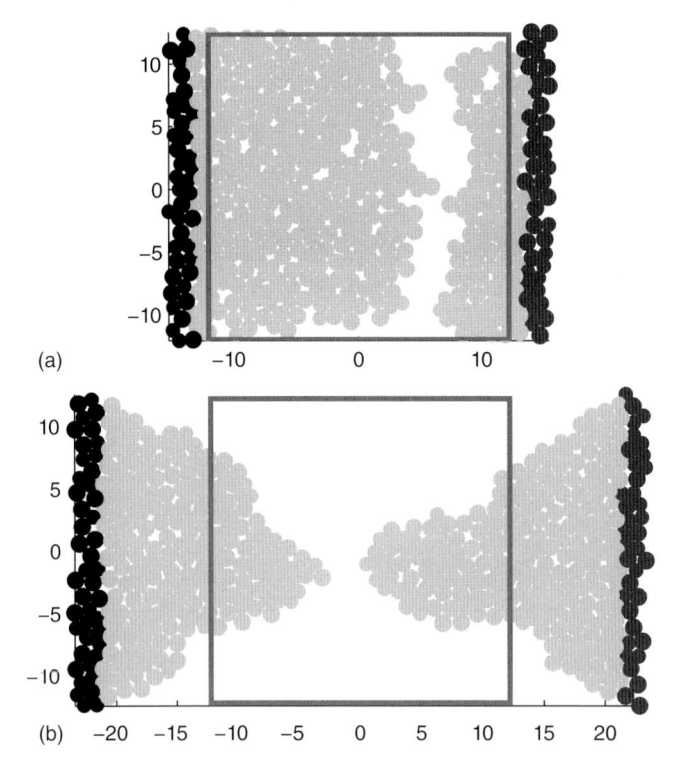

Figure 4.6 When the boundaries are pulled, the brittle fracture occurring for a shorter range of interatomic potential is shown in (a), where the shape of the solid is almost retained. The (b) shows the ductile fracture for larger range of interatomic potential, showing significant deformation in the shape in the form of plastic flow and necking. From Dauchot et al. (2011), with kind permission from American Physical Society.

grip boundary. In a recent work, Dauchot *et al.* (2011) claimed that by tuning the cut-off distance of the interatomic interaction, the ductile to brittle transition can be tuned (see Figure 4.6), and in the brittle regime, the grip boundary condition shows crack propagation. The mechanism of propagation of the crack shows formation of cavities ahead of crack tip, as was also sometimes seen experimentally (Bouchaud and Paun, 1999). This formation and subsequent coalescence of cavities ahead of crack tip also were considered as the mechanism behind roughness of fractured front (Procaccia and Zylberg, 2013), which we shall discuss later in Chapter 5.

4.5.2
Simulation of Highly Disordered Solids

The modeling of statistical physical properties arising out of fracture of highly disordered materials started with molecular dynamics simulations of such materials (Ray and Chakrabarti, 1985a, 1985b; Chakrabarti *et al.*, 1986). In two dimensions,

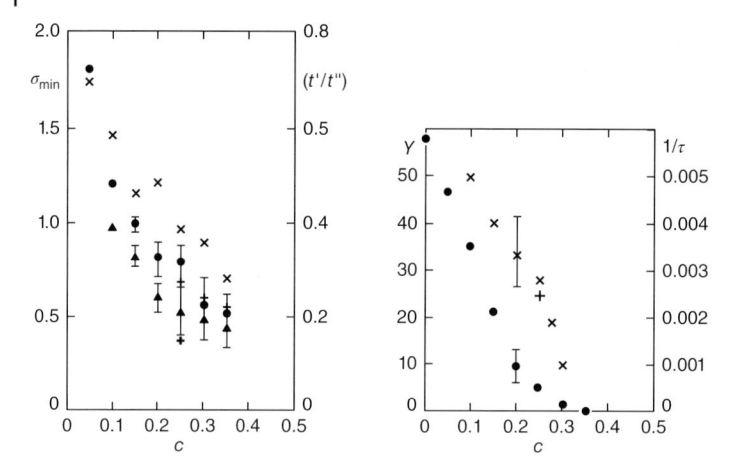

Figure 4.7 The variations of minimum fracture stress (σ_{min}) with concentration $c = 1 - p$ vanishing at the percolation threshold (a) and variation of Young's modulus (b) vanishing at $p = 1/3$. From Chakrabarti et al. (1986), with kind permission from Springer.

it was claimed (Ray and Chakrabarti, 1985a,b) that the minimum stress to initiate fracture vanishes at the percolation threshold as

$$\sigma_{min} \sim (p - p_c)^{T'}. \tag{4.24}$$

In Chakrabarti et al. (1986), a dilute triangular lattice was considered, where the atoms interact via Lennard-Jones (LJ) potential (Eq. (4.21)) which has a cut-off distance of $1.6d$, with d being the initial lattice constant and therefore the potential minimum. It was found that the average breaking stress σ_{min} vanishes at the percolation threshold (in this case $1/2$) (see Figure 4.7). The Young's modulus, however, vanishes at $p = 1/3$ (see Figure 4.7). The relaxation time can be calculated by fitting with the equilibrium strain as $l(t) - l(\infty) \sim \exp(-t/\tau)$, which approximately suggests that $1/\tau$ may vanish at the same concentration as the elastic modulus.

4.6
Summary

The fracture strengths of solids are not self-averaging quantities like the elastic constants (e.g., Y in Eq. (4.12)) where the effect of disorder enters through simple analytic formulas. Similar estimates for fracture strength (σ_f) yield results that are orders of magnitude higher than what are actually seen. The reason is that stress nucleates around defects and this leads to failure at a lower strength than predicted. The dependence of failure strength on defect size can be estimated using Griffith's formula in Sec. 4.1.1.

The failure strength of disordered solids depends on the system volume, and its distribution follows extreme statistics (Weibull or Gumbel) (Chakrabarti and Benguigui, 1997). In the limit of strong disorder (near percolation threshold), when the percolation correlation length exceeds the Lifshitz length, percolation statistics takes over (see Sec. 4.2). The average strength, given by Eq. (4.12), becomes precisely defined (even for infinite system size) and σ_f becomes volume independent (as in the equal load sharing fiber bundle model; see Alava et al. (2009a) for a review on size effect of failure strength).

We have also discussed the molecular dynamics route in simulating fracture. We have compared the simulations with Griffith's theory and have also discussed how by controlling the width of the interatomic force, the behavior of the sample changes from ductile to brittle.

Finally, we discuss the fiber bundle model, a limiting case of which shows stress nucleation and extreme statistics. In the particular case of local load sharing model, the failure strength decays as $\sigma_f \sim 1/(\ln N)^a$. Hence, for any finite load in a large system, size limit will find the weakest patch in the system and will get concentrated around it, leading to a catastrophic failure. A crossover between this extreme statistics for local load sharing case and self-averaging statistics of global load sharing case was also discussed. This can be done in various ways sec. 4.3.1.2. However, a common feature is that in all the cases, the models behave like either of the two extremes (local or global load sharing), but no intermediate, space dimension-dependent behavior was found.

5
Roughness of Fracture Surfaces

Fracture propagation is crucially dependent on the presence of defects, even at the microscopic scale. The global responses, namely the path of the crack propagation and the sudden release of energy while the crack propagates, are manifestations of the presence of defects. The macroscale nonuniformity in the path of the fracture brings the roughness in the fractured surface. For a long time, fracture surfaces are known to have scale- free structure, that is, there is no specific scale of unevenness in the morphology of the fracture surfaces. This suggests that the deviation in the path of crack propagation can be of significantly larger scale than that of the typical scale of the defects. In some sense, this scale-free nature is similar to critical fluctuations in a system, which can in principle be as large as the system size, in spite of the fact that the interaction length may just be nearest neighbors. This similarity with critical phenomena has attracted much attention to provide the description of fracture as a dynamical critical phenomenon (Herrmann and Roux, 1990; Chakrabarti and Benguigui, 1997; Bonamy and Bouchaud, 2011).

In this chapter, we will mainly focus on the "postmortem" analysis of fracture dynamics, that is, we will look into the "footprints" left by the crack front and try to find what information it can provide regarding the dynamics of fracture, and discuss the presence and absence of universality.

5.1
Roughness Properties in Fracture

Some understanding of the meandering nature of the path of propagation of fracture is obtained from the "postmortem" analysis of the roughness of the fractured surface. It was Mandelbrot *et al.* (1984) who first suggested that the roughness of the fracture surfaces is self-affine with a roughness exponent which is independent of the material (see Figure 5.1). This work and subsequent observation of this "universality" prompted active research in this line for several decades. In this chapter, we first describe the properties of self-affine surfaces and their scaling forms. We then describe the experimental findings in support of the universality and also other works that question it. Finally, we describe the

NATURE VOL. 308 19 APRIL 1984 ———————————————LETTERS

Fractal character of fracture surfaces of metals

Benoit B. Mandelbrot*, Dann E. Passoja†
& Alvin J. Paullay‡

* IBM Research Center, Yorktown Heights, New York 10598, USA
† Union Carbide Corporation, Tarrytown, New York 10591, USA
‡ Columbia University New York, Bronx Community College, Bronx, New York 10753, USA

When a piece of metal is fractured either by tensile or impact loading (pulling or hitting), the fracture surface that is formed is rough and irregular. Its shape is affected by the metal's microstructure (such as grains, inclusions and precipitates, whose characteristic length is large relative to the atomic scale), as well as by 'macrostructural' influences (such as the size, the shape of the specimen, and the notch from which the fracture begins). However, repeated observation at various magnifications also reveals a variety of additional structures that fall between the 'micro' and the 'macro' and have not yet been described satisfactorily in a systematic manner. The experiments reported here reveal the existence of broad and clearly distinct zone of intermediate scales in which the structure is modelled very well by a fractal surface. A new method, slit island analysis, is introduced to estimate the basic quantity called the fractal dimension, D. The estimate is shown to agree with the value obtained by fracture profile analysis, a spectral method. Finally, D is shown to be a measure of toughness in metals.

Figure 5.1 Benoit B. Mandelbrot (1924–2010) was a mathematician who developed the field of fractal geometry. He coined the term "fractal" for objects that are equally rough in all scales (having noninteger or fractional dimension). Among diverse areas, where "fractal" objects are found in nature, he showed that it also exists in the roughness properties of the fracture surfaces. He, along with his colleagues, first proposed the idea of a universal roughness exponent for materials that are very different from each other in their failure characteristics. [Abstract from, Mandelbrot et al. (1984); with kind permission from the Nature publishing group].

theoretical modeling of the propagation of fracture that attempts to explain the roughness properties of the fracture surfaces.

In a three-dimensional sample, the observations of fracture roughness can take place both in and out of plane of the propagation of fracture front (see Figure 5.2). The properties of roughness in both the cases have interesting features of their own, which we describe in the following.

5.1.1
Self-affine Scaling of Fractured Surfaces

The nature of a fracture surface is known to be self-affine (Bonamy, 2009; Bouchaud, 1997). Unlike a self-similar object, which is invariant under a global

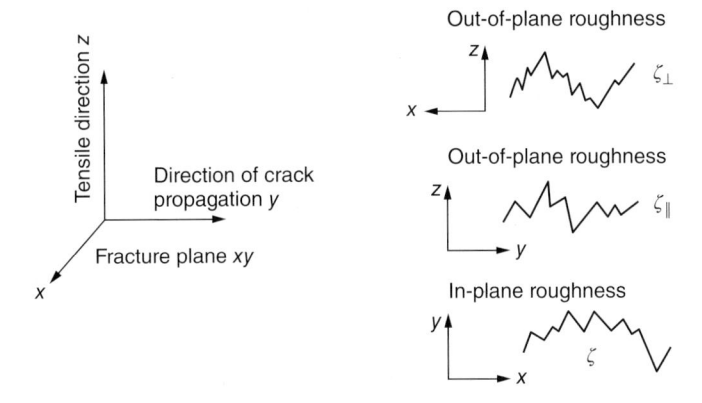

Figure 5.2 Schematic representation of in-plane and out-of-plane roughnesses. From Daguier et al. (1995) with kind permission from EDP sciences.

scale transformation in all directions, the scale transformation in self-affine objects is anisotropic in nature. Particularly, the self-affine object is invariant under the transformation $(x, y, z) \rightarrow (\lambda x, \lambda y, \lambda^\zeta z)$, where ζ is called the roughness exponent. There are many ways by which one can estimate the roughness exponent (see Schmittbuhl et al., 1995c for comparisons between different ways). One simple way to determine ζ is to measure the two-point correlation function, which follows

$$w(\delta) = \langle (h(r+\delta) - h(r))^2 \rangle_r^{1/2} \sim \delta^\zeta. \tag{5.1}$$

Here, the averaging $\langle .. \rangle_r$ is on different initial point r, and h is along z-direction. This relation is valid for self-affine surfaces; however, it does not guarantee self-affinity of a surface. There are other ways to determine the roughness exponent; one of them is the probability of return to zero values, that is, the probability that the surface returns to its initial z value after a distance δ. This scales as

$$p_0(\delta) \sim \delta^{-\zeta}. \tag{5.2}$$

These measurements are relevant for experimental observations of fracture roughness.

5.1.2
Out-of-plane Fracture Roughness

The first suggestion that the fractured surface might be a fractal came from Mandelbrot et al. (1984). This deals with out-of-plane roughness of the surfaces, that is, the height fluctuation in the direction perpendicular to the plane of fracture. They worked with metallic steel and observed the fractal properties. The basic procedure consisted of plating the fractured surface with a different material, then cutting it along the direction of average position of the fractured surface. Then, islands of nonplated regions appear, signifying the roughness of the surface. The area and boundaries of these islands were measured to determine the roughness exponent. The value of the roughness exponent was found to be close to 0.8.

Subsequent experiments were done to test these results. In one such early experiment, Bouchaud et al. (1990) studied roughness properties of aluminum alloys for different fracture modes and toughness values. In the commercial aluminum alloys used, two competing fracture modes exist. Either it happens by void formation at coarse constituents followed by transgranular rupture or ductile intergranular rupture. These modes were accessed by varying the temperature and quench rates. In addition, the toughness of these materials showed a variation from 28 to 45 MPa$\sqrt{m}$. As in the earlier case, the sample was then electroplated with Ni, and polished along the direction of average height of the fracture surface until islands of unplated regions appear. The fractal properties of these islands were measured by various correlation functions. The roughness exponents obtained for samples with different toughness as well as different fracture modes were all close to 0.8 with no systematic variation. This led the authors to conclude that the existence of universal roughness exponent is ~ 0.8 for fracture surfaces.

Subsequent experimental observations, too, supported this universal behavior. In Måløy et al. (1992), the authors measured the roughness of fracture surfaces in different materials such as aluminum alloy, steel, graphite, porcelain, Bakelite, and plaster of paris. Roughness exponent was estimated mainly by two methods: one is the power spectrum, which is the Fourier transform of the height correlation function and behaves as $P(f) \sim f^{-1-2\zeta}$ and the other was the zero return probability $p_0(r)$ mentioned earlier. In Figure 5.3, the power spectrum for the six samples and in Figure 5.4, the zero return probabilities of the same were plotted. As can be

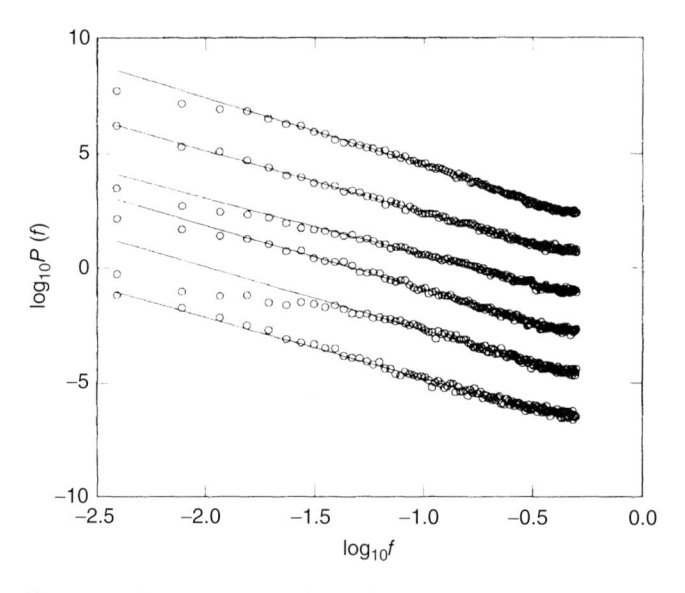

Figure 5.3 Power spectrum $P(f)$ as a function of frequency f for the six samples. From Måløy et al. (1992), with kind permission from American Physical Society.

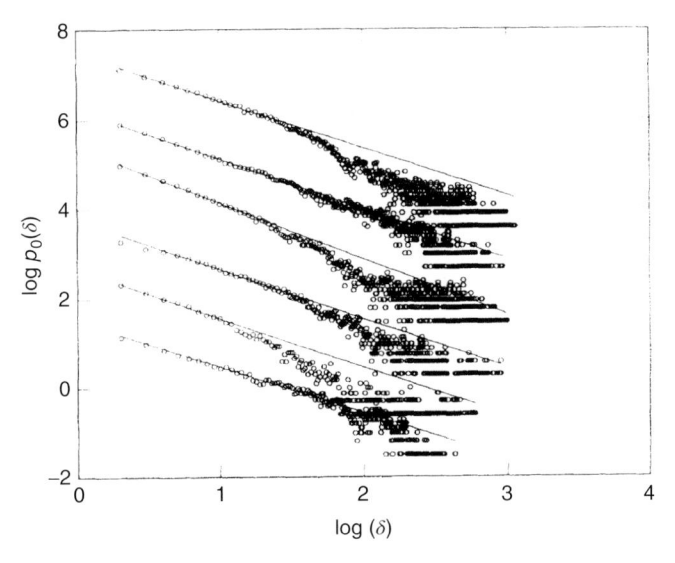

Figure 5.4 The zero return probabilities for the six samples. From Måløy et al. (1992), with kind permission from American Physical Society.

seen from the parallel nature in the log–log scale, there is no systematic variation in the roughness exponent values for the six samples. The average value of these estimates gives $\zeta = 0.87 \pm 0.07$, consistent with the earlier estimates. There have been many other studies, which support this claim of universality; for example, in Daguier et al. (1997) Ti$_3$ Al showed roughness scaling over five decades of length scale giving roughness exponent close to 0.8. Similar roughness behavior was found in rock fractures (Schmittbuhl et al., 1995b).

5.1.3
Distribution of Roughness: Mono- and Multi-affinity

The roughness properties can be much more richer than what is expressed in Eq. (5.1) (Bouchbinder et al., 2006). If one looks at the higher moments, for example, $\langle (h(r+\delta) - h(r))^k \rangle_r^{1/k} \sim \delta^{\zeta_p}$, how would the exponent change with k? To answer this question, one has to look beyond the pair correlation function and generally analyze the full distribution function of the roughness and its scaling properties. The distribution $P_\delta(\Delta h)$ of the height increments $\Delta h = h(r+\delta) - h(r)$ between two points separated by a distance δ ($r = \sqrt{x^2 + y^2}$) for self-affine surfaces scales as

$$P_\delta(\Delta h) = \delta^{-\zeta} f(\Delta h / \delta^\zeta). \tag{5.3}$$

For self-affine surfaces, $\zeta_k = k\zeta$. For multi-affine surfaces, ζ_k can be a complicated function of ζ, and the shape of the distribution function differs from the Gaussian shape. In Bouchbinder et al. (2006), such departures from mono-affinity were noted for two-dimensional fracture fronts. Later on, the problem was revisited in

Santucci et al. (2007), where various samples broken in various modes were considered both in two and three dimensions. These include fractured surfaces of granite block broken in mode-I, the propagation of quasi-two-dimensional fracture front in transparent sandblasted plexiglass (PMMA) plates, and fracture produced by tearing papers. The distributions of roughness for different materials have been found (Santucci et al., 2007) to be Gaussian and mono-affine above a certain length, that is, the distribution function can be fully characterized by a single roughness exponent.

Particularly, the authors aimed at finding the distribution function of the height fluctuation $P[h(x + \delta) - h(x)]$ and comparing it with a Gaussian distribution function. In the Comparison stage, one can define the ratio as

$$R_k(\delta) = \frac{\langle |h(x + \delta) - h(x)|^k \rangle_x^{1/k}}{\langle |h(x + \delta) - h(x)|^2 \rangle_x^{1/2}}. \tag{5.4}$$

In the case of Gaussian fluctuation (Δh) distribution, one has

$$P(\Delta h) = \frac{1}{\sqrt{2\pi w^2}} e^{-(\Delta h)^2/2w^2}, \tag{5.5}$$

where $w \sim \delta^\zeta$. Now, the ratio mentioned above takes a particular value for the Gaussian distribution, which is independent of δ and ζ. This ratio becomes

$$R_k^G = \sqrt{2} \left[\frac{\Gamma\left(\frac{k+1}{2}\right)}{\sqrt{\pi}} \right]^{1/k}. \tag{5.6}$$

One can now compute $R_k(\delta)/R_k^G$ for different fractured surfaces and compare its proximity to the Gaussian statistics. The comparison is expected to hold beyond a lower cut-off δ_0 related to the pixel size. It is seen, however, that the ratio converges to 1 beyond a crossover scale Λ. Again, at a much larger length scale, deviation from 1 is observed, which may be related to the finite size effect (see Figure 5.5). The crossover length Λ is related to the microstructures of the material and indeed scale with the respective microstructure sizes (see also Santucci et al. (2010); numerical simulations controlling the disorder length scale arrive at the same conclusion (Laurson and Zapperi, 2010)). Direct measurement of the fluctuation distribution function shows that it is Gaussian beyond a length scale Λ and below it, deviation from Gaussian behavior was noted (see Figure 5.6) suggesting multiscaling properties at small length scales, which crosses over to a single roughness exponent at large length scales.

5.1.3.1 Nonuniversal Cases
In contrast to the widely observed universal behavior of the roughness of fracture surfaces, there have been some incidences where the roughness exponent deviated from 0.8. Particularly, in sandstones, a much smaller value of the roughness exponent was observed (Boffa et al., 2000). The roughness was measured differently than the usual methods. The rough surface was illuminated in the grazing

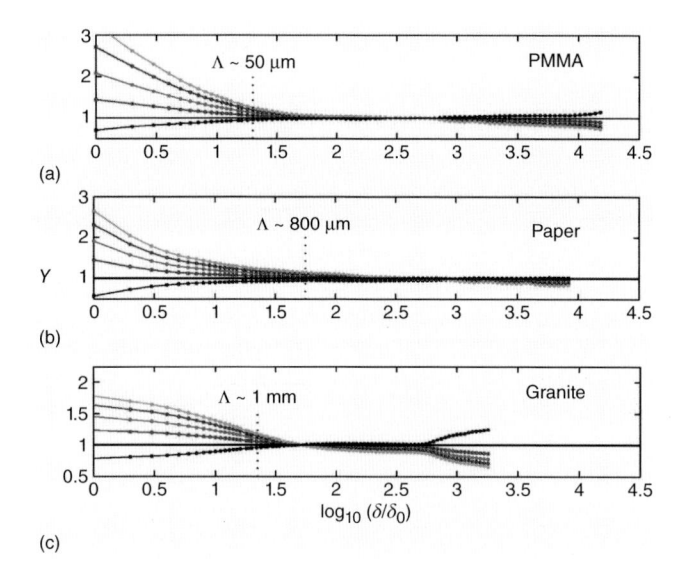

Figure 5.5 The convergence of moment ratios to unity beyond a length scale Λ, related to the microstructures of the respective materials. The fanning toward the end is due to finite size effect. From Santucci et al. (2007), with kind permission from American Physical Society.

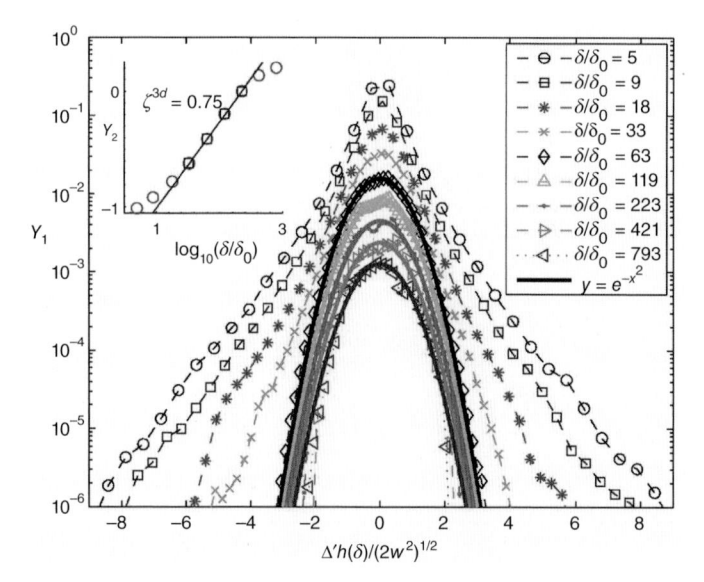

Figure 5.6 The scaled distribution function for the height fluctuations showing Gaussian behavior (parabolic shape) beyond a cut-off scale Λ. At a smaller scale, multiscaling is observed. From Santucci et al. (2007), with kind permission from American Physical Society.

angle and the distribution if the length of shadows were measured. It is known that (Hansen et al., 1995) the distribution function behaves as

$$P(l) \sim l^{-1-\zeta}, \tag{5.7}$$

where l is the length of the shadows, up to a cut-off length l^*, which is determined by the incidence angle of the light and the amplitude of roughness. These measurements show that the roughness exponent for sandstone is around 0.47, which is much smaller than the universally obtained (as well as in basalt checked in this method) value, 0.8.

Similar small exponent values were seen for sintered glass (Ponson et al., 2006a), where fracture surface roughnesses for different porosities, bead sizes, and crack speeds were measured. The roughness exponent was (0.40 ± 0.04) found to be independent of these porosities, bead sizes, and crack speeds (see Figure 5.7). However, the fracture mode showed transition from transgranular to intergranular propagation.

These two sets of exponents, observed consistently for different materials, suggest the existence of another universality class in terms of roughness of fractured surfaces. In Bonamy et al. (2006), it was proposed that the two types of exponents arise from the length scales in which one measures the roughness properties. Particularly, if one is within the fracture process zones (FPZ), where the crack propagates due to coalescence of damages, the elastic interface modeling (Fisher, 1985) of fracture front is no longer valid, and one gets a much higher roughness exponent value. Hence, damage screens the effect of elasticity in this region. On the other hand, if one is beyond the scale of FPZ or in the cases where no FPZ

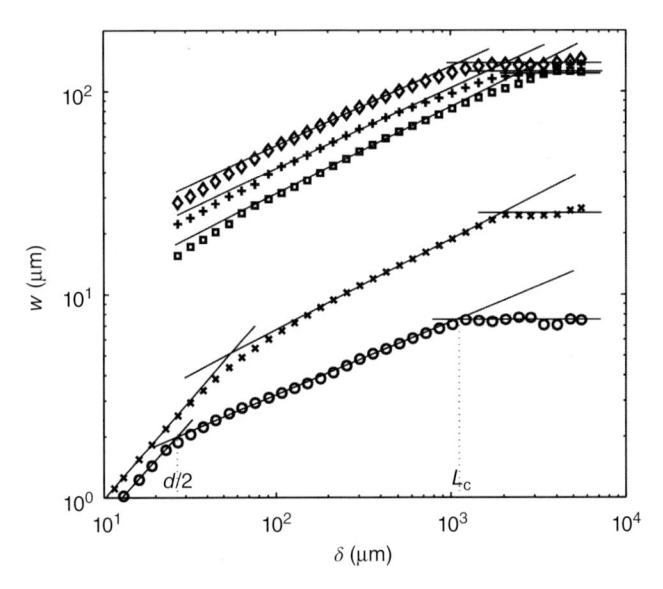

Figure 5.7 The roughness exponent for sintered glass for different bead sizes and velocities. The roughness exponent is close to 0.4. From Ponson et al. (2006a), with kind permission from American Physical Society.

5.1 Roughness Properties in Fracture

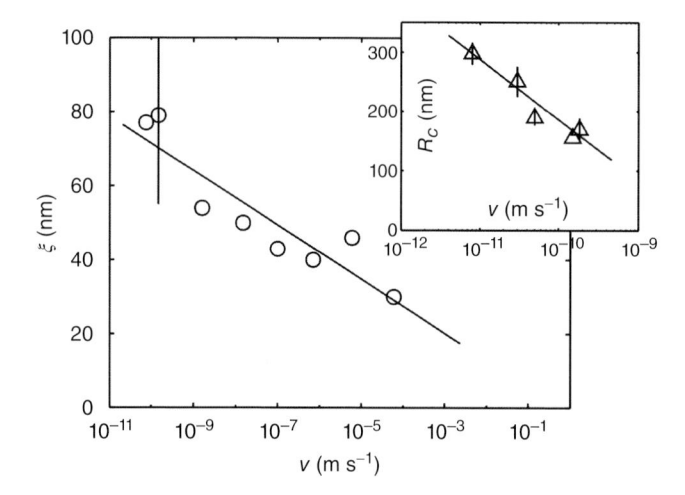

Figure 5.8 The variation of cut-off scales of the roughness exponent observed in silica glass. Power-law was obtained below ξ. The inset shows the variation of the FPZ size R_c measured along x-axis. Both the quantities vary as log (v), indicating that FPZ is the relevant cut-off scale. From Bonamy et al. (2006), with kind permission from American Physical Society.

develops, the elastic interface approach is valid and a much smaller roughness exponent is obtained.

The scaling of the height–height correlation functions was measured in homogeneous glass and glassy ceramics. It was found that the roughness exponent is close to 0.75 in homogeneous glass and 0.4 in glassy ceramics. In the case of homogeneous glass, the FPZ extends up to about 100 nm and varied logarithmically with the crack velocity. It was also seen that the cut-off scale, up to which the roughness exponent (0.75) was obtained also scaled logarithmically with the crack velocity (varied in the range 10^{-11} to 10^{-9} m s^{-1}) (see Figure 5.8). Indeed, FPZ was found to be the relevant cut-off scale.

In the case of glassy ceramics, however, the situation is different. The FPZ is again of the order of 100 nm, but the microstructure scale is extended up to the mean size of the beads, which is about 100 µm. The scaling of the roughness in this case, therefore, falls in the elastic range and has a value of 0.4, which is much smaller than the one obtained in the case of measurement within FPZ.

Similar observation was confirmed in Ponson et al. (2007), where the authors studied the roughness properties of Fontainebleau sandstones. The grain sizes of the material ranged from 100 µm to 500 µm. The roughness exponent showed a value about 0.43 up to the scale of 100 µm. Therefore, the FPZ, which is generally present near the fracture tip, did not develop here. The entire region, from the grain sizes to a scale limited by the system size, was dominated by brittle fracture and a smaller roughness exponent was found. The authors show further that the distribution of roughness is Gaussian in nature. Therefore, all moments $\langle [h(x+\delta) - h(x)]^k \rangle_x^{1/k}$ for the distribution function of the roughness should scale

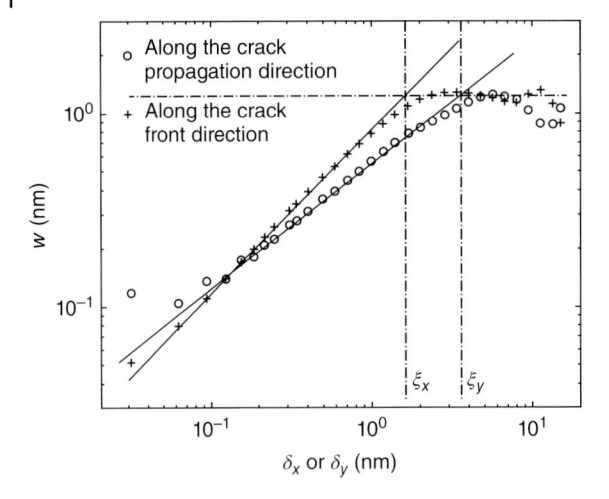

Figure 5.9 The one-dimensional height correlation function measured along and perpendicular to the crack propagation direction in quasi-crystals. From Ponson et al. (2006c), with kind permission from American Physical Society.

as δ^ζ, where ζ is the roughness exponent independent of the moment k. This absence of multi-affinity was another evidence of nonformation of process zones (see Section 5.1.3).

These studies, therefore, conclude that the value of the roughness exponent ζ indicates the failure mode of the material. For damage failure involving a process zone, the roughness exponent is 0.8 and for brittle failure, it is about 0.4 (which matches with predictions from elastic line depinning model, to be discussed in Chapter 8). The damage failure mode can be observed within a length scale from the grain size to the FPZ. The brittle failure will extend from the FPZ to the length scale of the system size and in the case of absence of FPZ, it will span the entire length scale from grain size to specimen size (see also Dalmas et al., 2008).

5.1.3.2 Anisotropic Scaling

In spite of the claim of universal roughness exponent for all materials (the two values mentioned above), there are observations that report a spread in the values of roughness exponents. In one way of explaining the spread, in Ponson et al. (2006d), the authors proposed that in two-dimensional samples, the roughness properties can be anisotropic. Particularly, if one measures roughness along the direction of the crack front and that of the crack propagation (i.e., perpendicular to the crack front) in both cases, self-affine roughness are seen, but with different roughness exponents (Bonamy, 2009; Ponson et al., 2006b,c), which could be the possible cause of the spread in the measures.

They had chosen silica glass and a metallic alloy and studied the mode-I fracture for a large velocity range. The crack is along x-direction and the direction of propagation is y. They measured the height–height correlation function in x-direction as $w(\delta_x) = \langle [h(x + \delta_x, y) - h(x, y)]^2 \rangle_{x,y}^{1/2}$ and that along y- direction as

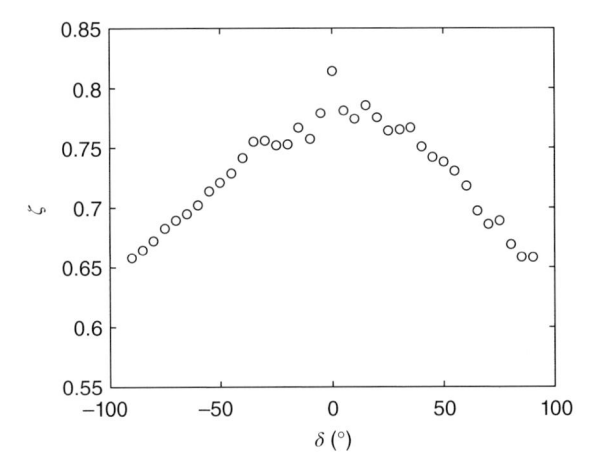

Figure 5.10 The variation of the roughness exponent for measurements along different angles from the crack front direction. From Ponson et al. (2006c), with kind permission from American Physical Society.

$w(\delta_y) = \langle [h(x, y + \delta_y) - h(x, y)]^2 \rangle_{x,y}^{1/2}$. Both of these showed self-affine behavior, but with different roughness exponents (see Figure 5.9 for this property in quasi-crystals). While along the crack front (perpendicular to the crack propagation direction) the roughness exponent was found to be close to 0.8, in the perpendicular direction to the crack front (parallel to crack front propagation direction) it is much smaller, 0.6. One can indeed measure the variation of the roughness exponent with the angle θ to the crack propagation direction. As can be seen from Figure 5.10, there is a continuous variation with the two extreme values $\zeta_\parallel \approx 0.6$ and $\zeta_\perp \approx 0.8$. The full characterization of the fracture roughness thus requires a two-dimensional height correlation function

$$w(\delta_x, \delta_y) = \langle (h(x + \delta_x, y + \delta_y) - h(x, y))^2 \rangle_{x,y}^{1/2}. \tag{5.8}$$

This quantity shows the Family–Vicsek scaling (Family and Vicsek, 1985; Ponson et al., 2006d)

$$w(\delta_x, \delta_y) \sim \delta_y^{\zeta_\parallel} f\left(\delta_x / \delta_y^{\zeta_\parallel/\zeta_\perp}\right), \tag{5.9}$$

where

$$f(u) \sim 1 \quad \text{if} \quad u \ll 1$$
$$\sim u^{\zeta_\perp} \quad \text{if} \quad u \gg 1.$$

The two exponents obtained in the process are found to be universal with respect to different materials, different modes of fracture, and also for a wide range of crack velocity.

The distances δ_y and δ_x can be scaled by topothesies l_y and l_x, respectively, defined as the scale at which the out-of-plane increment becomes equal to the in-plane increment, that is, $w(\delta_x = 0, \delta_y = l_y) = l_y$ and $w(\delta_x = l_x, \delta_y = 0) = l_x$.

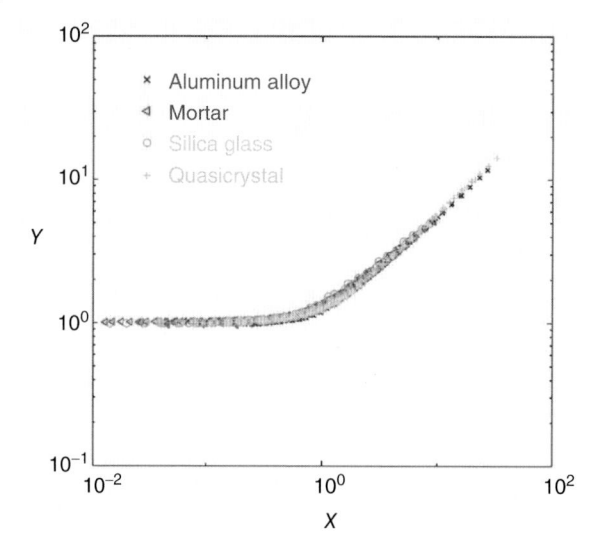

Figure 5.11 Data collapse for the dimension-less two-dimensional height correlation functions for different materials, showing universality (with $X = \left(\frac{l_x}{l_y}\right)^{1/\zeta_\perp} \frac{\delta_x/l_x}{(\delta_y/l_y)^{\zeta_\parallel/\zeta_\perp}}$ and $Y = \frac{w/l_y}{(\delta_y/l_y)^{\zeta_\parallel}}$). From Ponson et al. (2006c), with kind permission from American Physical Society.

The Family–Vicsek scaling then becomes

$$w(\delta_x, \delta_y) = l_y(\delta_y/l_y)^{\zeta_\parallel} g(u), \tag{5.10}$$

with

$$u = \frac{l_x}{l_y} \frac{\delta_x/l_x}{(\delta_y/l_y)^{\zeta_\parallel/\zeta_\perp}}, \tag{5.11}$$

where

$$g(u) \sim 1 \quad \text{if} \quad u \ll 1$$
$$\sim u^{\zeta_\perp} \quad \text{if} \quad u \gg 1.$$

The form of g is then found to be universal (see Figure 5.11).

5.1.4
In-plane Roughness of Fracture Surfaces

As discussed earlier, the fracture front propagates in three dimensions. The roughness properties of a fractured surface, therefore, are studied in both in- and out-of-plane directions. The in-plane roughness is the roughness of the crack front projected along the plane orthogonal to the direction of propagation (see Fig. 5.2). Here, we describe some experimental results measuring the in-plane roughness properties of fractured surfaces.

Daguier et al. (1995) studied the in-plane roughness properties of two materials: 8090 aluminum–lithium alloy and super α_2 Ti_3 Al-based material. While the former is highly anisotropic, the latter is not. In order to visualize the fracture front, China ink was impregnated under vacuum in the pinned condition. The crack front was observed after completion of the fracture process using the scanning electron microscope. The measured quantity was the average maximum height $h_{max}(\delta)$ (h is measured along y-direction, i.e., direction of crack propagation) within a window δ which scales as $h_{max}(\delta) \sim \delta^{\zeta}$, where ζ is the roughness exponent. For the two materials, the exponent values were found to be 0.6 and 0.54, respectively. These are significantly different from the theoretical prediction from the fluctuating line model.

Measuring the roughness of a crack front is difficult due to the fact that the front is not visible. The earlier experiment (Daguier et al., 1995) measured the crack front roughness after the fracture was completed. Schmittbuhl and Måløy (1997) did the first experiment, where a crack front was studied while it the crack was propagating (see also Måløy and Schmittbuhl, 2001). This was later refined in Delaplace et al. (1999). Here, two transparent polymethylmethacrylate GS (PMMA) plates were taken and one side of each plate was sandblasted. This introduces disorders of micrometer order. Then the two plates were annealed together to get a sample with a easy plane through which the fracture can propagate. Even though the sandblasting procedure destroys the transparency of the plates, it is recovered when the two plates are annealed at high temperature. However, when the fracture propagates (i.e., the two plates are separated), the plates become opaque again (due to difference in refractive indices), and therefore the boundary between the transparent and opaque regions is the fracture front. This front can now be viewed from the top and its statistics can be easily recorded. The images were treated (islands and overhangs removed) to get a single-valued fracture front (see Figure 5.12). The roughness exponent was measured in several different

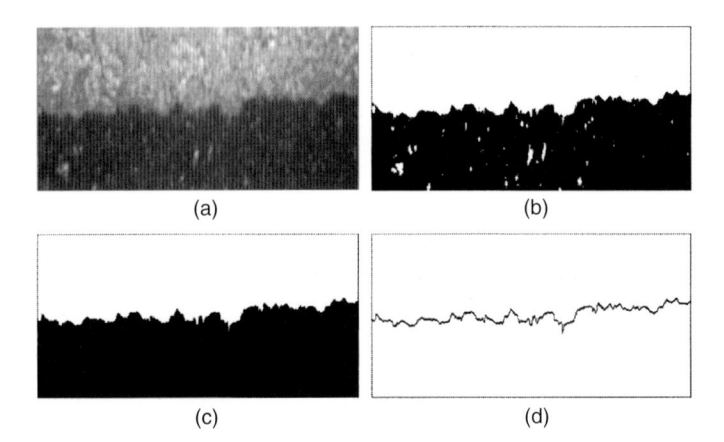

Figure 5.12 Experimental observation of fracture front. (a) raw image, (b) image after threshold, (c) image after removing the islands, and (d) the extracted fracture front. From Delaplace et al. (1999), with kind permission from American Physical Society.

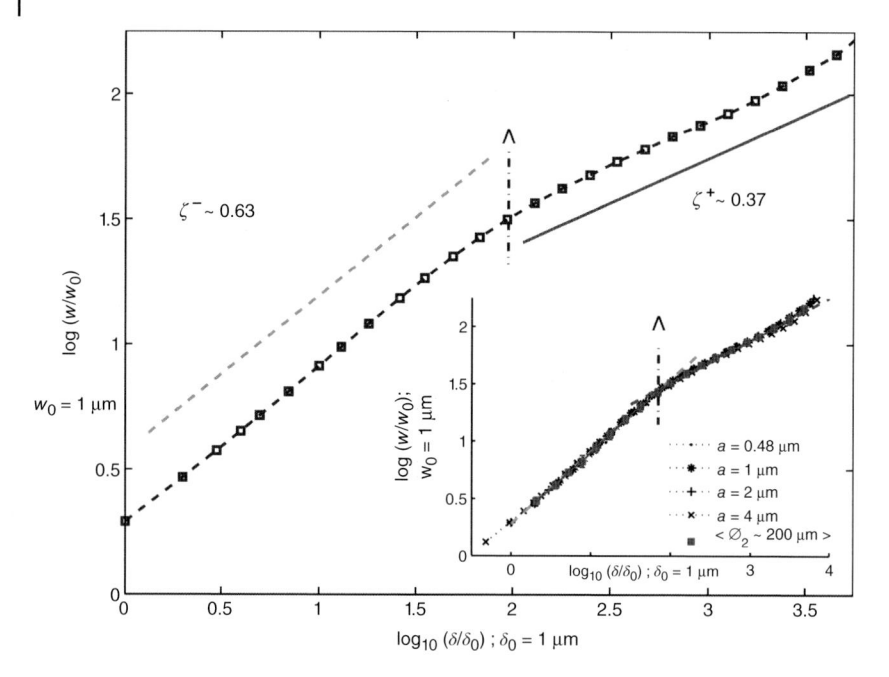

Figure 5.13 The scaling of the width of the fracture front showing two roughness exponents: ~ 0.63 below the characteristic length scale 100 μm and ~ 0.37 for larger length scales. The inset shows that the exponent values are independent of the image resolutions. From Santucci et al. (2010), with kind permission from EPL.

ways: (i) the variable bandwidth method, where a standard deviation w of the fluctuating surface was measured over a length segment l and was averaged over all positions of the segment. For a self-affine surface, $w \sim l^\zeta$, where ζ is the roughness exponent, which comes out to be 0.63 ± 0.03; (ii) multi-return probability method, where horizontal parallel lines were drawn a cross the fracture front, and the probability distribution of the distances d_s, where the line meets the front. It scales as $d_s^{1-\zeta}$. This method gives $\zeta = 0.55 \pm 0.05$; (iii) The power spectrum analysis method, where $S(f) \sim f^{1-2\zeta}$ and $\zeta = 0.64 \pm 0.06$; (iv) an average wavelet coefficient method, which gives 0.64 ± 0.03 (see Schmittbuhl et al., 1995c for different methods).

The in-plane roughness exponent obtained in these studies, however, is not the only value that is obtained. Particularly, a much lower exponent value (close to 0.35) was obtained in larger length scales. As we see, this value is closer to the theoretical estimates of roughness exponent using a fluctuating line model. It is argued that experimental observation of both the values (0.35 and 0.6) are possible, but they are valid in different length scales. While in the small length scales the coalescence-dominated propagation leads to an exponent value close to 0.6, in the large length scale limit, the elastic line depinning picture is valid and correspondingly a lower value of roughness exponent can be obtained.

5.1 Roughness Properties in Fracture

In Santucci et al. (2010), the authors studied fracture front propagation between two PMMA plates. The two plates were sandblasted, as mentioned earlier, to create an easy plane for fracture propagation. The size and type of the blasting particles were widely varied to check the universality of the observation. The loading procedure, propagation velocities, and resolution of the front were also varied.

The main measured quantity was the height fluctuation, $\Delta h(\delta) = h(x+\delta) - h(x)$. The scaling properties were extracted from the behavior of the width $w(\delta) = \langle \Delta h^2(r) \rangle^{1/2}$, which is expected to scale as $w(\delta) \sim \delta^\zeta$. In the particular case where the sample was prepared with 200 µm glass beads, the roughness exponent was found to be 0.60 ± 0.05 below the length scale 100 µm. However, in the length scale higher than this, the roughness exponent was found to be 0.35 ± 0.05. (see Figures 5.13 and 5.14)

Furthermore, the probability distribution $P(\Delta' h)$ of the height fluctuation $\Delta' h(\delta) = \Delta h(\delta) - \langle \Delta h \rangle$ was also studied by computing its structure functions $C_k(\delta) = \langle |\Delta h(\delta)|^k \rangle_x^{1/k}$. It is seen that the distribution function is Gaussian above the length scale $\Lambda \approx 100$ µm, signifying a mono-affine behavior. The structure function $C_k(\delta)$, when normalized by the set of values $R_k^G = \sqrt{2}(\Gamma(\frac{k+1}{2})/\sqrt{\pi})^{1/k}$ corresponding to the ratios $R_k^G = C_k^G(\delta)/C_2^G(\delta)$ obtained for a Gaussian and mono-affine fluctuation, collapses to a single curve, and the roughness exponent

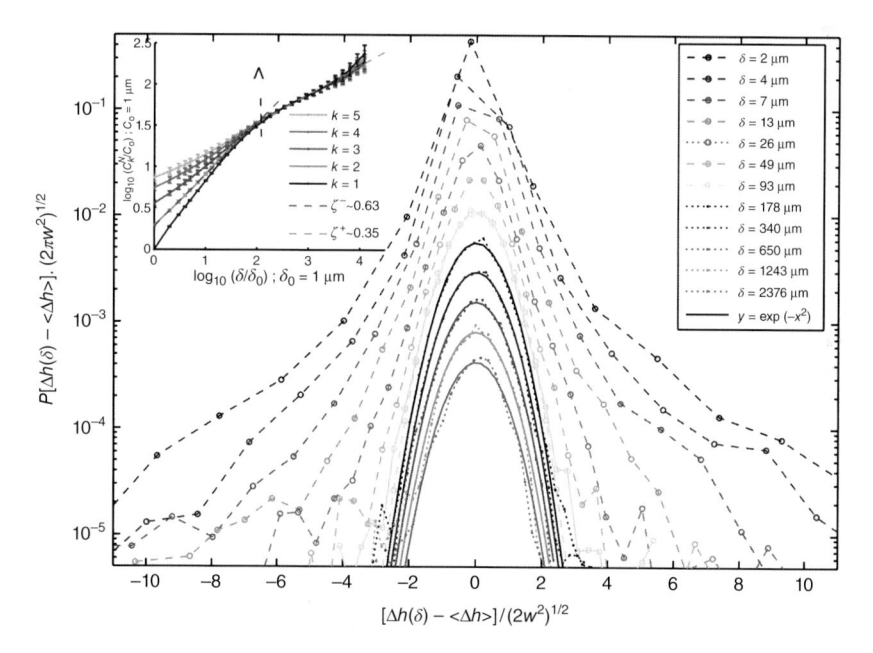

Figure 5.14 The normalized structure function shows Gaussian behavior for higher length scales and it deviates from Gaussian statistics for smaller length scales. The curves are arranged sequentially from lower (outside) to higher (inside) length scales. From Santucci et al. (2010), with kind permission from EPL.

is 0.35 ± 0.05. Below this length (~ 100 μm) scale, a clear deviation from the Gaussian statistics was noted. The fanning of the structure functions suggests an effective multi-affine surface. It was also suggested that the crossover length scale depends upon the ratio of the local stress drop to the local toughness disorder (see Section 5.1.3).

5.1.4.1 Waiting Time Distributions in Crack Propagation

In addition to the roughness properties of the fracture fronts, their intermittent dynamics are also of importance, and was studied both experimentally and theoretically. The local velocity fluctuations along the crack front and their spatial as well as temporal correlations were studied by two very different loading conditions (Måløy et al., 2006). In one case, a constant velocity drive was used, that is, the deflection between the two plates was a linear function of time $d(t) = v_p t$. In the second case, the creep motion was studied. Here, an initial displacement was given, which increased linearly with time up to some extent, and then the displacement was kept fixed. Force builds up initially and when driving stops, the crack keeps propagating by creeping and the force on the front gradually decreases.

To study the local fluctuations in velocities, a procedure called waiting time matrix (WTM) was introduced in Tallakstad et al. (2011). When one has a fluctuating propagating front represented by $h(x, t)$, one can define a matrix $H[x, h(x, t)] = 1$ and 0 elsewhere. The waiting time matrix W_T is then

$$W_T(x, y) = \sum_t H[x, h(x, t)], \tag{5.12}$$

where the sum is over all time steps.

The WMT can give a spatial structure of the local velocities. One can define a matrix $V(x, y)$, where the matrix element v represents the normal speed of the fracture front at the time it went through a particular position

$$v = \frac{a}{w_T \delta t}, \tag{5.13}$$

where w_T is the element of waiting time matrix. The local velocity along each front can now be obtained from

$$v(x, t) = V[x, h(x, t)]. \tag{5.14}$$

By computing $v(x, t)$ at every step, the spatiotemporal velocity map $V_t(x, t)$ can be formed. Then the average propagation velocity $\langle v \rangle$ of the front is defined as the average over all elements of $V_t(x, t)$. From the local velocities, one can now compute the normalized probability density function $P(v)$. When rescaled by the average velocity, all the curves from different experimental conditions collapse to a single one (see Figure 5.15). It was seen that

$$P(v/\langle v \rangle) \sim (v/\langle v \rangle)^\eta, \tag{5.15}$$

with $\eta = 2.55 \pm 0.15$ (see however Laurson et al., 2010). The data collapse in Figure 5.15 suggests quantitative equivalence between different loading conditions (constant velocity or creep). It is this fat-tailed distribution that leads to

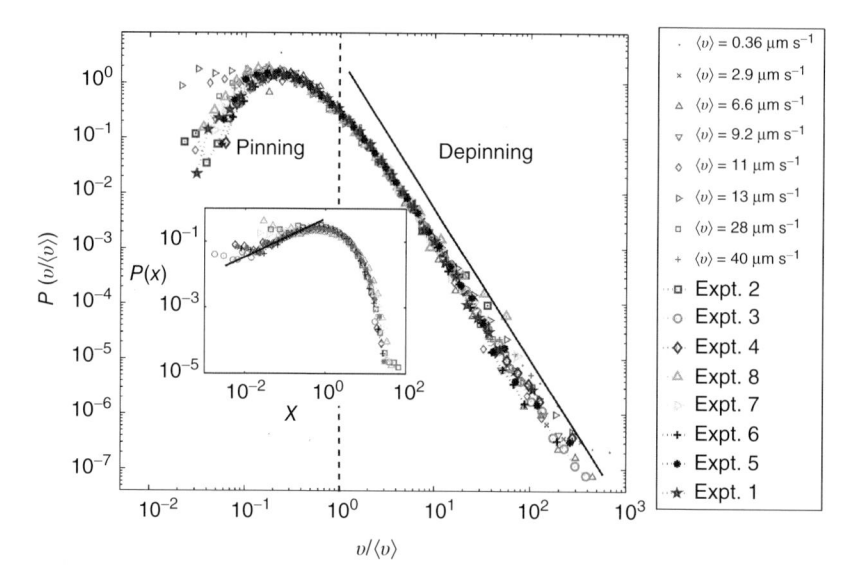

Figure 5.15 The scaled local velocity distribution function. The power-law decay is with exponent value close to 2.55. The results from different loading conditions match. From Tallakstad et al. (2011), with kind permission from American Physical Society.

breakdown of central limit theorem (CLT) while calculating the fluctuation of global velocity and predicts non-Gaussian statistics (Tallakstad et al., 2013). It was explained by invoking a generalized CLT, as the decay exponent is less than 3. The average velocity V tends to an alpha-stable Levy distribution defined by

$$\Psi(V; \alpha, \theta, c, \delta') = \frac{1}{2\pi} \int_{-\infty}^{\infty} \Phi(q) \exp(-iVq) dq, \tag{5.16}$$

with

$$\Phi(q) = \exp\left[iq\delta' - |cq|^\alpha \left\{1 + i\beta \frac{q}{|q|} \tan\left(\frac{\pi\alpha}{2}\right)\right\}\right] \tag{5.17}$$

for $\alpha \neq 1$ and $\eta = \alpha + 1$. From normalization, the parameters $\delta' = 0$ and $c = 1$ can be found. The skewness parameter has its extreme value 1 for nonnegative values of v. Thus, the distribution should converge to $\Psi(V; 1.7, 1, 1, 0)$, which it does.

To compare the spatiotemporal correlations of the local dynamics in the pinned and depinned regimes, a thresholding procedure was applied to the local velocity matrix $V(x, y)$, which gives a binary matrix V_C, which is defined as

$$\begin{aligned} V_C &= 1 \quad \text{for} \quad v \geq C\langle v \rangle \\ &= 0 \quad \text{for} \quad v < C\langle v \rangle \end{aligned} \tag{5.18}$$

for the depinned regime and

$$\begin{aligned} V_C &= 1 \quad \text{for} \quad v \leq \langle v \rangle / C \\ &= 0 \quad \text{for} \quad v > \langle v \rangle / C \end{aligned} \tag{5.19}$$

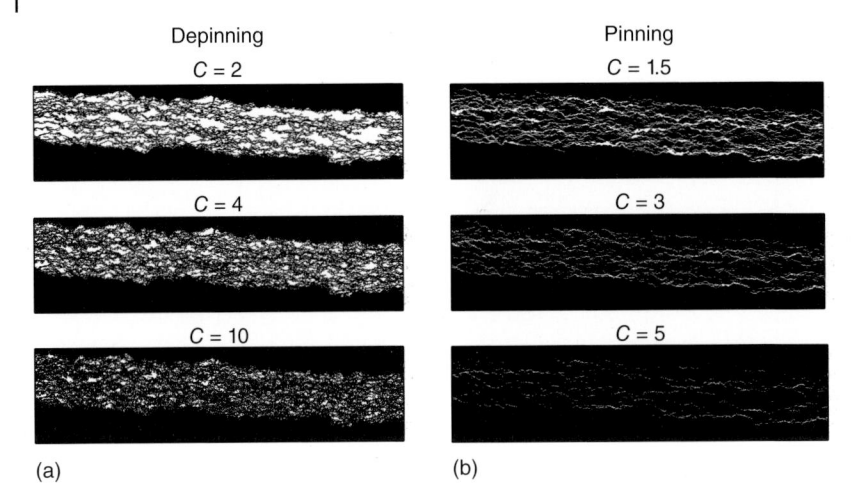

Figure 5.16 The thresholded matrix V_C for pinned (b) and depinned (a) regimes. From Tallakstad *et al.* (2011), with kind permission from American Physical Society.

for the pinned regime and C is the chosen threshold. Figure 5.16 shows an example of the thresholded matrix. It is seen that the clusters are longer along x-direction and rather narrow along y-direction. Indeed, the aspect ratio of the depinning clusters follows a power-law with exponent value 0.66. This shows that the clusters are anisotropic, and grow longer in the direction transverse to the direction of the crack propagation than along the direction of propagation.

The size distribution $P(S)$ of the clusters of size S was found. The basic form of the distribution is a power-law decay with an exponential cut-off. These are independent of the velocity as well as the loading conditions. This further indicates that the local dynamics are independent of the loading conditions. As is expected, the cut-off value decreases with the increase in the threshold C. The form of the size distribution was fitted with

$$P(S) \sim S^{-\beta} \exp(-S/S^*), \tag{5.20}$$

where S^* is the cut-off in the cluster size and $\beta = 1.56 \pm 0.04$. This is true for both the pinned and depinned regimes. As for the cut-offs, they seem to scale with the chosen threshold as $S^* \sim C^{-y_x}$, with $y_d = 1.77 \pm 0.16$ for the depinned regime and $y_p = 2.81 \pm 0.23$ for the pinned regime (see Figure 5.17).

5.1.5
Effect of Probe Size

Measurement of roughness properties of a surface is, of course, limited by the size and shape of the probe used (Lechenault *et al.*, 2010). These are particularly important while measuring crossover behaviors in smaller length scales. It was shown that given the shape and size of a probe, there exists a critical length scale

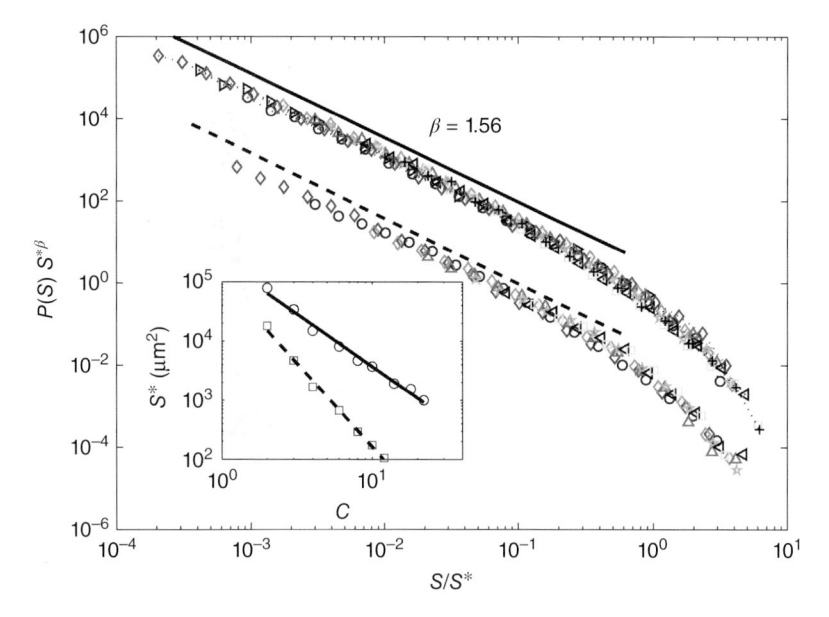

Figure 5.17 The collapsed cluster size distributions for different thresholds in pinned and depinned regimes. Inset shows the variation of cut-off in the cluster size with the threshold. From Tallakstad *et al.* (2011), with kind permission from American Physical Society.

below which one cannot accurately determine the roughness exponent. That critical length scale (δ_{x_c}) depends on the probe size (R), roughness exponent (ζ), and topothesy (l_t). Topothesy is the length scale for which the local slope of the profile becomes unity. For a self-affine surface,

$$\frac{w}{l_t} = \left(\frac{\delta_x}{l_t}\right)^\zeta. \tag{5.21}$$

Now assuming a probe shape of the form $\frac{|x|^\gamma}{2R^{\gamma-1}}$, the critical length up to which measurement is possible scales as

$$\delta_{x_c} \sim (R^{\gamma-1} l_t^{1-\zeta})^{1/(\gamma-\zeta)}. \tag{5.22}$$

The effect of the finite size of the probe is essentially to make the surface smoother. In Figure 5.18, such a surface and its changed version are shown in the inset, while the main figure shows the departure from the original roughness exponent for smaller length scale in case of the changed surface.

To see this effect experimentally, atomic force microscopy (AFM) measurements were performed on fracture surfaces in fused silica glass. The roughness exponent shows larger values of ~0.8 for smaller length scales and smaller values of ~0.2 for larger length scales. The image of the AFM tip was taken and fitted with the parameter $R = 40 \pm 5$ nm and $\gamma = 4.0 \pm 0.05$. A numerical profile with $\zeta = 0.13$ was taken and was modified using the given value of R. The profile

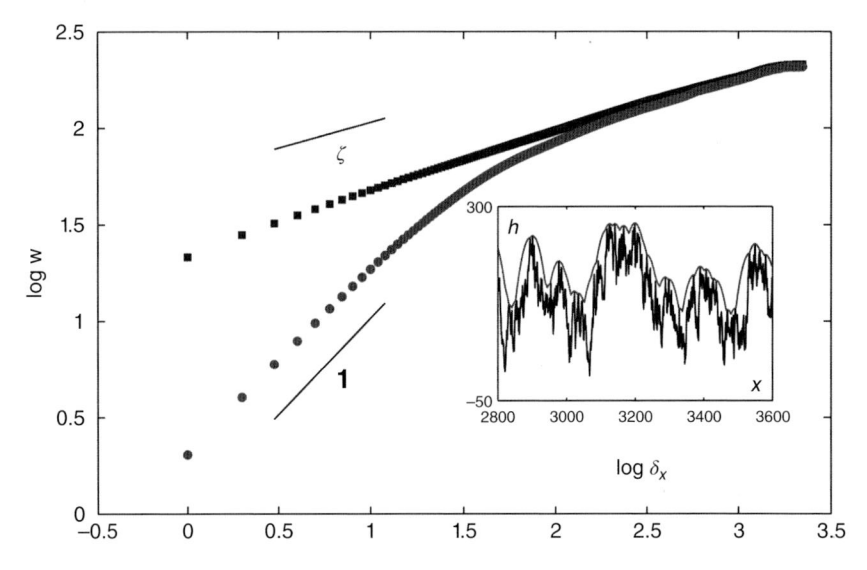

Figure 5.18 The height–height correlation function for a self-affine surface with $\zeta = 0.33$ and $l_t = 100$ and its smooth version for a finite probe size with $R = 7.8$. Inset shows the original and modified surfaces. From Lechenault et al. (2010), with kind permission from American Physical Society.

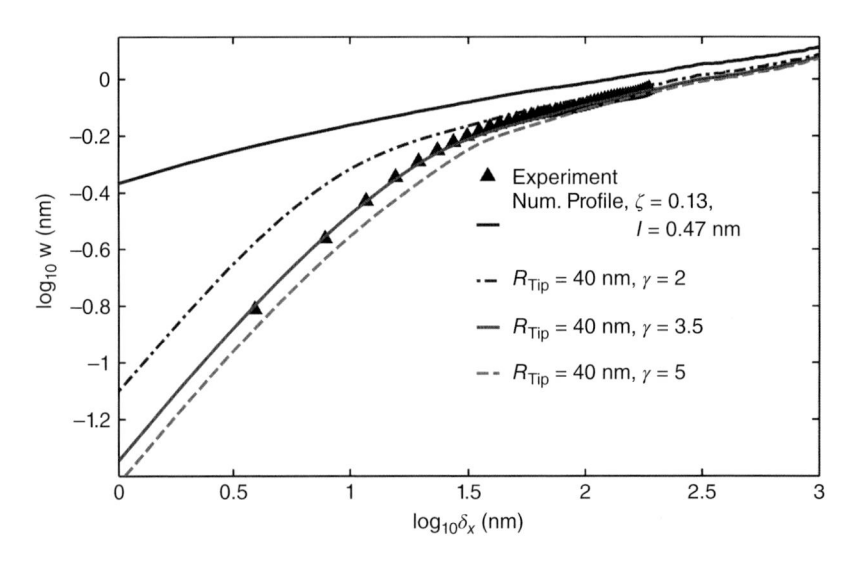

Figure 5.19 Height–height correlation function in experiment with probe size $R = 40$ nm. This is compared with a numerical profile with $\zeta = 0.13$ and its modified version with a 40 tip and different values of γ. From Lechenault et al. (2010), with kind permission from American Physical Society.

obtained was fitted with the experimental data for different γ. The best fit was for $\gamma = 3.5$ (see Figure 5.19), which is compatible with the measured value. This suggests that the small scale roughness exponent in silica glass may be due to finite size of the AFM tip. In general, large roughness exponents at small scales may suffer from this error in measurement, although in the large scale asymptotically the two results match.

5.1.6
Effect of Spatial Correlation and Anisotropy

Experimentally, anisotropy and spatial correlation of disorder can play an important role. It is, therefore, important to look into the effect of correlated as well as anisotropic disorders on the models of fracture. Particularly, the dependence of roughness exponent on such properties is of interest. In a random fuse model (Ansari-Rad et al., 2012), the effect of spatial correlation and anisotropy on the two-dimensional crack propagation and roughness exponent were studied. Because of the fact that for materials, the extended spatial correlations of many properties, such as porosity, hydraulic conductance, elastic moduli (Sahimi, 2002; Sahimi and Tajer, 2005; Knackstedt et al., 1998, 2001), and so on, are well described by fractional Brownian motion, the spatial correlation in conductances of the elements of the fuse model was also taken from fractional Brownian distribution. The power spectrum, in general, looks like

$$S(\omega) = \frac{a}{(\omega_x^2 + \omega_y^2)^{H+1}}, \tag{5.23}$$

where ω_i is the Fourier component in the ith direction and H is the Hurst index that indicates (anti)correlation for $(1/2 >)H > 1/2$, while the case $H = 1/2$ is purely random. Now, to account for anisotropy and a limiting length scale for spatial correlation, the power spectrum can be written as

$$S(\omega) = \frac{a}{(\omega_c^2 + \eta_x \omega_x^2 + \eta_y \omega_y^2)^{H+1}}, \tag{5.24}$$

where the constants η_x and η_y control the anisotropy. Particularly, setting $\eta_y = 1$ and changing $\eta_x < 1$ generate layers parallel to x-direction and vice versa. The cut-off frequency ω_c corresponds to a length scale $\xi = 1/\omega_C$ beyond which the conductances are uncorrelated.

The roughness exponents were estimated using standard methods. The crack propagation direction is, on average, parallel to y-axis. Figure 5.20 shows the effect of anisotropy on roughness exponent for different correlation lengths, with anisotropy along both x- and y-directions. As can be seen, when the anisotropy along y-axis was varied (η_y approaching zero), the roughness exponent changes from 0.7 to 0.55. On the other hand, when η_x was decreased, keeping $\eta_y = 1$, it decreases from 0.85 to 0.7. Therefore, the total range of variation in the roughness exponent is considerably high, implying a nonuniversal behavior with respect to anisotropy and spatial correlation in disorders.

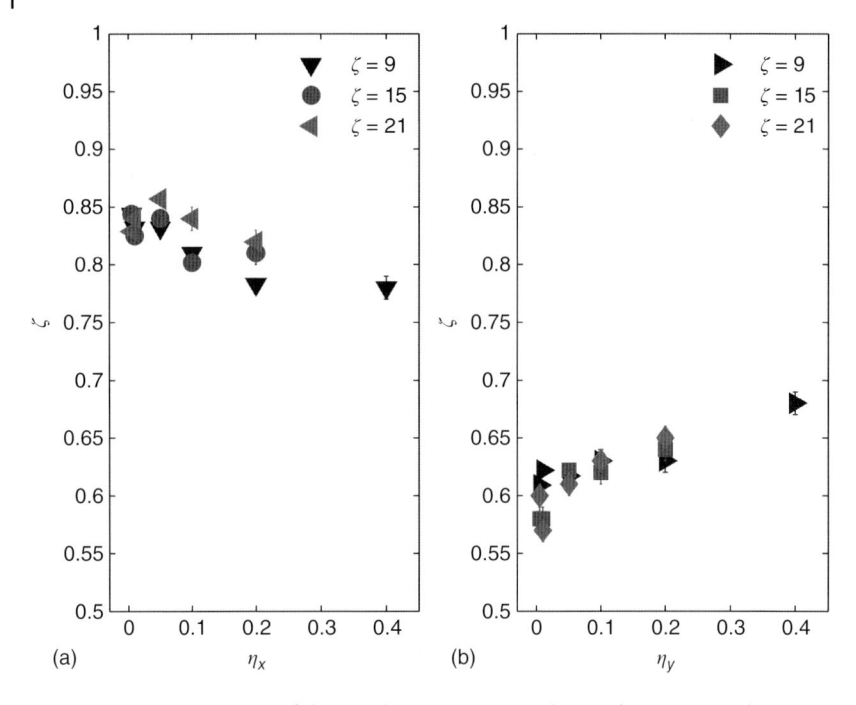

Figure 5.20 The variations of the roughness exponent values with anisotropy along y- (a) and x- (b) axes for different spatial correlation lengths (ξ). From Ansari-Rad et al. (2012), with kind permission from American Physical Society.

5.2
Molecular Dynamics Simulation of Fractured Surface

There have been many attempts to capture roughness exponent and other properties of fracture of two- and three-dimensional materials with much larger system sizes in molecular dynamics simulation studies (Kalia et al., 1997; Nakano et al., 1995; Chen et al., 2007, 2009). Mode-I fracture propagation with 130 million atoms in silicate glasses (SiO_2) has been studied. The propagation of fracture occurs through damage coalescence. Similar properties were found in other amorphous materials, such as $Si_3 N_4$ (Kalia et al., 1997). The measures of roughness exponents in these simulations of amorphous materials show interesting behaviors. The roughness exponent perpendicular to the direction of crack propagation ($\zeta_\perp$) shows a crossover behavior from 0.44 to 0.82 above a certain length scale, 25Å (Nakano et al., 1995). The roughness exponent parallel to the direction of crack propagation is $\zeta_\| = 0.75 \pm 0.08$, which is close to the experimental observations (Bonamy and Bouchaud, 2011) (Figure 5.21).

That the roughness of a fractured surface arise out of void formation ahead of crack and the tendency of the crack tip to move toward that have been proposed earlier from experimental observations (Bouchaud and Paun, 1999). Subsequent

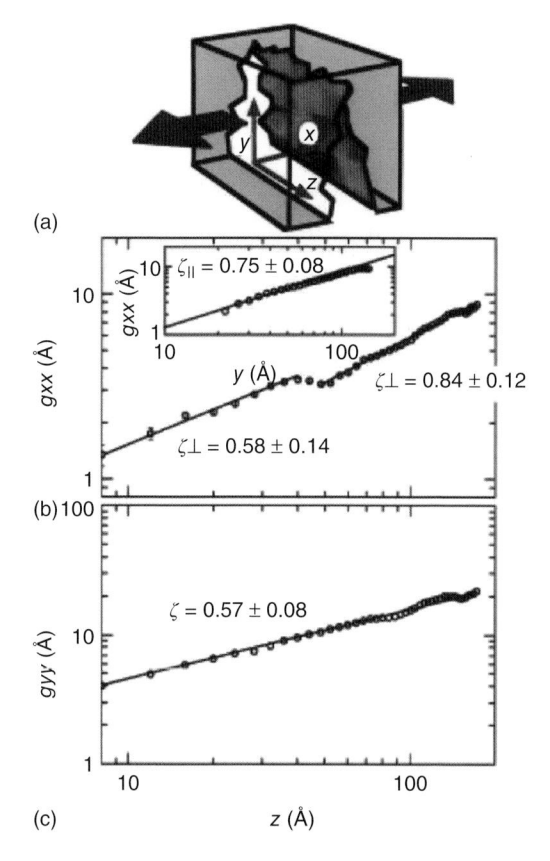

Figure 5.21 (a) Schematics of the mode-I crack propagation, (b) log–log plot of the height correlation for out-of-plane roughness in direction perpendicular to crack (inset showing that for parallel direction), and (c) the height correlation for in-plane roughness. From Kalia et al. (1997), with kind permission from American Physical Society)

attempts to explain the roughness exponent through plastic deformation (Bouchbinder et al., 2004) predicted roughness exponent higher than 0.5. This mechanism of crack propagation and subsequent roughness of the fractured surface was recently analyzed with molecular dynamics simulation (Procaccia and Zylberg, 2013) of binary mixture of amorphous material. The determination of pressure field clearly shows that the maximum pressure is formed ahead of the crack tip, thus leading to opening of voids (see Figure 5.22). As the crack propagates via coalescence of these voids, and since it can appear not only directly in front of the tip but also on either side of it, the crack is likely to be deviated, causing a rough broken surface. Furthermore, if the tip is deviated along one direction, it is more likely that a void will appear in that direction again, leading to a persistent random walk, for which the roughness exponent is greater than 0.5. Roughness exponent shows a value 0.66 in the small length scale, which crosses over to 0.5 in the longer length scale, where this effect of persistence is no longer present.

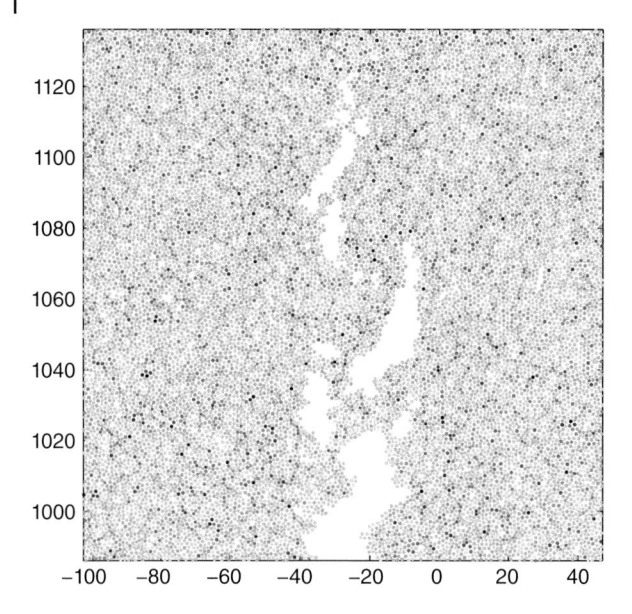

Figure 5.22 The snapshot of a crack tip propagating and voids opening in front of it. From Procaccia and Zylberg (2013), with kind permission from American Physical Society.

5.3
Summary

One of the most debated topics in fracture studies is the morphology of the fractured surfaces. Since the first suggestion by Mandelbrot *et al.* (1984), extensive studies were done both experimentally and theoretically (see Bouchaud, 1997; Bonamy, 2009; Bonamy and Bouchaud, 2011). The general observations in many different works led to a common feature of two different roughness exponent values for both in- and out-of-plane roughnesses. A crossover between a higher small length scale value (0.8) and a lower large length scale value (0.4) was seen. The crossover length scale was often related with the formation of a fracture process zone (plastic deformation) near the crack front. However, it was also suggested that the small-scale crossover could be related to the probe size used in the measurements (Lechenault *et al.*, 2010). While there is lack of agreement in the process of crack propagation, it was suggested (Schittbuhl *et al.*, 2003; Hansen and Schmittbuhl, 2003) that in the small scale, the propagation is percolation-like (coalescence) and in the large scale, the behavior is similar to elastic interface driven through a disordered medium (to be discussed in the next chapter). Finally, in view of the spread in the value of roughness exponent, the existence of self-affine scaling came under question recently (Katzav and Adda-Bedia, 2013). However, the spread could possibly come from spatial correlations and anisotropy as discussed in Section 5.1.6.

6
Avalanche Dynamics in Fracture

The scale-free characteristics of fracture are manifested in roughness of fracture front (as discussed in detail in Chapter 5) and also in avalanche statistics. In this chapter, we focus on the second topic.

When stress is increased slowly on a solid, its breaking dynamics proceeds through intermittent activities. The strength of a disordered solid may vary in different parts of it. The weak parts will break first, making the stress on the remaining parts higher, which may cause further breaking and so on, triggering what is called an avalanche. However, the avalanche may stop without breaking the solid completely. Then, the stress is to be further increased. The increase this time, however, may cause a much bigger response (in terms of breaking), since the solid is already stressed and many parts of it may already be on the verge of failure. Basically, the response of the solid may be very much nonuniform (in time) in spite of the fact that the applied stress is uniform (in time). The breaking process releases the elastic energy stored in the strained solid and it is detected mainly in the form of acoustic emissions.

We wish to distinguish two cases of avalanche dynamics: first, the gradual application of stress may lead to the complete failure of the solid (e.g., application of pressure on a rock sample) and second, where the fracture front is quasi-statically driven through the disordered solid and may in principle go on in the same manner until the boundaries of the sample are reached. The distinction is based on the time series of the response, that is, avalanche size and time. In the first case, the time series is not translationally (time) invariant and the average size of the avalanche increases as the failure is approached. In the second case, the time series is translationally (time) invariant leading to what people have often described as the self-organized dynamics. We will discuss that in Chapter 8. In this chapter, we focus on the first kind of avalanche dynamics, which leads to a failure of the solid.

The size distribution of the acoustic emissions from the solids, when stressed, follows a power-law distribution. This implies that there is no typical size of the response. Furthermore, the exponent value seems to be universal for many different materials and across widely different length and energy scales. These observations have led to developments of very simple discrete models, such as the fiber bundle model or the random spring model, to capture these features of failure

Statistical Physics of Fracture, Breakdown, and Earthquake: Effects of Disorder and Heterogeneity, First Edition.
Soumyajyoti Biswas, Purusattam Ray, and Bikas K. Chakrabarti.
© 2015 Wiley-VCH Verlag GmbH & Co. KGaA. Published 2015 by Wiley-VCH Verlag GmbH & Co. KGaA.

statistics. We describe the dynamics of these models in some detail and their predictions of the exponent values.

6.1
Probing Failure with Acoustic Emissions

A very important noninvasive probe to monitor failure properties of materials is to study the acoustic emissions from it (Petri et al., 1994; Zapperi et al., 1997b; Sethna et al., 2001). A fraction of the elastic energy released due to fracture is emitted as sound and is considered as a measure of strain (another fraction is emitted as light in mechanoluminescence of some inorganic scintillators (Tantot et al., 2013)). This process is used in failures in widely varying length scales, from laboratory scale, to earthquakes, landslides, and so on. The basic measured quantities are the size distribution of the emissions and the distribution of the waiting distribution of the acoustic emissions and duration of such events. Particularly, one has

$$P(E) \sim E^{-\beta} \tag{6.1}$$

and

$$Q(\tau) \sim \tau^{-\alpha}, \tag{6.2}$$

where E is the energy emitted in an avalanche and τ is the duration (see Figure 6.1). The estimates of the exponent values, however, vary in the ranges of $1.2 - 2$ for β and $1 - 1.3$ for α (Michlmayr et al., 2012). This may, in particular for the waiting time exponent, depend on the strain rate (for strain rate-controlled fracture), value of pressure (for creep), finite thresholding of emitted signals (Laurson et al., 2009), and so on. Even though there is no agreement as to the nature of this variation (Rosti et al., 2009), it is seen, in general, that the exponent β decreases before the catastrophic failure. This nature is seen for failures in various scales and materials, such as rock fracture, fiberglass, wood, and also in earthquakes. The precise

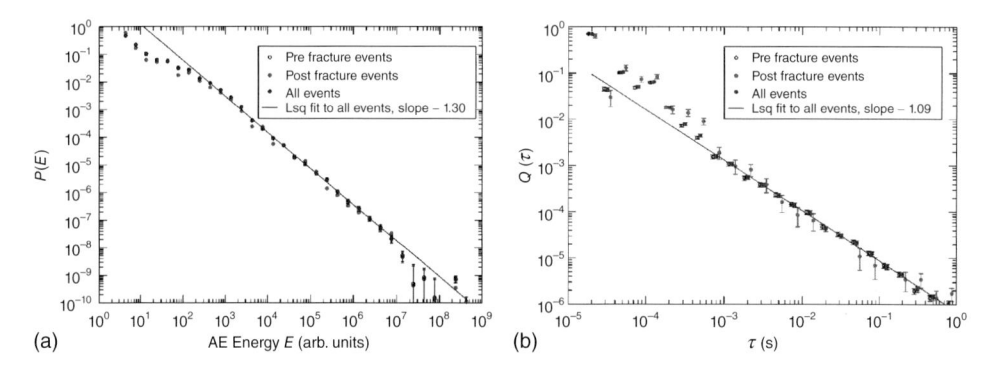

Figure 6.1 The size distribution of acoustic emission (a) and waiting time (b) showing scale-free behavior with exponent values indicated in the plot. From Rosti et al. (2009), with kind permission from IOP (UK).

analogy between acoustic emission statistics in a laboratory-scale failure experiments with brittle solids and that of earthquakes was shown in Baró *et al.* (2013). The avalanche size distribution exponent (β) value was found to be about 1.39.

There are numerous other examples of acoustic emission studies in experiments. One uses the change in the exponent value of the bursts size distribution to predict a catastrophic failure (Amitrano *et al.*, 2005). The particular field study in Amitrano *et al.* (2005) was conducted for a 1000 m^3 chalk cliff on the Normandie ocean shore in western France. Five probes were used, with a maximum distance of 50 m between them. Each probe had a geophone and an accelerometer. The measurements showed a peak of activities before the failure point was reached, much like the failure in statistical models of brittle fracture, for example, fiber bundle model. In addition, the exponent value of the size distribution of energy bursts showed significant decay before the catastrophic failure. Later on, it was shown (Cohen *et al.*, 2009) that the data fitted nicely with those taken from fiber bundle model (see Figure 6.2).

In Reiweger *et al.* (2009), the avalanche of dry snow slabs was modeled by fiber bundle model. In its simplest form, strain was developed in the system by shear rate. The fibers break beyond the failure thresholds. However, there is probability of reattachment of the fibers as well. By putting realistic model parameters, the real data were fitted with the model (see Figure 6.3).

The avalanche statistics and waiting time distributions are also widely studied for fracture front propagation (Bonamy and Bouchaud, 2011), which we shall discuss in Chapter 8.

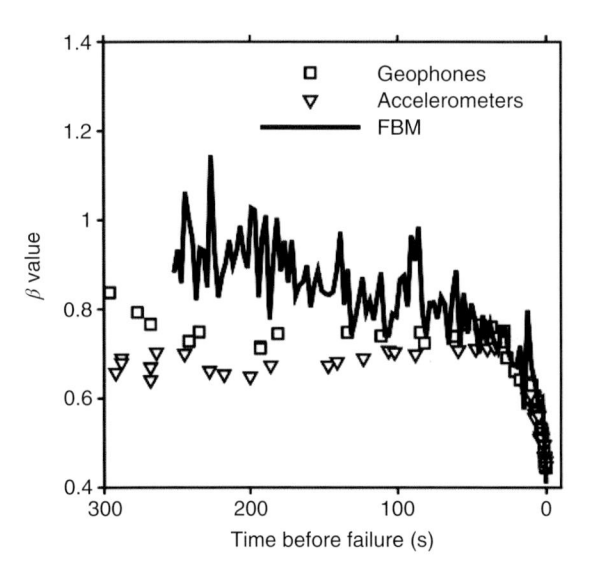

Figure 6.2 Fit of the change of energy bursts distribution exponent with time reported in Amitrano *et al.* (2005) with fiber bundle model. From Cohen *et al.* (2009), with kind permission from John Wiley and Sons.

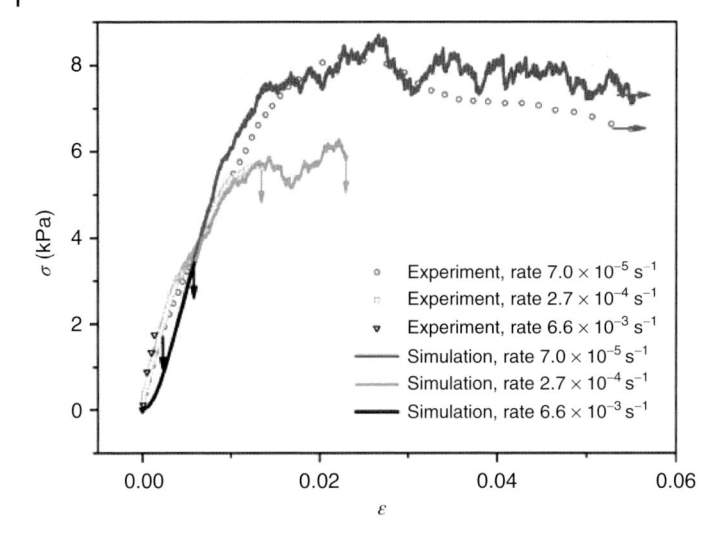

Figure 6.3 Comparison of the stress (σ)-strain (ϵ) curves of fiber bundle model for ice slab avalanche and experiment in Schweizer (1997). From Reiweger et al. (2009).

The scale-free distributions of burst sizes and waiting times (similar to Gutenberg–Richter and Omori laws of earthquakes) led to the speculation of fracture being a critical phenomenon. That in turn leads to the search of a diverging correlation length at the failure point. Let us discuss what happens to the correlation of damages. The most studied feature is the localization of damages (Lockner et al., 1991; Garcimartin et al., 1997). In particular, Garcimartin et al. (1997) and Guarino et al. (1998) measured the entropy of the positions of the damages in a quasi- two-dimensional sample (see Figure 6.4). Here, the surface of the sample is divided into grids and the pressure is increased at equal intervals up to failure point. If the number of microfractures in the ith grid is n_i, then the Shannon entropy is defined as

$$S = -\sum_i q_i \ln q_i, \qquad (6.3)$$

where $q_i = n_i/N$, with N being the total number of microfractures. The behavior of normalized entropy $s = S/S_e$ (S_e being the entropy for randomly placed N microcracks) with normalized pressure $p = P/P_f$ (P_f being the failure pressure) is shown in Figure 6.5. The linear decay quantifies the localization effect as can be seen from Figure 6.4. The decay further quantifies the proximity to failure (subject to the fact that the value of entropy is dependent on grid size logarithmically). Another quantity studied to quantify the localization is the spatial separation of microfracture events. The probability that two events are separated by a distance r decays as $r^{-1.4}$ (Maes et al., 1998).

6.1 Probing Failure with Acoustic Emissions | 73

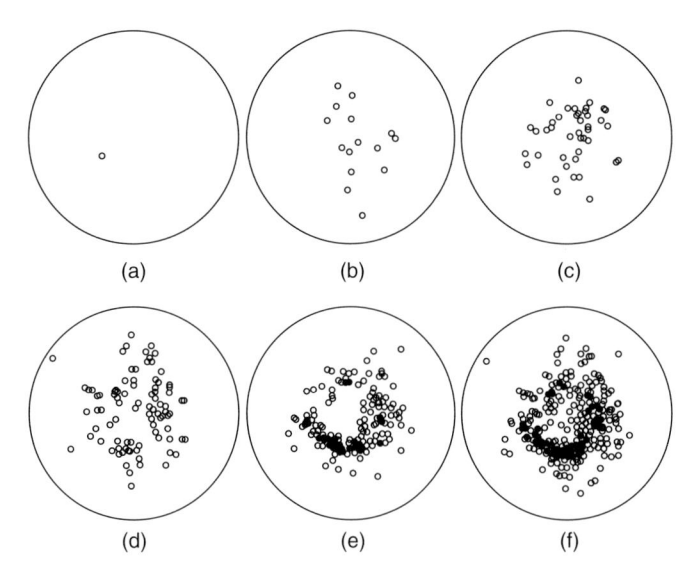

Figure 6.4 The localization of microfractures. Successive panels (a–e) show the localization as the pressure is increased by equal amount, and (f) shows all the microfractures. From Guarino et al. (1998), with kind permission of European Physical Journal (EPJ).

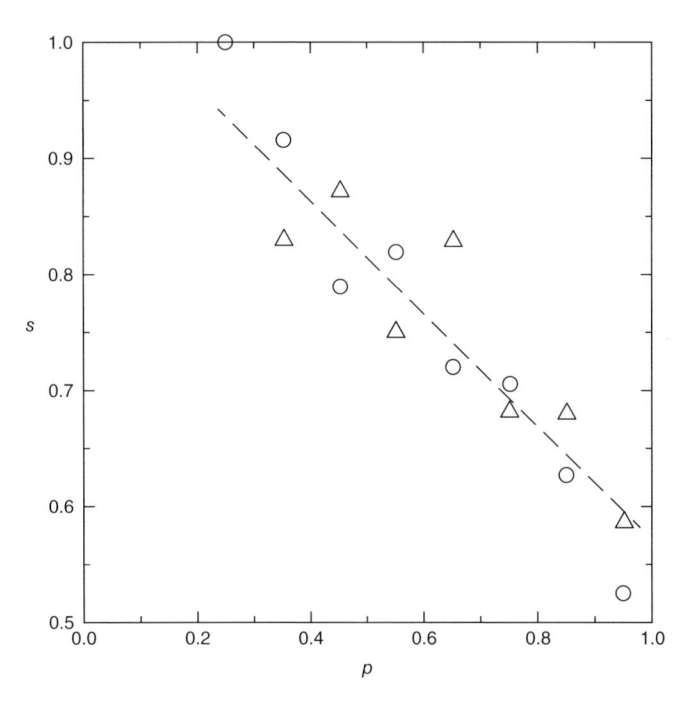

Figure 6.5 The variation of normalized entropy with normalized pressure showing an almost linear decay. From Guarino et al. (1998), with kind permission of European Physical Journal (EPJ).

6.2
Dynamics of Fiber Bundle Model

The observations of avalanche dynamics in fracture and related statistics in space and time have been attempted to be captured by simples models. Here, we discuss one such models, the fiber bundle model, already qualitatively introduced in Section 4.3.1.

The fiber bundle model, introduced by Peirce (1926) (see Figures 6.6 and 6.7) for testing the strengths of cotton yarns, shows different aspects of failure process

T355

32—X.—TENSILE TESTS FOR COTTON YARNS
v.—"THE WEAKEST LINK"
THEOREMS ON THE STRENGTH OF LONG AND OF COMPOSITE SPECIMENS

By FREDERICK THOMAS PEIRCE, B.Sc., F.Inst.P.
(British Cotton Industry Research Association).

INTRODUCTION AND SUMMARY

It is a truism, of which the mathematical implications are of no little interest, that the strength of a chain is that of its weakest link. It is equally true that the strength of a test specimen is that of its weakest element of length, whether it be a metal rod, a thread of yarn, or a cotton hair. This fact distinguishes the quantity, breaking load, from most other quantities, such as weight of which the value is determined by the average over all elements of the length. Tensile strength thus decreases with the length of specimen in a way which is definitely calculated from the distribution of strength of short specimens. The decrease in mean strength and in irregularity is directly proportional to the irregularity of the short specimens and to a factor, depending only on the multiple by which the length is increased and very simply calculated therefrom.

Variability along a specimen is shown necessarily to introduce negative skewness into all frequency curves of strength, counteracting or reinforcing any skewness that may arise from methods of production, and this must be taken into consideration when drawing conclusions from the shape of strength frequency curves. In cotton yarns, the method of production tends to produce positive skewness which is found with specimens of 3 inches. This is obliterated to yield a symmetrical curve by increasing the length to a foot or so, while leas show decided negative skewness. More generally, skewness is produced by irregularity in the frequency curves of any quantity "when the deviations of individual values are affected unequally by equal deviations of opposite sign among the constituent elements of a specimen," a criterion which fits most elastic measurements.

The relations between the strength of fibres, yarns, leas, and fabrics have often been studied empirically, but they are subject to so many disturbing factors that measurements do not lead to definite or simple conclusions, in the absence of a logical basis for comparing the results. In the present paper, five cases of specimens composed of parallel elements are analysed for a relation between the strength of the whole and of the parts. Actual specimens of all kinds of materials may reproduce the conditions more or less closely. The lea test can be brought under a simple case if modified as suggested in Paper I. of this series.

A correction is given for measurements of tendering when only a fraction of the length of a specimen is subjected to treatment, such as wear or exposure to light.

The present paper is mathematical throughout, but the conclusions are condensed into simple forms applicable to experimental results in this series or elsewhere.

Figure 6.6 Frederick Thomas Peirce (1896–1949) was born in Australia. He graduated from the University of Sydney in 1915 at the age of 19 with first class in physics and chemistry. After the first world war, he went to England and worked with W. H. Bragg. In 1921, he joined the British Cotton Industry Research Association. In 1926, he proposed the fiber bundle model (FBM) to model the strength of cotton cloths. This model, however, later proved to be very useful as a prototypical model of failure of disordered materials. Many variants of the model are still studied in modeling failure of disordered solids and fracture propagation. First page is shown from F. T. Peirce (1926).

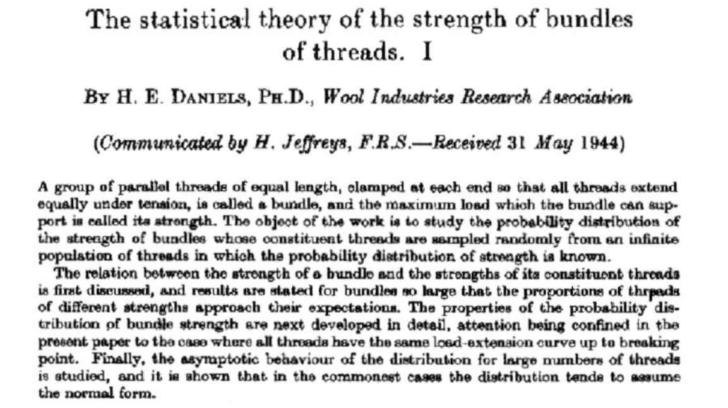

The statistical theory of the strength of bundles of threads. I

By H. E. Daniels, Ph.D., *Wool Industries Research Association*

(*Communicated by H. Jeffreys, F.R.S.—Received* 31 *May* 1944)

A group of parallel threads of equal length, clamped at each end so that all threads extend equally under tension, is called a bundle, and the maximum load which the bundle can support is called its strength. The object of the work is to study the probability distribution of the strength of bundles whose constituent threads are sampled randomly from an infinite population of threads in which the probability distribution of strength is known.

The relation between the strength of a bundle and the strengths of its constituent threads is first discussed, and results are stated for bundles so large that the proportions of threads of different strengths approach their expectations. The properties of the probability distribution of bundle strength are next developed in detail, attention being confined in the present paper to the case where all threads have the same load-extension curve up to breaking point. Finally, the asymptotic behaviour of the distribution for large numbers of threads is studied, and it is shown that in the commonest cases the distribution tends to assume the normal form.

Introduction

1. Consider an infinite population of threads of equal length whose probability distribution of strength is known. A number of threads are sampled randomly from this population, laid side by side and clamped at each end to form what we term a *bundle*. When a free load is applied to the bundle all threads extend by an equal amount, and the minimum load beyond which all the threads of the bundle break is called its *strength*. The present work is an investigation into the probability distribution of the strength of such bundles.

The problem presents itself naturally in the theory of strength testing of textile materials. For example, a test widely used in practice for estimating the strength of yarns is known as the 'hank' test for wool yarns, and the 'lea' test for cotton yarns. The method employed is to reel a hank of yarn containing a stated number of turns of specified circumference and to apply the breaking load to the hank by stretching it between two hooks. The word 'bundle' has been used here in preference to 'hank' or 'lea', since there is usually a small amount of slip at the hooks which makes the conditions of this test indeterminate. The testing of cloth samples for warp and weft strength is another instance of a practical test in which the specimens approximate to bundles, though in closely woven fabrics the cross-threads afford a measure of support which introduces complications. Other examples outside the field of textile testing could no doubt also be quoted.

2. The subject was first considered, as far as the author is aware, by F. T. Peirce (1926), relevant experimental work being published in previous papers in the same series. He deals exhaustively with the underlying physical considerations and derives useful formulae for the strength of large bundles.

Vol. 183. A. [405] 27

Figure 6.7 Henry Ellis Daniels (1912–2000) was a British statistician. He graduated from the University of Edinburgh. He worked as a professor of mathematical statistics in the University of Birmingham. He was the President of the Royal Statistical Society between 1974 and 1975. He had significant contribution on the study of dynamical properties of the fiber bundle model. First page of Daniels (1945) is shown in the figure.

in disordered systems (for reviews, see Pradhan *et al.* (2010), Kawamura *et al.* (2012)). The fiber bundle (see Figure 6.8) is made up of N fibers or linear (Hookean) springs, each having identical spring constant κ. The bundle carries a load $W = N\sigma$, and the failure thresholds $(\sigma_{\text{th}})_i$ are different for each fiber (i). For the equal (global) load sharing model considered here, the lower platform is absolutely rigid, therefore no local deformation occurs and consequently there can be no stress concentration in the vicinity of the failed fibers. This ensures equal load sharing process, that is, all the intact fibers share the load W equally and the load per fiber increases with the failure of fibers. The strength of each fiber $(\sigma_{\text{th}})_i$ in the system is given by the load per fiber value it can withstand, and beyond which it fails. The strength of the individual fibers are drawn randomly from a normalized density

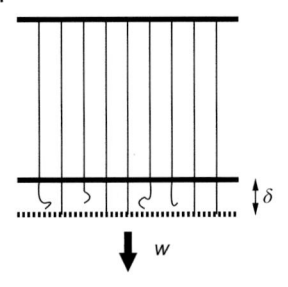

Figure 6.8 The fiber bundle consists initially of N fibers attached in parallel to a fixed, rigid plate at the top and a downwardly movable platform from which a load W is suspended at the bottom. In the equal load sharing model considered here, the platform is absolutely rigid and the load W is consequently shared equally by all the intact fibers.

$p(\sigma_{th})$ within the interval 0 and 1 such that

$$\int_0^1 p(\sigma_{th})d\sigma_{th} = 1.$$

The equal load sharing assumption does not consider "local" fluctuations in stress (and its redistribution and nucleation), and renders the model as a mean-field one.

The breaking dynamics starts when an initial stress σ (load per fiber) is applied on the bundle. The fibers having strength less than σ fail instantly. It is because of this rupture, the total number of intact fibers decreases and rest of the (intact) fibers have to bear the applied load on the bundle. Hence, effective stress on the fibers increases and this compels some more fibers to break. These two sequential operations, namely the stress redistribution and further breaking of fibers, continue until a static point is reached, where either the surviving fibers are strong enough to bear the applied load on the bundle or all fibers fail.

This breaking dynamics can be represented by recursion relations in discrete time steps. For this, let us consider a very simple model of fiber bundles where the fibers (having the same spring constant κ) have a uniform strength distribution $p(\sigma_{th})$ up to a cut-off strength normalized to unity, as shown in Figure 6.9: $p(\sigma_{th}) = 1$ for $0 \leq \sigma_{th} \leq 1$ and $p(\sigma_{th}) = 0$ for $\sigma_{th} > 1$. Let us also define $U_t(\sigma)$ to be the fraction of fibers in the bundle that survives after (discrete) time step t, counted from time $t = 0$ when the load is applied (time step indicates the number of stress redistributions). As such, $U_t(\sigma = 0) = 1$ for all t and $U_t(\sigma) = 1$ for $t = 0$ for any σ. If σ_f is the failure threshold of the model, then in the long time limit $U_t(\sigma) = U^*(\sigma) \neq 0$ if $\sigma < \sigma_f$, and $U_t(\sigma) = 0$ if $\sigma > \sigma_f$.

Therefore, the simple recursion relation for $U_t(\sigma)$ is given by (see Figure 6.9)

$$U_{t+1} = 1 - \sigma_t; \qquad \sigma_t = \frac{W}{U_t N}$$

or, $\qquad U_{t+1} = 1 - \dfrac{\sigma}{U_t}.$ \hfill (6.4)

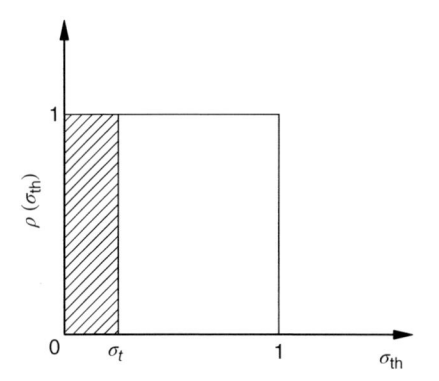

Figure 6.9 The simple model considered here assumes uniform density $\rho(\sigma_{th})$ of the fiber strength distribution up to a cut-off strength (normalized to unity). At any load per fiber level σ_t at time t, the fraction σ_t fails and $1 - \sigma_t$ survives.

At the stable state $(U_{t+1} = U_t = U^*)$, the aformentioned relation gives

$$U^{*^2} - U^* + \sigma = 0.$$

The solution is

$$U^*(\sigma) = \frac{1}{2} \pm (\sigma_f - \sigma)^{1/2}; \sigma_f = \frac{1}{4}.$$

Here, σ_f is the critical value of initial load per fiber value beyond which the system fails completely. The solution with (+) sign is the stable one, whereas the one with (−) sign is unstable (and not physically relevant) (Pradhan and Chakrabarti, 2001; Pradhan et al., 2002; Bhattacharyya et al., 2003). The quantity $U^*(\sigma)$ must have a real value since it is a physical quantity: the fraction of the original system that survives under an applied stress σ for $0 \leq \sigma \leq \sigma_f$. Clearly, $U^*(0) = 1$. Therefore the stable solution is

$$U^*(\sigma) = U^*(\sigma_f) + (\sigma_f - \sigma)^{1/2}; \quad U^*(\sigma_f) = \frac{1}{2} \quad \text{and} \quad \sigma_f = \frac{1}{4}. \tag{6.5}$$

For $\sigma > \sigma_f$, the dynamics never stops until $U_t = 0$ when all the fibers are broken (no stable fixed point).

6.2.1
Dynamics Around Critical Load

It may be noted that the quantity $U^*(\sigma) - U^*(\sigma_f)$ behaves like an order parameter that determines a transition from a state of partial failure ($\sigma \leq \sigma_f$) to a state of total failure ($\sigma > \sigma_f$):

$$0 \equiv U^*(\sigma) - U^*(\sigma_f) = (\sigma_f - \sigma)^{\beta'}; \beta' = \frac{1}{2}. \tag{6.6}$$

6 Avalanche Dynamics in Fracture

To study the dynamics away from criticality ($\sigma \to \sigma_f$ from below), we replace the recursion relation (6.4) with a differential equation

$$-\frac{dU}{dt} = \frac{U^2 - U + \sigma}{U}.$$

Close to the fixed point, we write $U_t(\sigma) = U^*(\sigma) + \epsilon$ (where $\epsilon \to 0$). This, following Eq. (6.6), gives

$$\epsilon = U_t(\sigma) - U^*(\sigma) \approx \exp(-t/\tau), \tag{6.7}$$

where $\tau = \frac{1}{2}\left[\frac{1}{2}(\sigma_f - \sigma)^{-1/2} + 1\right]$. Near the critical point, we can write

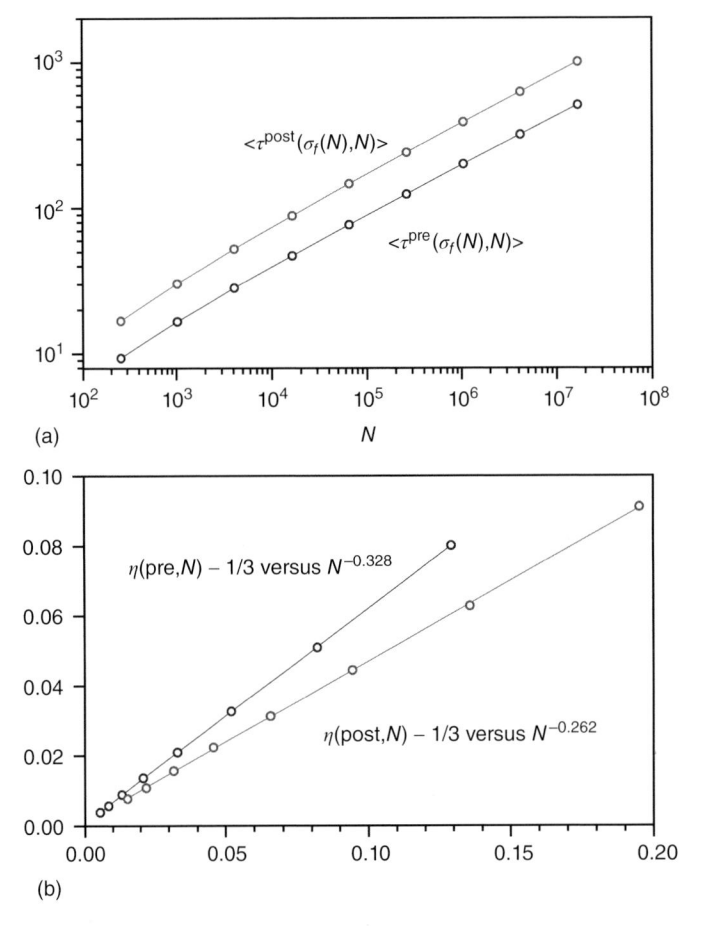

Figure 6.10 (a) The log–log plot of the average relaxation time at the critical point with system size. The slope is close to 1/3 in both precritical and postcritical regimes. (b) The difference between the slope $\eta(N)$ and 1/3 is seen to decay with system size in a power-law as indicated. From Roy et al. (2013), with kind permission from American Physical Society.

$$\tau \propto (\sigma_f - \sigma)^{-z}; z = \frac{1}{2}. \tag{6.8}$$

Therefore, the relaxation time diverges following a power-law as $\sigma \to \sigma_f$ from below.

The relaxation dynamics in this model was extensively studied recently in Roy et al. (2013). For a given bundle (with uniform failure threshold distribution), the critical load σ_f^i was first determined. Then the relaxation time was measured for a given applied load, different from the critical point. Keeping the deviation from critical load fixed, the process was repeated for many independent realizations. It was seen that as the deviation from critical point tends to zero, the relaxation time in the precritical and postcritical regions tends to two distinct values, say $\langle \tau^{\text{pre}} \rangle$ and $\langle \tau^{\text{post}} \rangle$ and $\langle \tau^{\text{post}} \rangle / \langle \tau^{\text{pre}} \rangle \to 2$ as $N \to \infty$. The system size dependences of $\langle \tau^{\text{pre}}(\sigma_f(N), N) \rangle$ and $\langle \tau^{\text{post}}(\sigma_f(N), N) \rangle$ are shown in Figure 6.10(a). The slopes (η) of the curves in fact approach $1/3$. This can be seen from Figure 6.10(b), where the differences of the slopes from $1/3$ are shown to be going to zero with increase of system size. It was therefore concluded that

$$\langle \tau(\sigma_f(N), N) \rangle \sim N^\eta, \tag{6.9}$$

with $\eta = 0.333 \pm 0.001$.

As for the scaling away from critical point, the average relaxation times $\langle \tau(\sigma, N) \rangle$ against $\sigma_f(N) - \sigma$ are plotted for three system sizes in Figure 6.11(a). These lines can be made to collapse by finite-size scaling, as shown in Figure 6.11(b). These scalings suggest a general form for the relaxation time as follows

$$\langle \tau(\sigma, N) \rangle / N^\eta \sim \mathcal{F}\left[(\sigma_f(N) - \sigma)N^\theta\right]. \tag{6.10}$$

The best-fit values are reported to be $\eta = 0.336$ and $\theta = 0.666$. The scaling function has two regimes: for small argument value, it is a constant and for large argument, it is a power-law. The constant regime is approximately extended up to $[\sigma_f(N) - \sigma]N^\theta \approx 1$. This means

$$\sigma_f(N) - \sigma \sim N^{-\theta}. \tag{6.11}$$

Furthermore, the power-law part would imply

$$\langle \tau(\sigma) \rangle \sim (\sigma_f - \sigma)^{-z}. \tag{6.12}$$

From Eq. (6.10), one then gets

$$z\theta = \eta \tag{6.13}$$

giving $z = 0.50 \pm 0.01$ as can be seen from Eq. (6.8). The same set of exponent values is seen for the postcritical region as well. Of course, these values do not depend on the distribution of thresholds chosen. These observations also rule out a logarithmic system size dependence on the prefactor of the relaxation time as was earlier claimed (Pradhan and Hemmer, 2007).

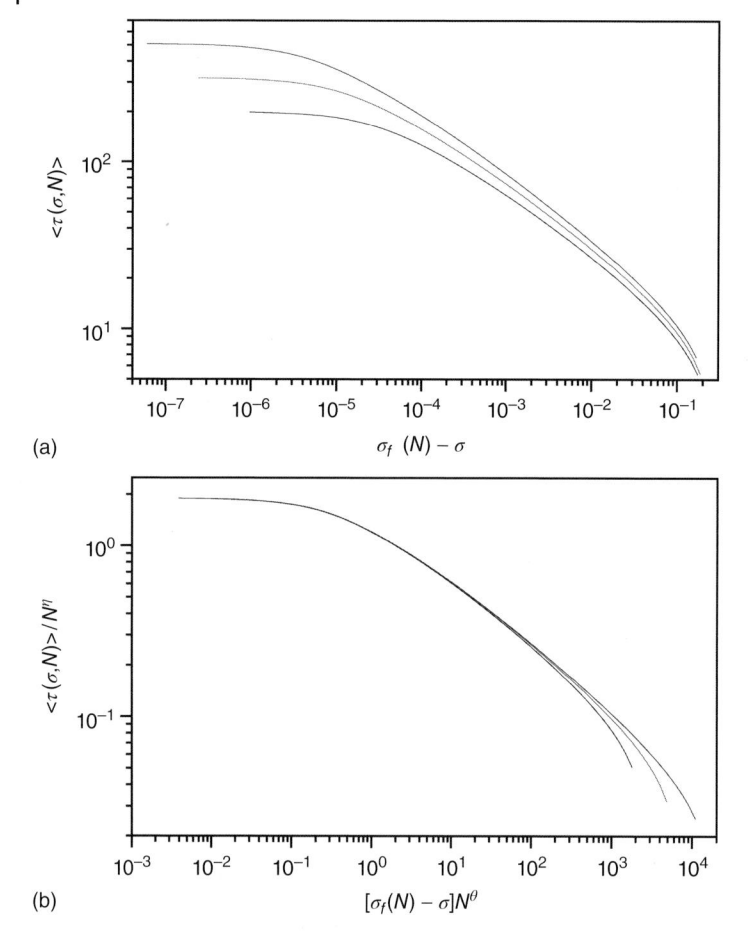

Figure 6.11 (a) The variation of relaxation time with distance from critical load for three different system sizes ($N = 2^{20}$, 2^{22}, 2^{24}, bottom to top). (b) The finite-size scaling plot with exponent values $\eta = 0.336$ and $\theta = 0.666$. The scaling function is a constant for small value of argument and then decays in a power-law (with exponent value close to 1/2). From Roy et al. (2013), with kind permission from American Physical Society.

One can also consider the breakdown susceptibility χ, defined as the change of $U^*(\sigma)$ due to an infinitesimal increment of the applied stress σ

$$\chi = \left| \frac{dU^*(\sigma)}{d\sigma} \right| = \frac{1}{2}(\sigma_f - \sigma)^{-\gamma}; \gamma = \frac{1}{2} \tag{6.14}$$

from Eq. 6.6. Hence, the susceptibility diverges as the applied stress σ approaches the critical value $\sigma_f = \frac{1}{4}$. Such a divergence in χ had already been observed in the numerical studies.

6.2.2
Dynamics at Critical Load

At the critical point ($\sigma = \sigma_f$), a different dynamic critical behavior in the relaxation of the failure process was observed. From the recursion relation (6.4), it can be shown that decay of the fraction $U_t(\sigma_f)$ of unbroken fibers that remain intact at time t follows a simple power-law decay:

$$U_t = \frac{1}{2}\left(1 + \frac{1}{t+1}\right), \tag{6.15}$$

starting from $U_0 = 1$. For large t ($t \to \infty$), this reduces to $U_t - 1/2 \propto t^{-x}; x = 1$; a strict power-law which is a robust characterization of the critical state (see, however, Zapperi et al., 1997a).

6.2.3
Avalanche Statistics of Energy Emission

The avalanche statistics in the fiber bundle model are well known in terms of the number of fibers breaking in an avalanche. In the same way, one can also study the avalanches of energies, which is a more relevant quantity in terms of experimental realizations.

In Pradhan and Hemmer (2008), the energy avalanches of global load sharing fiber bundle model were studied. Let there be N fibers breaking in an avalanche. If the threshold of the weakest fiber breaking in the avalanche is x_{min} and that of the strongest fiber is x_{max}, then $x_{max} \approx x_{min} + \frac{n}{Np(x)}$, where $p(x)$ is the probability density function of the failure threshold distribution of the fibers. Since the second term decays as $1/N$, in the large system size limit, one would expect the difference to be small. Hence, the energy released in a burst of size N can be approximated by $E = nx^2/2$ (assuming for elongation x, the energy stored is $x^2/2$ for Hookean springs). Now, the number of bursts of size N beginning with a threshold value in x and $x + dx$ is known (Hemmer and Hansen, 1992) to be

$$f(n,x)dx = N\frac{n^{n-1}}{n!}\frac{1 - P(x) - xp(x)}{x}X(x)^n e^{-nX(x)}dx, \tag{6.16}$$

where

$$X(x) = \frac{xp(x)}{1 - P(x)}. \tag{6.17}$$

Therefore, the number of bursts with energies less than E is

$$G(E) = \sum_n \int_0^{\sqrt{2E/n}} f(n,x)dx \tag{6.18}$$

with the energy density being

$$P(E) = \frac{dG}{dE} = \sum_n (2En)^{-1/2} f(n, \sqrt{2E/n}). \tag{6.19}$$

The high energy limit of this function can be evaluated to be

$$P(E) \approx N \frac{C}{E^{5/2}}, \qquad (6.20)$$

where

$$C \frac{x_c^4 p^2(x_c)}{4\pi^{1/2}[2p(x_c) + x_c p'(x_c)]}. \qquad (6.21)$$

Therefore, the high energy limit of the burst size distribution is a universal power-law with exponent value 5/2, independent of the threshold distribution (see Figure 6.12). The low energy limit, however, is not universal.

6.2.4
Precursors of Global Failure in the Model

In the case of failure phenomena, the knowledge of precursor is very important using which the failure can be predicted early. For this model, there exist several such precursors. The susceptibility χ with σ grows following Eq. 6.14 suggesting one such possibility: $\chi^{-1/2}$ decreases linearly with σ to 0 at $\sigma = \sigma_f$ from below. In addition, Pradhan and Hemmer (2009) studied the rate $R(t) (\equiv -\frac{dU_t}{dt})$ of failure (see Figure 6.14) of fibers following the dynamics as in Eq. 6.14 for $\sigma > \sigma_f$ and found that the rate is minimum at a time t_0 that is half of the total failure time t_f of the system. This relation has been shown to be independent of the breaking threshold distribution of the fibers.

The failure time of a bundle of fiber can also be predicted by noting the rate of energy release (Pradhan and Hemmer, 2011) in the breaking process. Particularly,

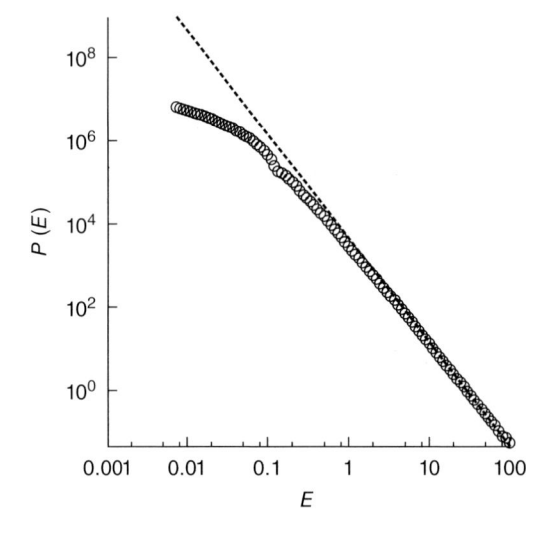

Figure 6.12 The distribution of the energy bursts with uniform threshold distribution of global load sharing fiber bundle model. From Pradhan and Hemmer (2008), with kind permission from American Physical Society.

for an overloaded system, the total failure time of the system is almost twice the time at which the energy release reaches a minimum (similar to the case of breaking rate mentioned earlier). Since the number of fibers within the threshold values x and $x + dx$ is $Np(x)dx$, and the fibers are Hookean springs up to the breaking point, the energy released at time t is

$$E_t = \int_{\sigma/U_{t-1}}^{\sigma/U_t} \frac{x^2}{2} Np(x)dx. \tag{6.22}$$

The applied load is to be taken slightly more than the global failure threshold $\sigma = \sigma_f + \delta\sigma$. Now, minimum of emission rate can be located by equating the derivative of E_t to zero. The energy emission is minimum for

$$1 - U_t = P\left[\sqrt{\sigma^3 p(\sigma/U_t) U_t^{-4}}\right]. \tag{6.23}$$

Now, in the particular case of uniform threshold distribution $[0:1]$,

$$\tau_E = \frac{\tau_f}{2} - \frac{3}{2}, \tag{6.24}$$

where τ_E is the time of energy emission minimum and τ_f is the failure time of the system. A similar relation holds for Weibull (and for that matter other similar) threshold distribution, except for the last constant. Noting that τ_f and τ_E are large, effectively it can be said that the energy release minimum occurs at half the total failure time, independent of the threshold distributions (see Figure 6.13).

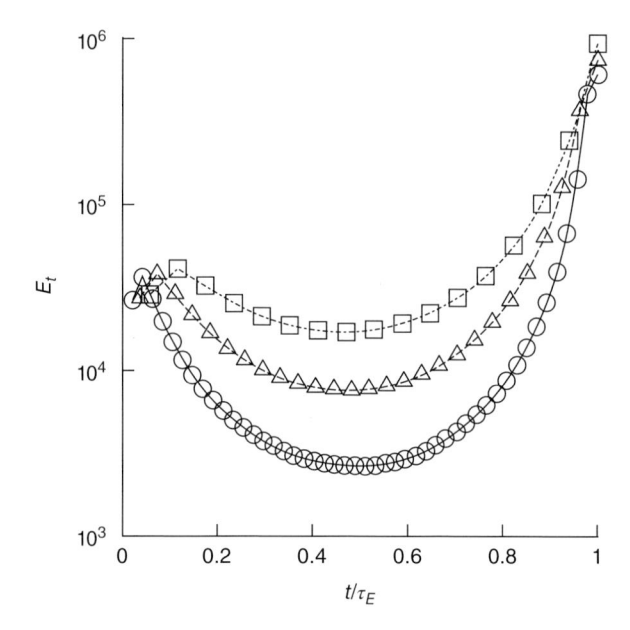

Figure 6.13 The energy emission in a scaled time. The minimum occurs near the half of the total failure (Eq. 6.24). The different curves are for different excess loads: $\sigma - \sigma_f = 0.001$ (circles), 0.003 (triangles), and 0.007 (squares). From Pradhan and Hemmer (2011), with kind permission from American Physical Society.

6.2.5
Burst Distribution: Crossover Behavior

For the fiber bundle model, if the load is increased until a new failure takes place, a burst is defined as the number (Δ) of failures following that initial failure. The size distribution of such bursts ($P(\Delta)$) is a power-law. It was shown for a generic case (independent of threshold distribution) that the form of this distribution (for continuous loading) is

$$P(\Delta)/N = C\Delta^{-\beta} \qquad (6.25)$$

in the limit $N \to \infty$. The burst exponent β has the value 5/2 for average over all $\sigma(= 0$ to $\sigma_f)$, and it is universal (Hemmer and Hansen, 1992). However, the burst exponent value can depend, for example, on the details of loading process and also from which point of the loading the burst statistics are recorded. If the burst distribution is recorded for only for the bursts near the critical point ($\sigma \lesssim \sigma_f$), then the exponent (β) value becomes 3/2 (Pradhan et al., 2005b). For equal load sharing model with uniform strength distribution, the burst distribution is shown (Figure 6.15) for recording statistics that starts from different points of effective loading, denoted by x_0, with $x_t = \sigma/U_t$ being the elongation or the effective loading (for linear elastic behavior) at any point t). The crossover behavior is clearly seen as the failure point is approached. In these studies, the load

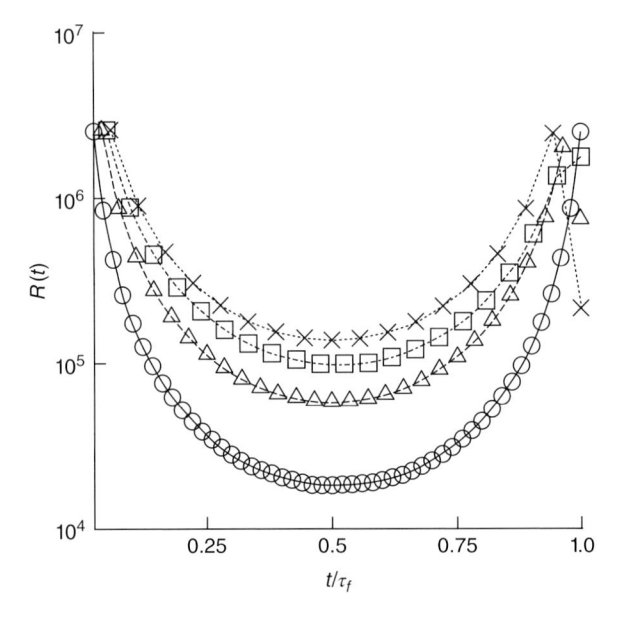

Figure 6.14 The breaking rate $R(t)$ versus the rescaled step variable t_f/t for the uniform threshold distribution for a bundle of $N = 10^7$ fibers. Different symbols are for different excess stress levels $\sigma - \sigma_f$: 0.001 (circles), 0.003 (triangles), 0.005 (squares), and 0.007 (crosses). From Pradhan and Hemmer (2009), with kind permission from American Physical Society.

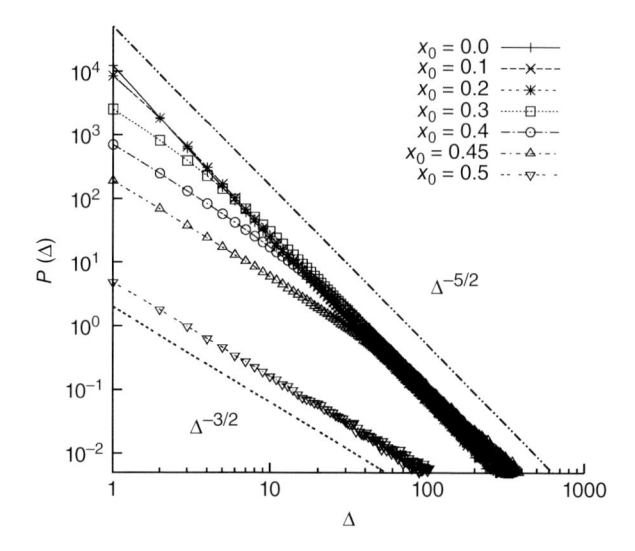

Figure 6.15 The burst size distribution for different values of x_0 in the equal load sharing model with uniform threshold distribution. The number of fibers is $N = 50\,000$. From Pradhan et al. (2010), with kind permission from American Physical Society.

increase rate is extremely slow and the increase is assumed to stop once a fiber fails (redistribution timescale is much faster than load increase, hence a separation of timescales can be assumed). The consequent avalanches are studied at that load. Once an avalanche stops, the load is increased again. This process is realistic in the case of earthquakes where stress accumulation takes place over years but the failure or earthquake events are very fast. However, if the increase in load is fixed ($d\sigma$), then the aforementioned exponent value of β becomes 3: $\Delta \sim \frac{d(1-U^*)}{d\sigma}$, giving $\Delta^{-2} = \sigma - \sigma_f$ (from Eq. 6.5) and since $P(\Delta)d\Delta \sim d\sigma$, $P(\Delta) \sim \frac{d\sigma}{d\Delta} \sim \Delta^{-\beta}$, $\beta = 3$ (Pradhan et al., 2002). In fact, the earthquake frequency statistics may indeed show a crossover behavior similar to the one discussed above: If event frequency is denoted by $P(M)$, then it is known that (GR law) $P(M) \sim M^{-\beta}$, where M denotes the magnitude (may be assumed to be related to avalanche size Δ in the models) and β value is found (Kawamura, 2006) to be more ($\beta \approx 0.9$) for statistics over a smaller interval (before the main shock), compared to the long time average value ($\beta \approx 0.6$); see Figure 6.16.

6.2.6
Abrupt Rupture and Tricritical Point

In the GLS fiber bundle model, there exists a trivial limit when all the fibers have equal failure thresholds, where the system fails abruptly without any precursory damage. In other words, there is no stable configuration with partially damaged state. This is, of course, the limit of brittle failure, when the stress–strain relationship remains linear up to the failure point. In the case of broader disorder,

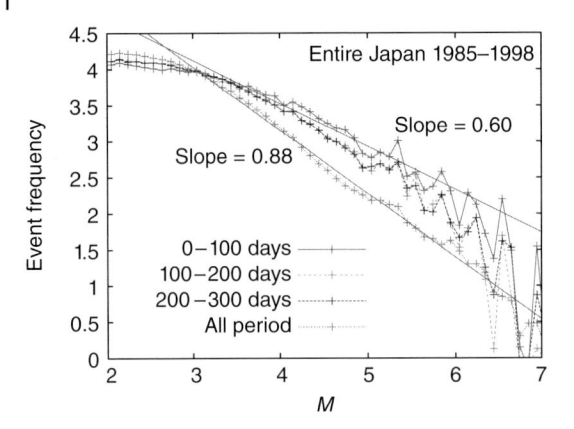

Figure 6.16 Crossover signature in the local magnitude distribution of earthquakes in Japan. During the 100 days before main shock, the exponent is 0.60; much smaller than the average value 0.88. From Kawamura (2006), with kind permission from Springer.

as is expected, there will be partially damaged stable states and hence precursory events before catastrophic rupture. The stress–strain relationship will consequently become nonlinear and the system behaves as a quasi-brittle one. It is interesting to note (Andersen et al., 1997) that this transition from brittle to quasi-brittle behavior happens for a finite width of the disorder and not merely a trivial point where all thresholds are equal.

Consider a rectangular threshold distribution having a lower cut-off strength at σ_L and upper cut-off strength at σ_R. Mathematically, it can be represented as (Bhattacharyya et al., 2003)

$$\rho(\sigma_{\text{th}}) = 0 \quad \text{for} \quad 0 \leq \sigma_{\text{th}} < \sigma_L$$
$$= \frac{1}{\sigma_R - \sigma_L} \quad \text{for} \quad \sigma_L \leq \sigma_{\text{th}} \leq \sigma_R$$
$$= 1 \quad \text{for} \quad \sigma_R < \sigma_{\text{th}}. \tag{6.26}$$

The physically relevant fixed point of the dynamics is (see Appendix C for details)

$$U^* = \frac{\sigma_R}{2(\sigma_R - \sigma_L)} + \frac{1}{(\sigma_R - \sigma_L)^{1/2}} \left[\frac{\sigma_R^2}{4(\sigma_R - \sigma_L)} - \sigma_0 \right]^{1/2}. \tag{6.27}$$

The usual critical behavior (when the distribution is broad enough) is the same as that discussed earlier. However, when the distribution becomes narrow, a transition to brittle behavior takes place (Roy and Ray, 2014; Roy, 2012). The condition for abrupt failure is when $\frac{\sigma_R}{2(\sigma_R - \sigma_L)} = 1$, that is, $\sigma_R = 2\sigma_L$. Also, the condition for rupture is $\frac{\sigma_R^2}{4(\sigma_R - \sigma_L)} - \sigma_0 = 0$, where we consider σ_R to be the driving parameter for the transition, which is a measure of the strength of the disorder. The point where the abrupt failure sets in is a tricritical point, separating the continuous failure (with suitably defined order parameter as before in the quasi-brittle region) and abrupt failure. At the tricritical point, the order parameter behaves as (again considering strength of disorder as the driving parameter)

$$O \sim \left(\sigma_R - \sigma_{R,f}^{TCP}\right). \tag{6.28}$$

Also, regarding the timescale divergence

$$\tau \sim \left(\sigma_R - \sigma_{R,f}^{TCP}\right)^{-1}, \tag{6.29}$$

see Appendix C for details. The numerical observation of this divergence was noted in Roy and Ray (2014), Roy (2012). This set of exponent values is different from that seen for the usual nonabrupt case.

6.2.7
Disorder in Elastic Modulus

Usually, the disorders in fiber bundles are introduced in terms of the failure thresholds. However, as real materials are often made up of different components having different stiffness, it is important to note the features of such mixtures in the framework of fiber bundle model (Karpas and Kun, 2011). In the simplest limit, the system can be a mixture of two components, each having elastic modulus Y_1 and Y_2 (with $Y_1 < Y_2$) with respective fractions of fibers p_1 and p_2 and no disorder in failure thresholds. The relevant parameters are $r = Y_2/Y_1$ and $k = p_2/p_1$. It was shown that if these parameters satisfy a relation $k < 1 - 1/r$, then the critical load of the system is determined by the second subset, otherwise the critical load is determined by the first subset. Furthermore, it was shown that the single-component system can always carry a higher load than the two-component mixture. In the case of strong heterogeneity, where a fraction of fibers do not break at all (Hidalgo et al., 2008a), it was shown that the avalanche exponent shifts to a value of 9/4 from the mean-field prediction 5/2 beyond a value of the fraction.

A more interesting feature is seen when the distribution is continuous. When there is a substantial fraction of fibers in the immediate vicinity of the maximum Young's modulus (which will break first) they will break immediately following the first one, and a catastrophic breakdown will follow. The same is not true for a Weibull distribution due to the exponential tail. This effect is more clearly seen in a power-law distribution $p(Y) \sim Y^{-a}$, where one can vary a to see the different regimes.

The constitutive relations were shown to be (Karpas and Kun, 2011)

$$\sigma_a(\epsilon) = \left[\frac{1-a}{2-a}\frac{1}{Y_{max}^{1-a} - Y_{min}^{1-a}}\right](\sigma_{th}^{2-a}\epsilon^{a-1} - Y_{min}^{2-a}\epsilon)$$

$$\sigma_1(\epsilon) = \frac{1}{\ln\frac{Y_{max}}{Y_{min}}}(\sigma_{th} - Y_{min}\epsilon),$$

$$\sigma_2(\epsilon) = \frac{Y_{max}Y_{min}}{Y_{max} - Y_{min}}\left[\ln\left(\frac{\sigma_{th}}{Y_{min}}\right)\epsilon - \ln(\epsilon)\epsilon\right], \tag{6.30}$$

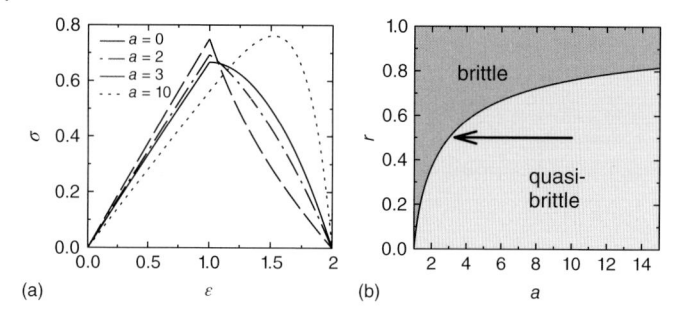

Figure 6.17 (a) The stress–strain relationship for different a's, showing brittle (sharp maximum) and quasi-brittle behavior (rounded maximum). (b) The phase diagram following Eq. (6.31). From Karpas and Kun (2011), with kind permission from EPL.

where $\sigma_a(\epsilon)$, $\sigma_1(\epsilon)$, $\sigma_2(\epsilon)$ correspond to exponent value a, and special cases of the exponent value 1 and 2 respectively. The two limits of elastic modulus are $Y_{\min} = 0.5$ and $Y_{\max} = 1$. As can be seen from Figure 6.17(a), for low values of a, there is a sharp maximum after which a catastrophic failure occurs. This is the brittle region. For larger values of a, a quadratic maximum is seen, signifying that precursory avalanches occur before the catastrophic one, thus a quasi-brittle behavior is seen. It was shown that the critical fraction for a given value of a for which the brittle to quasi-brittle transition is seen is given by

$$r_c(a) = (a-1)^{1/(2-a)}. \tag{6.31}$$

The resulting phase diagram is shown in Figure 6.17(b). The statistics of avalanche size distribution changes from the usual value 5/2 to a value 3/2 as the transition is approached.

6.3
Interpolations of Global and Local Load Sharing Fiber Bundle Models

The simplicity of the GLS fiber bundle model allows for analytical treatment and it also captures the qualitative features of threshold-activated dynamical failures, namely the scale-free avalanche statistics due to intermittent failure. However, in real systems, the load transfer mechanism will be more complicated than the global load sharing mechanism due to local deformations. In this context, therefore, several attempts were made to generalize the fiber bundle model. Here we discuss those studies, which essentially interpolate between the two extreme cases of load sharing, namely the GLS and LLS versions. While the GLS version has a finite critical failure threshold, the LLS version has no nonzero failure threshold in the thermodynamic limit, Particularly, $\sigma_f \sim 1/\ln N$, where N is the system size. Therefore, another important aspect of these studies is to capture any dimension-dependent critical behavior in fiber bundle models, which of course is absent for GLS scheme. Very recently a general criterion demarcating GLS and LLS schemes in terms of range of load redistribution was proposed in Biswas et al. (2014).

6.3.1
Power-law Load Sharing

Hidalgo et al. (2002b) considered the case where once a fiber fails, its load is shared by all the intact fibers, but the sharing fraction decays in a power-law with the distance of an intact fiber from the failed fiber. The fibers are arranged in a square lattice and periodic boundary conditions are taken. The load transfer mechanism is as follows: When a fiber breaks, all the remaining fibers share its load. If a failure takes place at (x_j, y_j), then an intact fiber at a point (x_i, y_i) receives a load proportional to (apart from the load of the failed fiber) $r_{ij}^{-\gamma}$, where $r_{ij} = \sqrt{(x_i - x_j)^2 + (y_i - y_j)^2}$. Clearly, by changing γ, one can change the "effective" range of interaction. The limits $\gamma \to 0$ and $\gamma \to \infty$ correspond to GLS and LLS schemes, respectively.

This model was studied using Monte Carlo simulations. The failure threshold σ_f, in general, depends on both system size and γ. However, for very low γ values, σ_f is practically independent of both γ and system size N ($= L \times L$). This is, of course, the limit of GLS scheme. But, as the γ value is increased, σ_f starts depending on both γ and N (see Figure 6.18). The value of γ for which the failure threshold starts depending on N is close to $\gamma_c \approx 2$. The $\gamma > \gamma_c$ regime has two parts. When γ is much greater than 2, the failure threshold σ_f depends on N, but is practically independent of γ. There is an intermediate region, where the dependence on both γ and N is prominent. The most important feature of this model is that, in the thermodynamic limit, for $\gamma > \gamma_c$, the failure threshold σ_f goes to zero. As can be seen from Figure 6.19, the failure threshold $\sigma_f \sim \frac{1}{\ln N}$. This is the LLS-type behavior.

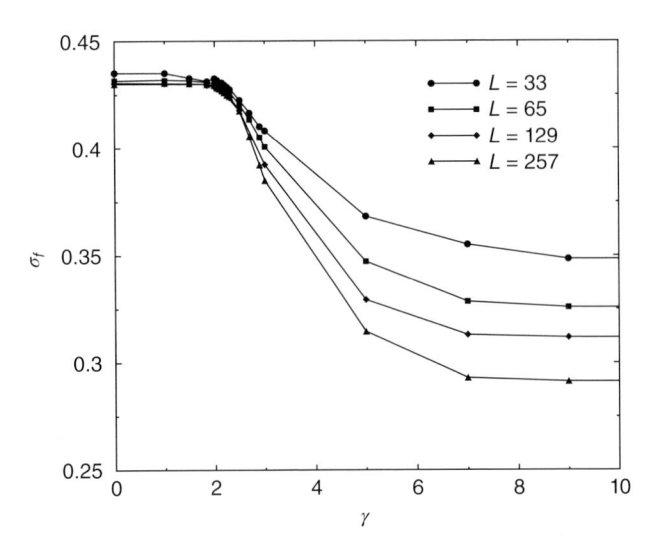

Figure 6.18 Failure thresholds for different values of γ and for different system sizes. For $\gamma > \gamma_c \approx 2$, the failure threshold decreases with system size. From Hidalgo et al. (2002b), with kind permission from American Physical Society.

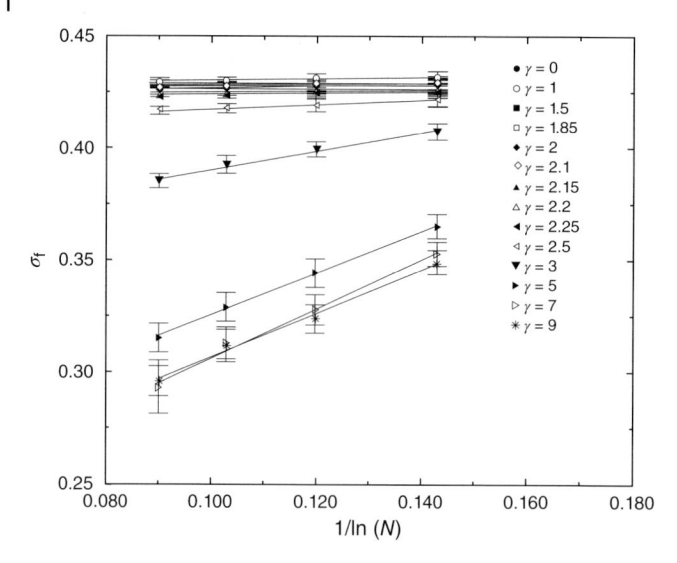

Figure 6.19 Dependence of failure threshold on system size. For $\gamma > \gamma_c$, the failure threshold goes to zero in the thermodynamic limit. From Hidalgo et al. (2002b), with kind permission from American Physical Society.

Hence, it is seen that as γ is increased, the critical behavior shifts from completely global to completely local load sharing mechanism. This behavior is also observed in the avalanche size distribution. When $\gamma < \gamma_c$, the avalanche size distribution follows a power-law with exponent value 5/2 as in the GLS case. But, for $\gamma > \gamma_c$ no power-law distribution is seen.

The spatial structure of the failed fibers also reflects this "localization." When $\gamma < \gamma_c$, the fibers break at random positions, and this is essentially equivalent to the random percolation problem (although complete failure occurs before reaching the percolation threshold). However, when $\gamma > \gamma_c$, the failures are often correlated and stress concentration takes place. Therefore, a large patch of broken fibers grow in size and complete failure takes place at small stress.

6.3.2
Mixed-mode Load Sharing

Another form of interpolation was proposed by Pradhan et al. (2005a). In this case, when a fiber fails, g fraction of its load is distributed among the nearest surviving neighbors and the rest $1 - g$ fraction is distributed equally among the remaining fibers. Clearly, the two extreme limits, where GLS and LLS models are retrieved, are $g = 0$ and $g = 1$, respectively. As before, it is of interest to see how the failure threshold depends on this load sharing fraction.

Numerical analysis of this model was carried out in one dimension. In this case, the number of nearest surviving neighbors is always two. As the value of g is increased, the failure threshold of the system decreases due to local load

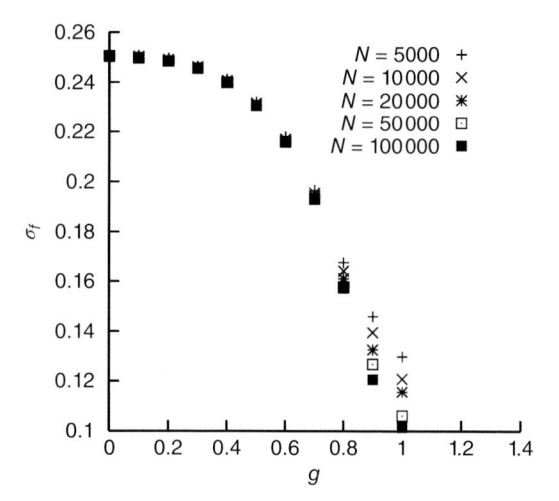

Figure 6.20 Dependence of failure threshold on the fraction of local load share g. Beyond $g = 0.8$, the failure threshold decreases with system size. From Pradhan et al. (2005a), with kind permission from American Physical Society.

concentration around failures. However, up to a significant fraction (≈ 0.7), the failure threshold remains practically independent of the system size (Figure 6.20). Beyond $g \approx 0.8$, the failure threshold is clearly dependent on system size. If σ_f is plotted against $1/\ln N$, this behavior is clearly seen (Figure 6.21).

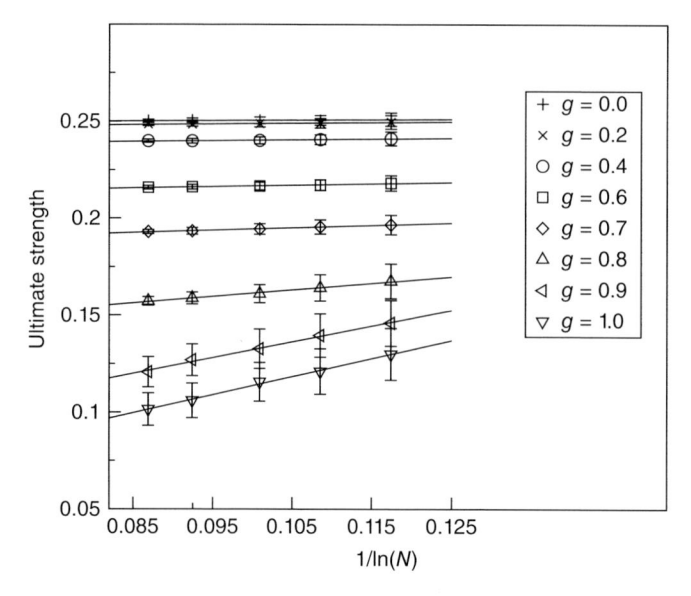

Figure 6.21 Dependence of failure threshold on system size for different values of g. The failure threshold of the system goes to zero in the thermodynamic limit for $g > 0.8$. From Pradhan et al. (2005a), with kind permission from American Physical Society.

The crossover in the behavior of failure threshold is also reflected in the avalanche size distributions. For $g < g_c \approx 0.8$, the avalanche size distribution follows a power-law with exponent 5/2 as is seen for GLS model. However, when $g > g_c$, a clear departure from the power-law behavior has been observed. It also indicates that not only there is a finite threshold for the crossover, the critical behavior in the global load sharing-dominated region is identical with the completely GLS scheme. This point is further clarified with the behavior of susceptibility and relaxation time. The susceptibility χ is defined as the number of fibers broken due to small increase in the total load. And the relaxation time τ is the time taken (number of scans) by the system to reach a steady state (no further failure) when load is applied. In the GLS scheme, both these quantities diverge near failure threshold as $\chi \sim (\sigma_f - \sigma)^{-1/2}$ and $\tau \sim (\sigma_f - \sigma)^{-1/2}$. It is seen that for $g < 0.7$, both these quantities follow similar divergence and for $g > 0.8$, no such divergence is seen (Figures 6.22 and 6.23).

6.3.3
Heterogeneous Load Sharing

As pointed out earlier, the load sharing process in the system is a manifestation of the elastic properties of the base plate. Hence, local disorders in that material may lead to heterogeneous load sharing processes. In order to see some effects of that, the system of a bundle of fibers was divided into two parts: a fraction p of the fibers follows completely local load sharing process and the rest $1-p$ fraction follows global load sharing process (Biswas and Chakrabarti, 2013a). In increasing the load, the direct loading mechanism was followed, that is, a load was

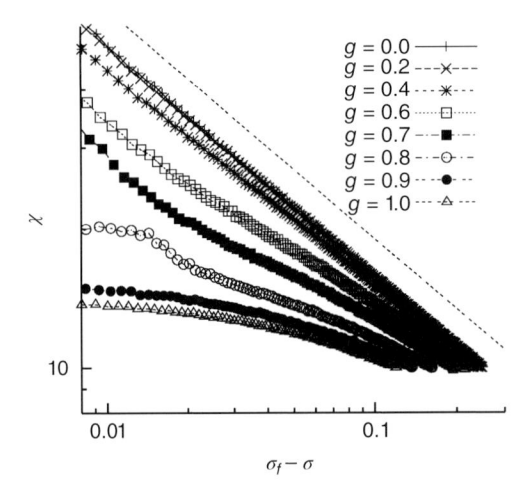

Figure 6.22 The susceptibility χ is plotted against $(\sigma_f - \sigma)$. The power-law divergence with exponent 1/2 is seen up to $g = 0.7$, beyond which no such divergence is noted. From Pradhan et al. (2005a), with kind permission from American Physical Society.

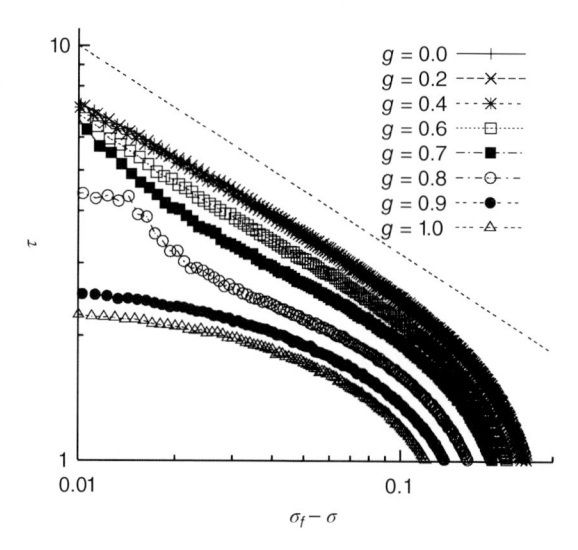

Figure 6.23 The relaxation time τ is plotted against $(\sigma_f - \sigma)$. The power-law divergence with exponent 1/2 is seen up to $g = 0.7$, beyond which no such divergence is noted. From Pradhan et al. (2005a), with kind permission from American Physical Society.

directly applied in the intact fiber. Comparisons with other loading mechanisms are discussed in the later part.

This model was studied in both one- and two-dimensional (square lattice) geometries numerically. Since this is also an interpolation mechanism ($p = 0$ corresponds to GLS and $p = 1$ corresponds to LLS), the results are to be compared with those of Pradhan et al. (2005a), where a fraction g of the load of the broken fiber is shared locally and the rest are distributed globally. In the following, we discuss the simulation results in one and two dimensions for this model.

6.3.3.1 Dependence on Loading Process

As mentioned earlier, there can be several ways in which external load can be applied into the system. The two extreme cases are (1) increasing the load up to the point when the weakest fiber breaks and (2) applying the desired load directly on the intact fiber. In addition to these two ways, one can also adopt a third possibility when the load is increased by equal amount in every step (after an avalanche stops). In GLS scheme, the fraction of surviving fibers and for that matter the critical point would not depend on how one increases the load (no memory effect). However, the situation can change when there is local load sharing mechanism present (even partially).

In the case of this model, the three ways of increasing the load were studied and it was found that the critical point differs for finite p values. In Figure 6.24, the fraction of surviving fibers for equal load increment and direct loading for $p = 0.5$ are shown. The results show that the critical point is significantly different for the

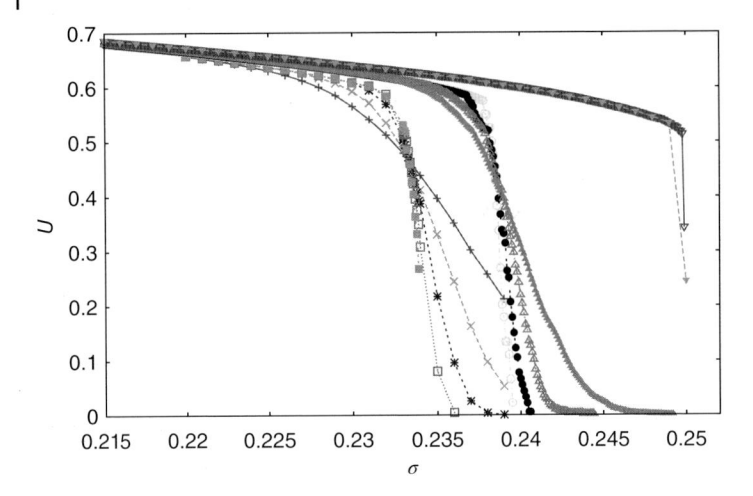

Figure 6.24 The fraction of surviving fibers is plotted against external load for two loading processes: equal load increment with $\delta\sigma = 0.0001$ and direct loading, for $p = 0.5$ and 0.0. For $p = 0.0$, the two curves for two loading processes exactly fall on top of each other, while those for $p = 0.5$ are significantly different. From Biswas and Chakrabarti (2013a), with kind permission of European Physical Journal (EPJ).

two cases. The same case for $p = 0$ (GLS model) is also shown, but it is clear that the load increment mechanisms have no effect on the GLS version, as expected. This is due to the fact that spatial correlations in avalanches are present in the local load sharing, which depends on the load increment method. But for GLS, there is no spatial structure as such (no notion of distance).

6.3.3.2 Results in One Dimension

For one-dimensional case, the fibers are arranged in a linear lattice. After applying load on the intact fiber, the entire lattice was scanned once and the fibers having threshold below the applied load were broken. This would create patches of broken fibers. Now, among them, the fraction of load carried by the local load sharing fibers, is distributed to the nearest surviving neighbors and the rest is distributed globally among the remaining fibers. This may lead to further failures and so on.

While the fraction of surviving fibers with external load was plotted, to accurately determine the critical point, the reduced fourth-order Binder cumulant (Binder and Heermann, 1988) was also studied. It has the form

$$B = 1 - \frac{\langle U^4 \rangle}{3\langle U^2 \rangle^2}, \tag{6.32}$$

and has the property that its value is independent of system size at the critical point. This helps in determining the critical point very accurately. In Figure 6.25, the fraction of surviving fibers with external load and for different system sizes (at $p = 0.3$ and 0.0) are shown. Then, in Figure 6.26, the Binder cumulants for

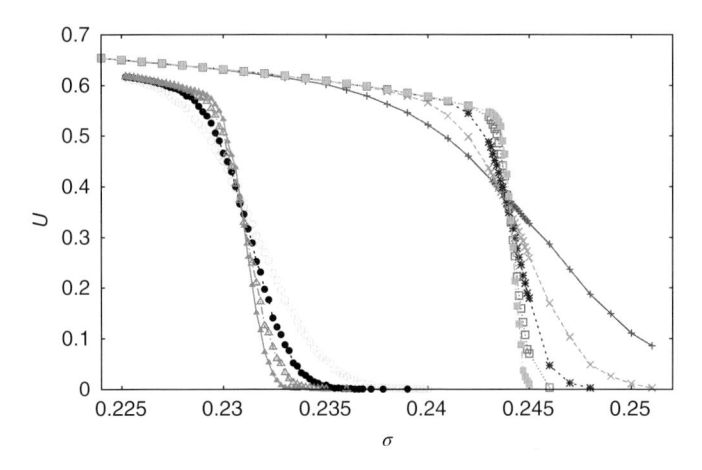

Figure 6.25 The fraction of surviving fibers is plotted with load per fiber for $p = 0.3$ in both one (left curves; $L = 10000, 25000, 50000$, and 100000) and two dimensions (right curves; $L = 50, 100, 200, 400$, and 700, $N = L \times L$). From Biswas and Chakrabarti (2013a), with kind permission of European Physical Journal (EPJ).

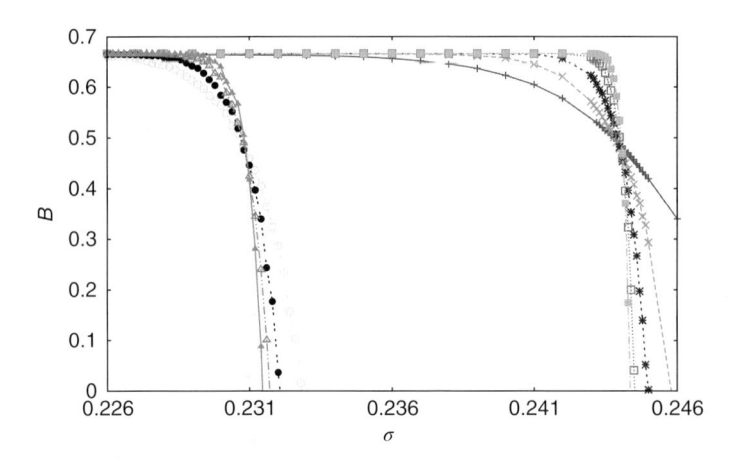

Figure 6.26 The Binder cumulants are plotted with load per fiber for $p = 0.3$ in both one (left curves; $L = 10000, 25000, 50000$, and 100000) and two dimensions (right curves; $L = 50, 100, 200, 400$, and 700, $N = L \times L$). From Biswas and Chakrabarti (2013a), with kind permission of European Physical Journal (EPJ).

different system sizes (also for $p = 0.3$ and 0.0) are plotted. The common crossing point, which signifies size-independent value, is the critical point. In this way, the phase boundary was determined.

To determine the critical behavior, the relaxation time is shown in Figure 6.27. The relaxation time is expected to diverge near the critical point as $\tau \sim (\sigma_f - \sigma)^{-z}$. From the plots, it can be seen that $z = 0.50 \pm 0.01$, which is in agreement with the GLS exact result $z = 1/2$. It was checked that these exponent values remained

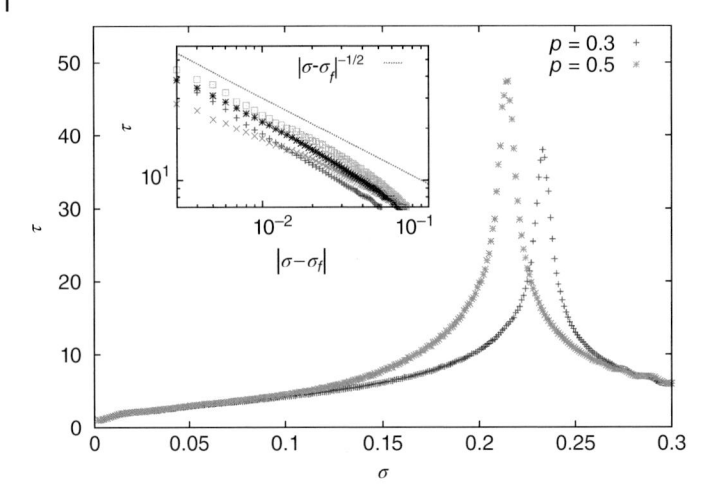

Figure 6.27 Relaxation time (τ) is plotted for one-dimensional heterogeneous model with $p = 0.3$ and 0.5. A power-law variation of the form $\tau \sim |\sigma - \sigma_f|^{-z}$ was observed (for both sides of the critical point), with $z = 1/2$, similar to the global load sharing model. From Biswas and Chakrabarti (2013a), with kind permission of European Physical Journal (EPJ).

unchanged up to $p = 0.95$ and beyond that no power-law variation was observed. Of course $p = 0.95$ is the crossover point in the phase boundary, beyond which the system fails as soon as any finite load is applied and there is no common crossing point of the Binder cumulant. This indicates that the mean-field critical behavior is present up to the crossover point and beyond that it goes over to local load sharing behavior, where of course critical behavior is not present. Of note the crossover point is almost up to the point of percolation. In one dimension, if the disorder has to percolate, its concentration will have to be unity. In this case, that value is almost reached.

6.3.3.3 Results in Two Dimensions

Similar to the case in one dimension, this heterogeneous load sharing mechanism can be followed in the more realistic case of two dimensions as well. In this case, after each scan of the lattice, the clusters of the broken fibers are identified. Then, the fraction of the load to be shared locally is distributed along the cluster boundary and the rest is redistributed globally. The fraction of surviving fibers with external load for various p values is shown in Figure 6.28.

As before, one can determine the critical point by the common crossing point of the Binder cumulant for different system sizes (see Figure 6.29). The critical point decreases with increasing fraction of the local load sharing fibers. But above a threshold value $p_c = 0.53 \pm 0.01$, there is no common crossing point of the Binder cumulant (see Figure 6.30). This suggests a crossover to local load sharing.

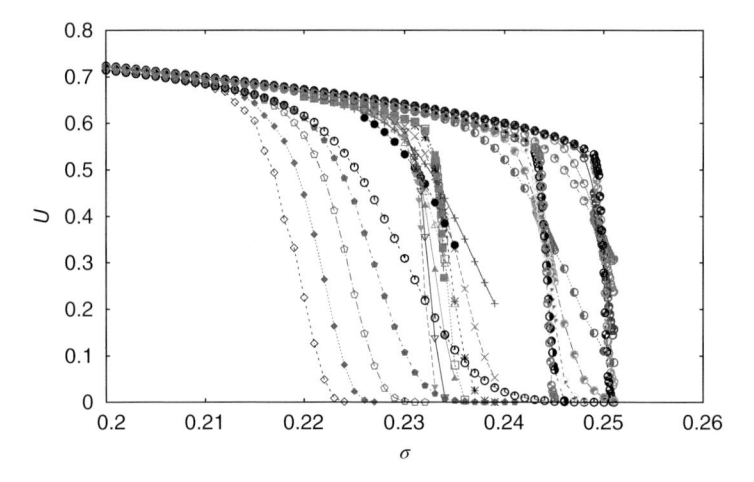

Figure 6.28 The fraction of surviving fibers is plotted against load per fiber for (left to right) $p = 0.0$, 0.3, 0.5, 0.52, and 0.6 and for $L = 50$, 100, 200, 400, and 700 for each p in two dimensions. Where the transition point became sharper with system size for $p < p_c$, the critical load goes to zero for $p > p_c$, signifying a crossover to LLS behavior. From Biswas and Chakrabarti (2013a), with kind permission of European Physical Journal (EPJ).

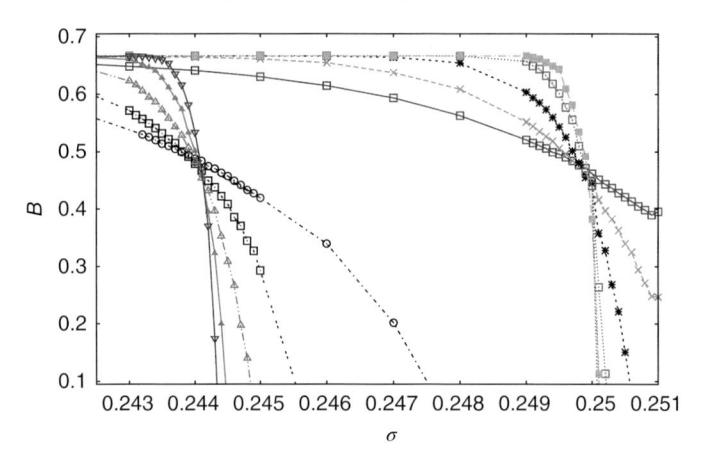

Figure 6.29 The Binder cumulants are plotted with load per fiber for $p = 0.0$ and 0.3 and different system sizes ($L = 50$, 100, 200, 400, and 700) in two dimensions. The clear common crossing point gives σ_f in each case. From Biswas and Chakrabarti (2013a), with kind permission of European Physical Journal (EPJ).

This point is also slightly below the site percolation threshold ($p_c = 0.5927\ldots$) (Stauffer and Aharony, 1994). The critical behavior, determined from the avalanche size distribution (Figure 6.31) and divergence of relaxation time (Figure 6.32), is same as that of GLS model. The phase diagrams for one and two dimensions for this model are shown in Figure 6.33.

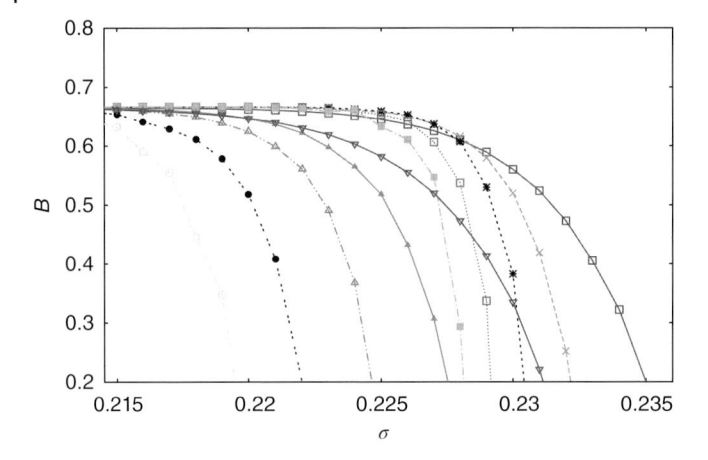

Figure 6.30 The Binder cumulants are plotted with load per fiber for $p = 0.55$ and 0.60 and different system sizes in two dimensions. By sharp contrast, with the case of $p < p_c$, here we do not find any common crossing point for different sizes. From Biswas and Chakrabarti (2013a), with kind permission of European Physical Journal (EPJ).

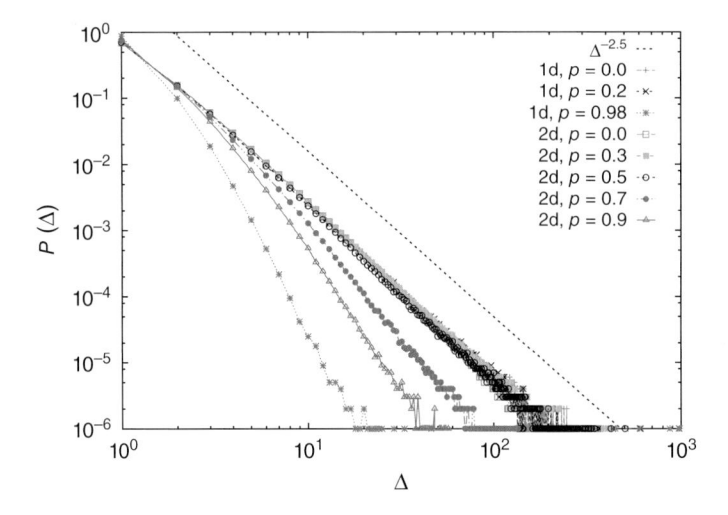

Figure 6.31 The avalanche size distribution is plotted for $p = 0.0$, 0.5, and 0.7. It is seen that for $p = 0.0$ and 0.5 the size distribution follows the same power-law with exponent 2.5 ± 0.01 and for $p = 0.7 > p_c$, there is no scale-free behavior. From Biswas and Chakrabarti (2013a), with kind permission of European Physical Journal (EPJ).

The stress (σ)-strain (σ^*) (with $\sigma^* = \sigma/U$) relation for the model (see Figure 6.34) was also studied. While individual fibers behave linearly, the collective stress–strain curve is nonlinear but approaches linearity as $\sigma \to 0$. The behaviors for both one and two dimensions and for different values of p, which is a measure of weakness of the material, were shown and the breakdown occurs for lower stress as p is increased.

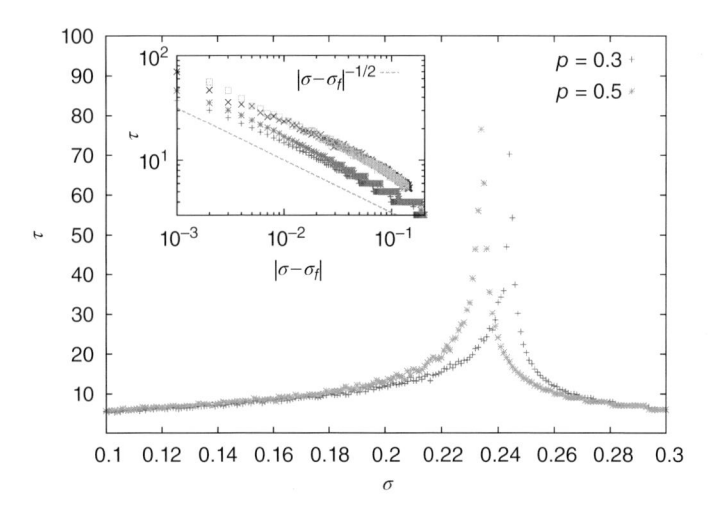

Figure 6.32 Relaxation time (τ) is plotted for two-dimensional heterogeneous model with $p = 0.3$ and 0.5. A power-law variation of the form $\tau \sim |\sigma - \sigma_f|^{-z}$ was observed (for both sides of the critical point), with $z = 1/2$, similar to the global load sharing model. From Biswas and Chakrabarti (2013a), with kind permission of European Physical Journal (EPJ).

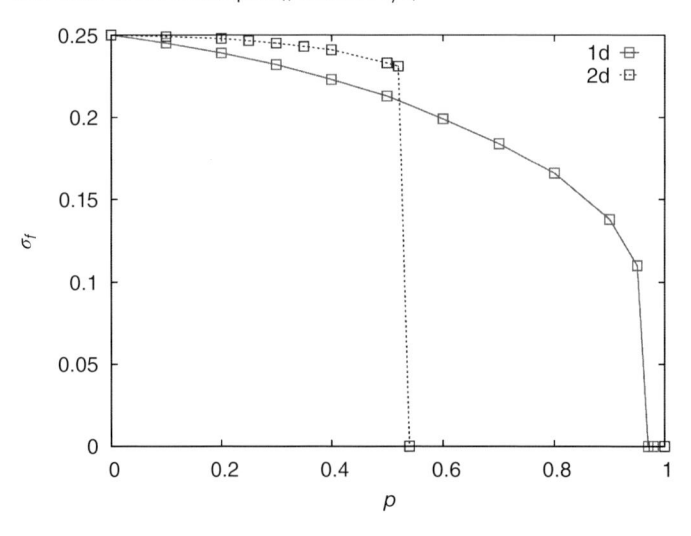

Figure 6.33 The phase diagrams for heterogeneous load sharing in one and two dimensions. The crossover points are at $p_c = 0.95 \pm 0.01$ and $p_c = 0.53 \pm 0.01$, respectively. From Biswas and Chakrabarti (2013a), with kind permission of European Physical Journal (EPJ).

In Figure 6.35, the fraction of surviving fibers for different loading methods is shown for a given system size and p value. It was found that the weakest fiber breaking and the direct loading are two limits, and equal load increment method interpolates between the two. For very small value of the fixed increment rate ($\delta\sigma = 0.0001$), the curve is almost identical with that in the weakest fiber breaking.

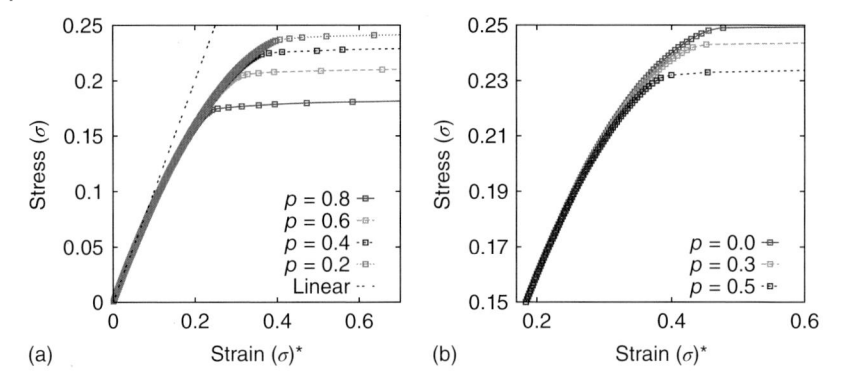

Figure 6.34 (a) shows the stress (σ)–strain (σ^*) relation for different p values for an one-dimensional model. The linear relation is plotted to show that the relation converges to linearity when $\sigma \to 0$. Similar behaviors are shown for two-dimensional version of the model on (b). From Biswas and Chakrabarti (2013a), with kind permission o European Physical Journal (EPJ).

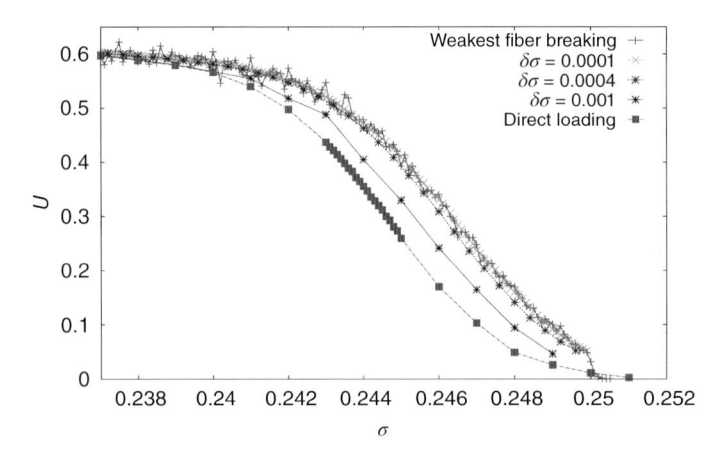

Figure 6.35 The fraction of surviving fibers is plotted against external load for different loading processes when $p = 0.3$. For equal load increment, $\delta\sigma = 0.0001$ almost matches the curve obtained for loading up to the weakest fiber breaking. The equal step load increment with $\delta\sigma = 0.0004$ and 0.001 is also shown. They are in between the $\delta\sigma = 0.0001$ curve and that due to direct loading. From Biswas and Chakrabarti (2013a), with kind permission of European Physical Journal (EPJ).

This is because, for $\delta\sigma \sim 1/N$, the equal load increment method should be effectively same as that of the weakest fiber breaking.

Finally, note that even though the critical point is dependent on the load increment mechanisms for finite p values, the crossover point and the critical behavior below the crossover point are unaffected by the load sharing mechanisms. The results in the aforementioned discussion are for direct loading, except for the avalanche size distribution, which is for the weakest fiber breaking.

6.3.3.4 Comparison with Mixed Load Sharing Model

This model is also an interpolation scheme between LLS and GLS. In fact this is the heterogeneous version of the mixed load sharing model (Pradhan et al., 2005a) (we refer to this as 'g' model). This model was studied in one dimension in Pradhan et al. (2005a). This model has been studied in two dimensions as well. Even though the amount of load shared locally on average is same for the two models, the local fluctuations introduced in the present version make it significantly different from the earlier one.

The relaxation time divergence of this model in two dimensions (Figure 6.36) shows the same critical behavior as that of the one-dimensional version (and GLS model). In Figure 6.37, the phase diagrams of the two models in one and two dimensions are compared, which shows the effect of Heterogeneity, as the phase boundaries do not coincide.

6.4
Random Threshold Spring Model

The fiber bundle model discussed so far is essentially a mean-field model. The attempts to consider local effects led it to the situation where it cannot support any finite load in the large system size limit. The random threshold spring model mimics the elastic deformation of the solids a little more realistically, and takes somewhat the local deformations into account. We describe this model and its relation to avalanche dynamics here.

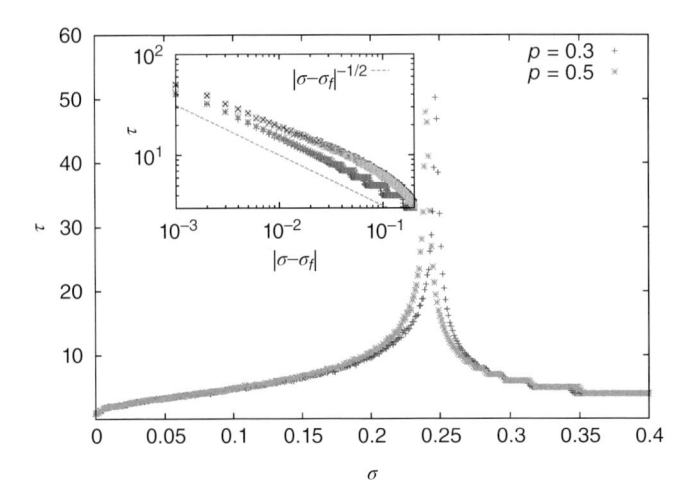

Figure 6.36 Relaxation time (τ) is plotted for two-dimensional homogeneous model (Pradhan et al., 2005a) with $p = 0.3$ and 0.5. A power-law variation of the form $\tau \sim |\sigma - \sigma_f|^{-z}$ was observed (for both sides of the critical point), with $z = 1/2$, similar to the global load sharing model. From Biswas and Chakrabarti (2013a), with kind permission of European Physical Journal (EPJ).

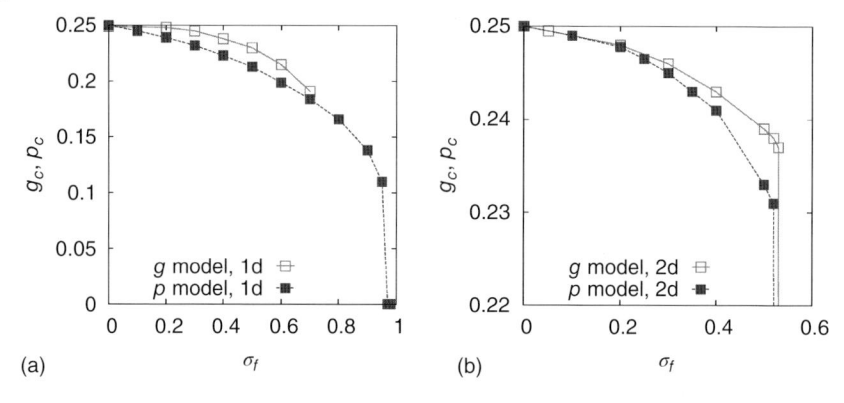

Figure 6.37 (a) shows the comparisons of the phase diagram of the heterogeneous and the homogeneous models in one dimension and (b) shows the comparisons of the phase diagram of the heterogeneous and the homogeneous models in two dimensions. From Biswas and Chakrabarti (2013a), with kind permission of European Physical Journal (EPJ).

Let us discuss (Ray, 2006) a simplified picture of the heterogeneous systems and do not take into account the full details of the defects or their effects on the elastic response of the system as the defects grow. We consider a discrete two-dimensional lattice where the bonds are Hookean springs (of identical spring constant) and mimic the heterogeneity by assigning a random breaking threshold σ_{th} drawn from a distribution $p(\sigma_{th})$ to each of the springs. The dynamics progresses in much the same way as the fiber bundle model described earlier, except for the fact that the springs are interconnected, in contrast with the parallel connections of the fibers in the fiber bundle model.

The network is subjected to a tensile stress in both x- and y-directions. A spring behaves like a Hookean object except that it can be stretched till the threshold value when it ruptures irreversibly (like the longitudinal fibers in the fiber bundle model). Subsequent to a rupture, the stress is redistributed over the remaining intact part of the network. Breaking up of a spring mimics the nucleation or onset of fracture. It can lead to further breaking of the springs, and the breaking process continues or the breaking event may stop where the stress level on the network is to be increased to induce further breaking. We discuss how the breakdown properties of this random spring network model give rise to power-laws in breakdown events and compare the results with those of experimental findings.

The study of fracture in random spring network is carried out by molecular dynamics simulation. The system consists of a $L \times L$ ($L = 50$, 100, and 200) square network with central and rotationally invariant bond-bending forces. The potential energy of the network is (Zapperi et al., 1997a, 1999b)

$$V = \frac{a}{2} \sum_{\langle ij \rangle} (\delta r_{ij})^2 g_{ij} + \frac{b}{2} \sum_{\langle ijk \rangle} (\delta \theta_{ijk})^2 g_{ij} g_{jk}, \qquad (6.33)$$

where δr_{ij} is the change in the length of the spring between the nearest neighbor sites $<ij>$ from its equilibrium value (which is the lattice spacing in the starting unstretched condition and is taken to be unity) and $\delta\theta_{ijk}$ is the change in the angle between the adjacent springs ij and jk from its equilibrium value, which is taken to be $\pi/2$ to ensure the square lattice structure of the unstretched starting configuration of the network (see Figure 6.38). If the spring ij is present, $g_{ij} = 1$, otherwise (when the spring is broken) $g_{ij} = 1$. Force constants of the central and the bond-bending force terms are a and b, respectively. The dimensionless equation of motion

$$\frac{d^2 r_i}{dt^2} = \gamma_1 \sum_{<j>}(\delta r_{ij})g_{ij} + \gamma_2 \sum_{<jk>}(\delta\theta_{ijk})\frac{\partial\theta_{ijk}}{\partial r_i}g_{ij}g_{jk}, \qquad (6.34)$$

involves two parameters $\gamma_1 = at_0^2/m$ and $\gamma_2 = bt_0^2/ml_0^2$ in terms of the mass m associated with the lattice sites, an arbitrary length scale l_0 and an arbitrary timescale t_0. The ratio $\gamma_1/\gamma_2 = a/l_0 b$ is a characteristic of the system under consideration. This suggests that the dynamical features of the network as described by the equation of motion do not depend on the choice of the scale of mass or time. The obvious choice for l_0 is unity, which is the lattice spacing of the lattice at the unstretched condition. $\gamma_1 = 1.0$ and $\gamma_2 = 0.1$ were chosen. A small value of γ_2, much less than the value of γ_1, allows the fracture to develop without much deformation of the network. The dynamics starts with all the springs intact so that $g_{ij} = 1$ for all neighboring ij's and with each spring, a random breaking threshold $(\sigma_{th})_{ij}$ is associated, chosen from a uniform distribution $\rho(\sigma_{th}) \in [0, 2]$.

A constant external force F is imposed on the sites of the boundary and the system is allowed to evolve dynamically using Verlet algorithm (Allen and Tildesley, 1987),

$$\vec{r}_i(t + \Delta t) = 2\vec{r}_i(t) - \vec{r}_i(t - \Delta t) + \vec{F}_i(t)(\Delta t)^2. \qquad (6.35)$$

Here, $\vec{F}_i(t)$ is the force (as determined from the potential energy and boundary condition) and $\vec{r}_i(t)$ is the position vector of the site i at time t. The simulation involves discrete time t in steps of Δt. After N iterations, the time elapsed is $n\Delta t$ while the real time elapsed is $nt_0\Delta t$. In order to speed up the computation, one

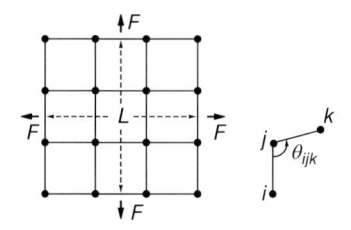

Figure 6.38 Schematic diagram of the elastic network used in the simulation. This is the network before the application of the force and with all the bonds intact. The bonds are Hookean springs and there is an angular force between any two adjacent bonds. The deformations are measured from the square configuration of the network. From Ray (2006), with kind permission from Elsevier.

would wish to choose a large value of Δt. However, there is an upper limit to this value given by the convergence time for the fastest developing components of the stress distribution, which is generally very small in disordered systems. In this case, $\Delta t = 0.01$ was chosen. Also, a small viscous damping was added to the evolution to avoid excessive oscillations and to achieve equilibrium for a given applied force faster. For a given applied force, once the system reaches equilibrium, it was checked if any spring ij is stretched beyond its cut-off value $(\sigma_{th})_{ij}$ and if this happens, the spring was snapped irreversibly (g_{ij} for that spring is set to zero). Once the springs are broken, the system is again brought to equilibrium and the springs are checked again to see if the initial set of breaking initiates further rupturing of springs. When no more breaking of springs takes place, the external force F is increased in small steps. At each step, the number of broken bonds is counted, which constitute an avalanche. In order to average over disorder, the simulation was repeated for 50 different configurations of threshold values σ_{th}.

These simulations show that the fracture in the spring network develops in a series of bursts of spring-rupturing processes. In one such burst, bonds rupture from different parts of the network in a random fashion. Figure 6.39 shows the

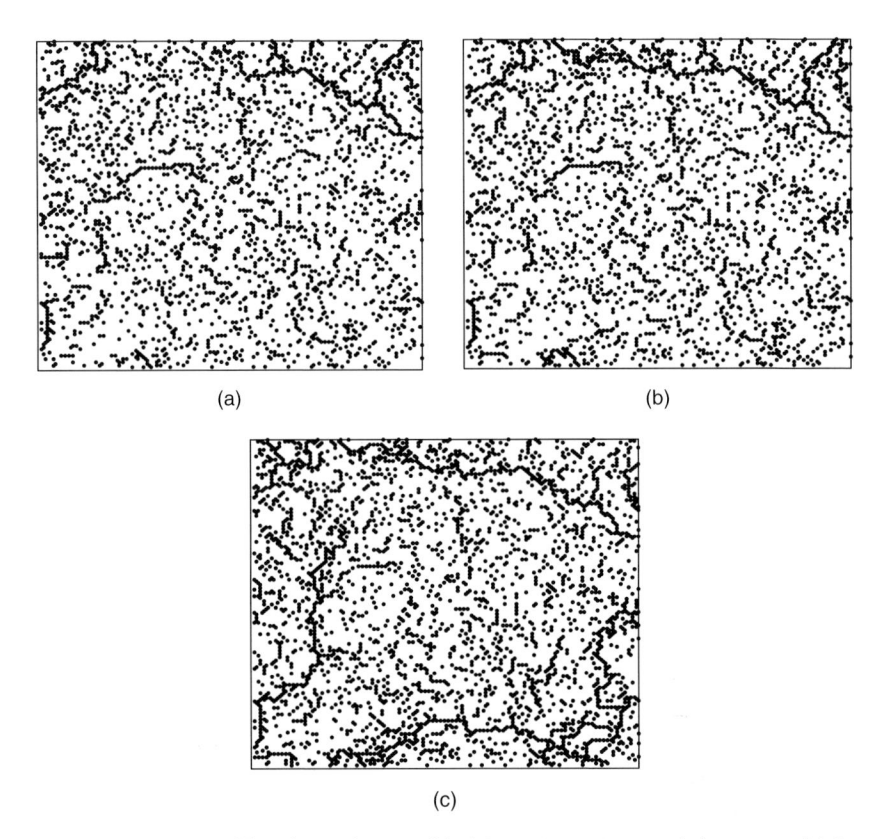

Figure 6.39 Ruptured bonds are shown in black in a 100×100 network for stresses (a) $F = 0.10$, (b) $F = 0.20$, and (c) $F = 0.25$. From Ray (2006), with kind permission from Elsevier.

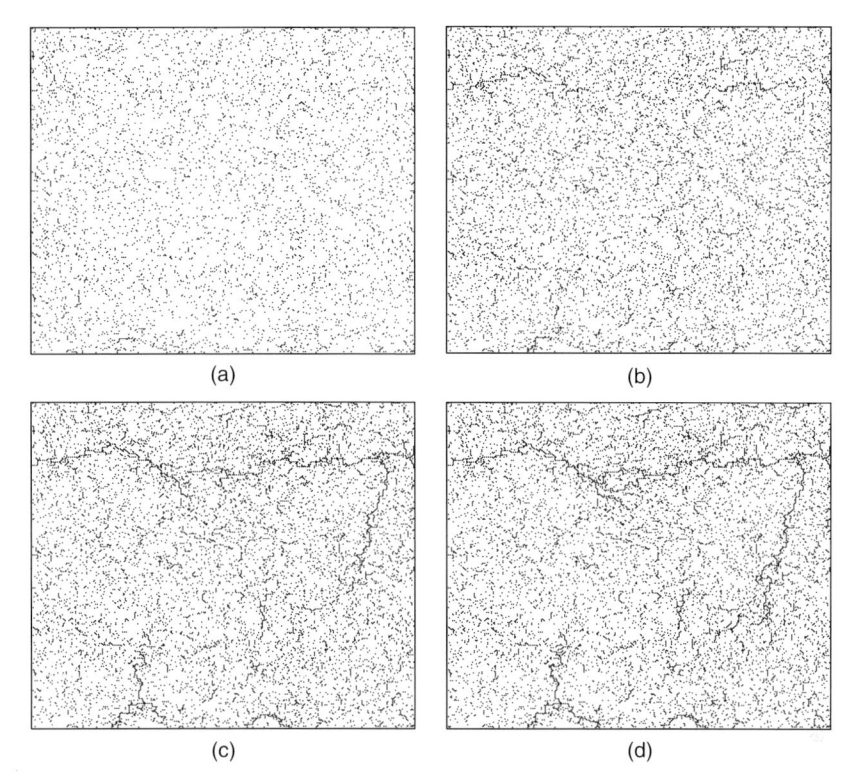

Figure 6.40 Ruptured bonds are shown in black in a 200 × 200 network subjected to the stress $F = 0.20$ at molecular dynamics time steps (a) $t = 120\,000$, (b) $t = 240\,000$, (c) $t = 360\,000$, and $t = 480\,000$. From Ray (2006), with kind permission from Elsevier.

ruptured bonds in a 100 × 100 lattice for $F = 0.10$, 0.20, and 0.25. A well-defined macroscopic fracture across the network was found at $F = 0.20$. Below this critical value of F, there is no crack that spans the network and the bonds rupture randomly and uniformly over the network. This phenomenon has also been observed in the experiment (Garcimartin et al., 1997) as mentioned earlier. Figure 6.40 shows the development of fracture in the network with time. At early times, the bonds are broken randomly over the network. At later times, the microcracks start coalescing and a large crack develops, which wins over the others and engulfs nearby microcracks to form a crack that spans the system. One keeps track of the clusters formed by the adjacent broken bonds (Zapperi et al., 1999a). In this respect, an isolated single broken bond form a cluster of size one. The number n_c of such clusters grows with the stress F following the relation $n_c = L^2 g(F)$ (see Figure 6.41), where $g(F)$ is a scaling function of F. This relation remains valid till the breakdown point indicating that the final crack results from sudden coalescence of few large microcracks without any drastic change in the number n_c. The final breakdown resembles a first-order transition, and the scaling form of n_c further strengthens this point of view. Figure 6.42 shows the variation of the average

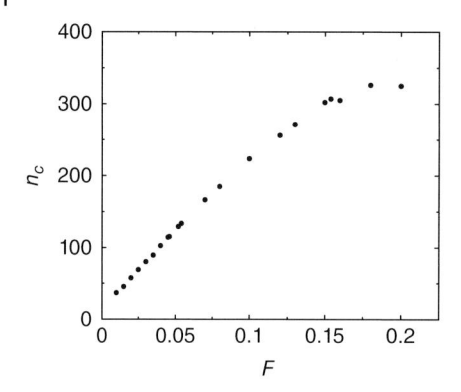

Figure 6.41 The number of clusters n_c of ruptured bonds is plotted against the stress F in a network of size $L = 50 \times 50$. From Ray (2006), with kind permission from Elsevier.

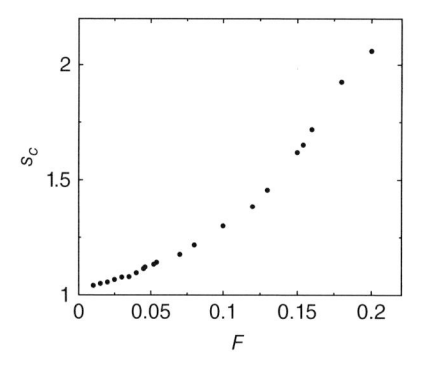

Figure 6.42 The average size s_c of the clusters of ruptured bonds is plotted against the stress F in a $L = 50 \times 50$ network. From Ray (2006), with kind permission from Elsevier.

size s_c of the clusters of ruptured bonds with the stress F. No evidence of divergence of s_c was found, which is a strong indication that it is not a second-order phase transition. In fact, the average size s_c remains finite and quite small, which suggests that the final breakdown is a highly correlated phenomenon involving coalescence of very few microcracks. Next, the distribution $P(\Delta)$ of the size Δ of burst or avalanche (number of bonds that snap in a burst) integrated over all the values of stress F up to the breakdown point F_c was considered. It was found that $P(\Delta) \sim \Delta^{-b}$ with $b = 5/2$ as is shown in Figure 6.43. This amplitude distribution can be transformed into an energy distribution (the energy is proportional to the square of the amplitude) giving the exponent $\beta = \frac{1+b}{2} = 1.75$. This is closer to the experimental results. In Figure 6.44, $<\Delta>^{-2}$ is plotted against F and a linear behavior is seen, which suggests $<\Delta> \sim (F_c - F)^{-1/2}$ so that the exponent $\lambda = 0.5$ in this model.

Therefore, we see that the dynamical response of a simple elastic network in the presence of a threshold and extreme dynamical rules (assigning a random breaking threshold with each bond and specifying the extreme dynamical rule in the

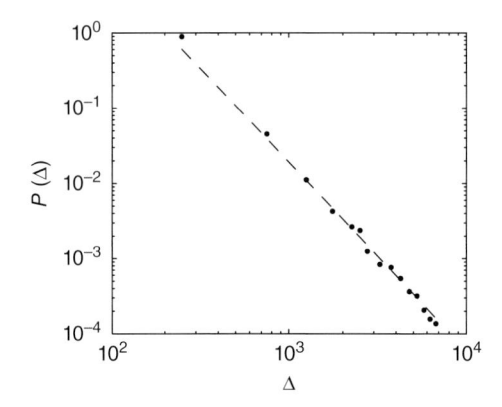

Figure 6.43 The size distribution $P(\Delta)$ is plotted against the size Δ of avalanche of ruptured bonds integrated over all the values of stress up to the breakdown point F_c in a $L = 50 \times 50$ network. From Ray (2006), with kind permission from Elsevier.

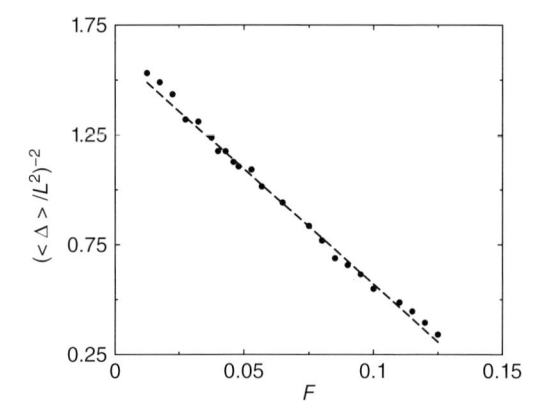

Figure 6.44 $(<\Delta>/L^2)^{-2}$ is plotted against F, where $<\Delta>$ is the average size of the avalanche of ruptured bonds in $L = 50 \times 50$ network integrated up to the stress F, and F_c is the stress at which the network fails completely. From Ray (2006), wih kind permission from Elsevier.

bond breaking process) produces several features characteristic of fracture in heterogeneous materials. It gives the power-law behaviors of the avalanche statistics, which are observed in the experiments. The simulation shows the right trend of development of fracture with time and with stress as is observed in experiments.

6.5 Summary

The presence of heterogeneity affects the dynamics of crack propagation drastically. While it causes deflection of crack tip due to inhomogeneity in pressure and plastic deformation, crack tip can also get arrested due to the presence of defects (Kierfeld and Vinokur, 2006) (see however, Ramos et al., 2013). The spatial nonuniformity in crack propagation of heterogeneous solids (to be discussed in Chapter 8) is also reflected in the temporal sequence, leading to punctuated equilibrium and scale-free size distributions of avalanche statistics.

The experimental observations of avalanche dynamics are also complimented by theoretical models. We have mainly discussed the fiber bundle model and also the random threshold spring model. The inherent mean-field nature of the equal load sharing fiber bundle models (discussion in Section 6.2) enables us to construct recursion relations (Eq. (6.14) for example), which captures essentially all the intriguing features of the failure dynamics. The "type" of phase transition in such models has been a controversial issue. Earlier, it was suggested to be a first-order phase transition, because the surviving fraction of fibers has a discontinuity at the breakdown point of the bundles. However, as the susceptibility shows divergence ($\chi \propto (\sigma_f - \sigma)^{-\gamma}; \gamma = 1/2$) at the breakdown point, the transition has been later identified to be of second order (Pradhan and Chakrabarti, 2001). The dynamic critical behavior of these models and the universality of the exponent values are straightforward. Here, divergence of relaxation time (τ) at the critical point ($\tau \propto (\sigma_f - \sigma)^{-z}; z = 1/2$) indicates "critical slowing" of the dynamics, which is characteristic of conventional critical phenomena. At the critical point, one observes power-law decay of the surviving fraction in time ($U_t(\sigma_f) \propto t^{-x}$; $x = 1$). Furthermore, the exponent values are universal for different distributions of failure thresholds.

In addition to the critical behavior around the failure point, the model shows the intermittent dynamics or avalanche statistics when slowly driven (external load increased slowly). The mean-field nature of the model enables one to calculate the avalanche size exponents. The value of the exponent, however, depends on (5/2 or 3) the way the load is increased. Also, the size distribution exponent changes from 5/2 to 3/2 as the failure point is approached (see section 6.2.3).

The other extreme of the model, that is, the local load sharing version, does not show avalanche dynamics. As discussed in Chapter 4, its failure properties are governed by extreme statistics and the critical strength of the system vanishes in the large system size limit. However, since the GLS version has no notion of length and the effect of fluctuations cannot be seen there, there have been many attempts to see the effect of spatial fluctuation in the model by interpolating it between the GLS and LLS schemes (Hidalgo et al., 2002b, 2008b; Pradhan et al., 2005a; Raischel et al., 2006). The basic result of the interpolation schemes is that up to a certain value of the tuning parameter that interpolates between the two regimes, the avalanche statistics (are other related quantities) behave as GLS case and beyond that value, the behavior abruptly shifts to LLS case. Often one can interpret the threshold value of the parameter physically. For the power-law load sharing case (Hidalgo et al., 2002b), that is, when the load of a broken fiber is shared by the intact fibers in some inverse power of the distance from the broken fiber, the threshold value for the power-law exponent is 2 in a square lattice (up to which the function remains normalizable). For the mixed-mode load sharing case (fraction of load shared locally and the rest distributed globally), however, the meaning of the threshold value is not immediately clear. But for the heterogeneous case, where a fraction of the fibers is global load sharing and the other fraction is local load sharing, an upper limit of the threshold can be argued to be the random percolation threshold and indeed the simulation results were consistent with it (Biswas and Chakrabarti, 2013a).

We have also discussed the random spring model for failure. The size distribution of avalanche gives an exponent value 5/2, and when translated to the energy distribution, it gives the exponent 1.75. However, the nature of transition in this model is first-order.

7
Subcritical Failure of Heterogeneous Materials

We have discussed so far the failure properties when the applied load on a solid is gradually increased until it breaks completely. However, even if a load smaller than the breaking threshold of a solid is applied on it, it can eventually break after sufficient time due to microcracks and their nucleations driven by the external noise, mainly temperature. This so called creep rupture has applications in various practical matters concerning stability of structures in different scales.

In this chapter, we discuss creep rupture events in heterogeneous solids and also failure due to fatigue due to repeated loading. The heterogeneity plays a crucial role in determining the time to failure in case of a rupture event by renormalizing the temperature. Here, we mainly discuss experimental observations and theoretical modeling attempts of different aspects of creep failure, namely the time to failure, the dynamics of strain rate and size distributions of avalanches (e.g., acoustic emissions), and waiting time. In addition to that, we briefly discuss localization of damage in creep.

7.1
Time of Failure Due to Creep

We begin by extending the framework of Griffith's criterion of crack propagation to creep dynamics (Pradhan and Chakrabarti, 2003a). One can consider the crack propagation for $\sigma_0 < \sigma_f$ at a finite temperature to capture the fatigue behaviors of the material. Consider a crack (of length a) in three dimensions. It can propagate at a finite temperature T, with a probability $\exp(-E(a)/k_B T)$, with

$$E(a) = E_e + E_s \sim \Gamma a^2 - \frac{\sigma_0^2}{Y} a^3, \tag{7.1}$$

where E_e and E_s are elastic and surface energies, respectively, as obtained in 4.1.1 One can now define failure time as the inverse of the rate

$$\tau \sim \exp[E(a)/k_B T] = \exp\left[A\left(1 - \frac{\sigma_0^2}{\sigma_f^2}\right)\right], \tag{7.2}$$

Statistical Physics of Fracture, Breakdown, and Earthquake: Effects of Disorder and Heterogeneity, First Edition.
Soumyajyoti Biswas, Purusattam Ray, and Bikas K. Chakrabarti.
© 2015 Wiley-VCH Verlag GmbH & Co. KGaA. Published 2015 by Wiley-VCH Verlag GmbH & Co. KGaA.

with $A \sim a^3 \sigma_f^2/(Yk_BT)$. This shows that failure time will increase very fast when σ_0 is gradually decreased below σ_f and becomes vanishingly small when $\sigma_0 > \sigma_f$.

However, it was later argued by Pomeau (1992) (see also Rabinovitch et al., 2004) that the barrier height that a nucleation process must overcome should be calculated by maximizing $E(a)$ with respect to crack length. Clearly, $E_m \sim 1/\sigma_0^4$. Hence, the time of failure should go as

$$\tau \sim \exp\left[\sigma_f^4/\sigma_0^4\right], \tag{7.3}$$

for three-dimensional homogeneous material without preexisting voids. In general, for a d-dimensional solid, the time of failure takes the form (Guarino et al., 1999b)

$$\tau \sim \exp\left(C\frac{\Gamma^d T^{d-1}}{T_{\text{eff}} \sigma^{2d-2}}\right), \tag{7.4}$$

where C is a constant depending on the geometry of the material and T_{eff} is the effective temperature considering the enhancement by internal disorder (to be discussed later).

The above calculation is valid when the crack is formed in a homogeneous solid. However, the presence of flaw distribution in the solid can change the failure time considerably. If there are preexisting cracks (Rabinovitch et al., 2004), then if one assumes that a crack has to increase from length a_0 to a_m, then it will have to overcome a reduced barrier given by $E = E_m - E_0$, where

$$E_0 \sim \Gamma a_0^2 - \frac{\sigma_0^2}{Y}a_0^3. \tag{7.5}$$

The time of failure, therefore, scales as

$$\tau \sim \exp\left((A/\sigma_0^4 + B\sigma_0^2 a_0^3 - Ca_0^2)\right), \tag{7.6}$$

where A, B, and C are constants depending upon geometry, elastic modulus, and surface energy. Now, for the distribution of the preexisting flaws, one uses extreme value statistics and can consider the Gumbel distribution:

$$p(a_l) = \frac{1}{q}\exp\left(-(a_l - \mu)/q\right)\exp\left(-\frac{a_l - \mu}{q}\right), \tag{7.7}$$

where μ and q are free parameters. These parameters can be varied to match the experimental results.

In addition to the macroscopic models mentioned earlier, there have been many attempts to theoretically model the creep dynamics and slow ruptures from microscopic discrete elements having heterogeneities in the form of failure thresholds (see Vanel et al., 2009 for a review). In the following, we discuss those attempts mainly from the fiber bundle model studies.

7.1.1
Fluctuating Load

As mentioned earlier, the failure of a sample can take place before its critical load, when thermal noise helps in overcoming the nucleation barrier. It was also argued that heterogeneity of the sample enhances the effect of temperature, hence facilitating the failure. The arguments were based on the Griffith's criterion of crack propagation. A model-based approach is, however, more difficult to tackle analytically, particularly in low dimensions. Here, we discuss the subcritical failure of fiber bundle models, where some exact and approximate calculations are possible, since it is generally a mean-field model.

To take into account the creep test in disordered material, Guarino et al. (1999b) introduced the fiber bundle model with thermal noise as well as internal disorder. Approximate analytical calculations of this model were done later by Roux (2000). Following that, let us first discuss the case of homogeneous bundle, where the breaking threshold is same ($\sigma_{th} = 1$) for every fiber, but the force on each fiber has a fluctuating part depending on temperature, but is temporally and spatially uncorrelated:

$$\sigma = \sigma_0 + \eta, \tag{7.8}$$

where the noise η is taken from the following distribution function

$$p(\eta) = \frac{1}{\sqrt{2\pi k_B T}} \exp\left(-\frac{\eta^2}{2k_B T}\right). \tag{7.9}$$

Now, the breaking probability at each time is the probability that the noise must be greater than $1 - \sigma_0$, that is, $\int_{1-\sigma_0}^{\infty} p(x)dx = P(1 - \sigma_0)$, say, where $P(\eta)$ is the cumulative distribution function. Now, since the noise is uncorrelated in time, the surviving probability up to time t is

$$p_1(t) = [1 - P(1 - \sigma_0)]^t. \tag{7.10}$$

Therefore, the probability that the entire bundle survives is

$$p_N(t) = [1 - P(1 - \sigma_0)]^{Nt}. \tag{7.11}$$

This is an exponential decay, where the typical timescale can be interpreted as the average time of failure of the bundle (i.e., the time when surviving probability has decayed substantially). Clearly, that timescale (or, average failure time for the first fiber) is given by

$$\langle \tau_1 \rangle = -\frac{1}{N \ln(1 - P(1 - \sigma_0))}. \tag{7.12}$$

The structure remains the same after each failure of fiber, and only the number of fibers is reduced. Therefore, the total failure time of the bundle will be

$$\langle \tau_f \rangle = \sum_{i=1}^{N} \frac{-1}{(N-i)\ln(1 - P(1 - N\sigma_0/(N-i)))}. \tag{7.13}$$

For large N, it can be written as an integral as follows:

$$\langle \tau_f \rangle = \frac{1}{N}\int_0^N \frac{-N}{(N-i)\ln(1 - P(1 - N\sigma_0/(N-x)))}dx = \int_{\sigma_0}^{\infty} \frac{-1}{\ln(1 - P(1 - y))} \frac{dy}{y}. \tag{7.14}$$

This cannot be integrated to give a closed-form solution for the case of Gaussian noise. However, when the load $\sigma_0 \ll 1$ and $k_B T \ll 1$, the failure time is dominated by the first fiber's breaking time. Then, P is much smaller than 1. One can now expand the derivative of τ_f with respect to σ_0 for small P and integrate that. Now,

$$\frac{\partial \langle \tau_f \rangle}{\partial \sigma_0} = \frac{1}{\sigma_0 \ln(1 - P(1 - \sigma_0))} \approx \frac{1}{\sigma_0 P(1 - \sigma_0)}. \tag{7.15}$$

For $(1 - \sigma_0)^2 \gg k_B T$, one gets by expanding the error function

$$P(1 - \sigma_0) = \frac{\sqrt{k_B T} \exp\left(-\frac{(1-\sigma_0)^2}{2k_B T}\right)}{\sqrt{2\pi}(1 - \sigma_0)} [1 + \mathcal{O}(k_B T)]. \tag{7.16}$$

Keeping the leading order terms in $k_B T$, one gets

$$\langle \tau_f \rangle \approx \frac{\sqrt{2\pi k_B T}}{\sigma_0} \exp\left(\frac{(1 - \sigma_0)^2}{2k_B T}\right). \tag{7.17}$$

This result compares well (see Figure 7.1) with numerical results for different values of $k_B T$.

Let us now consider the effect of quenched disorder in the failure thresholds of the fibers. Let the failure thresholds for each fiber σ_{th}^i are Gaussian distributed with mean 1 and variance $k_B \Theta$, where Θ is an equivalent temperature to measure the variance. As before, the survival probability of the full bundle up to time t can be written as

$$s(t) = \prod_{i=1}^{N} [1 - P(\sigma_{th}^i - \sigma_0)]^t. \tag{7.18}$$

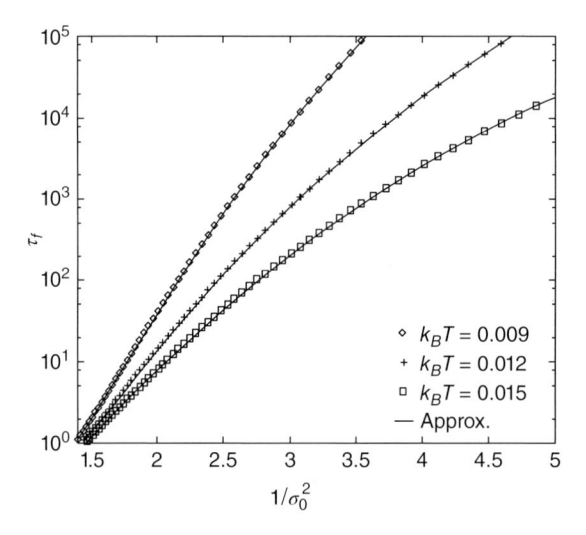

Figure 7.1 The dependence of failure time (τ_f) on external load (σ_0) for different temperatures. The numerical points match well with approximate expression Eq. (7.17). From Roux (2000), with kind permission from American Physical Society.

In the limit of large N, this can be written as

$$s(t) = \exp(-t/\tau_1), \qquad (7.19)$$

with

$$\frac{1}{\tau_1} = -\int_{-\infty}^{+\infty} \ln[1 + x - \sigma_0] g(x) dx, \qquad (7.20)$$

where $g(x)$ is a Gaussian distribution as mentioned earlier. Now, in the limit $k_B T \ll (1 - \sigma_0)^2$ and $k_B \Theta \ll (1 - f_0)^2$, for small P, one gets

$$\frac{1}{\tau_1} = N \sqrt{\frac{T}{2\pi\sqrt{\Theta}}} \exp\left(-\frac{(1-\sigma_0)^2}{2k_B T}\right)$$

$$\int_{-\infty}^{+\infty} \frac{1}{1+x-\sigma_0} \times [1 - \mathcal{O}(k_B T)] \exp\left(-\frac{(T+\Theta)x^2}{2k_B T\Theta}\right)$$

$$\times \exp\left(-\frac{x(1-\sigma_0)}{k_B T}\right) dx. \qquad (7.21)$$

Further reducing it in the limit $(1 - \sigma_0)^2 \gg k_B \Theta(T + \Theta)/T$ gives

$$\tau_1 = \frac{\sqrt{2\pi}}{N} \frac{(1-\sigma_0)}{\sqrt{k_B(T+\Theta)}} \exp\left(\frac{(1-\sigma_0)^2}{2k_B(T+\Theta)}\right). \qquad (7.22)$$

One can in fact compare this with the previous expression without quenched disorder Eq. (7.17), and see that the role of disorder in the failure threshold is to increase the effective temperature of of the system, as one can write the effective temperature as

$$T_{\text{eff}} = T + \Theta. \qquad (7.23)$$

This is a step toward explaining the experimental fact (Pauchard and Meunier, 1993; Guarino et al., 1999a) that for fitting the data, one has to assume a much higher effective temperature than the actual room temperature. However, this is only the time up to the first failure, which is a small fraction of the total time of failure (discussed later).

For the disordered case (Scorretti et al., 2001; Ciliberto et al., 2001), the failure thresholds (σ_f) are random variables taken from the distribution $G(\sigma_f)$. This threshold distribution will now evolve with time. If one denotes the fraction of surviving fibers at time t as $U(t)$, then

$$\frac{\partial G(\sigma_f, t)}{\partial t} = -G(\sigma_f, t) P\left(\sigma_f - \frac{\sigma_0}{U(t)}\right)$$

$$\int_0^\infty G(\sigma_f, t) d\sigma_f = U(t). \qquad (7.24)$$

The initial distribution, $G(\sigma_f, 0)$ is a Gaussian with mean 1 and variance $k_B \Theta$. In Figure 7.2, the time evolution of $G(\sigma_f, t)$ is shown. As can be seen, the distribution is eroded from the left-hand side whereas the right-hand side remains almost unchanged. Thus, it is safe to assume that the dynamics is controlled by

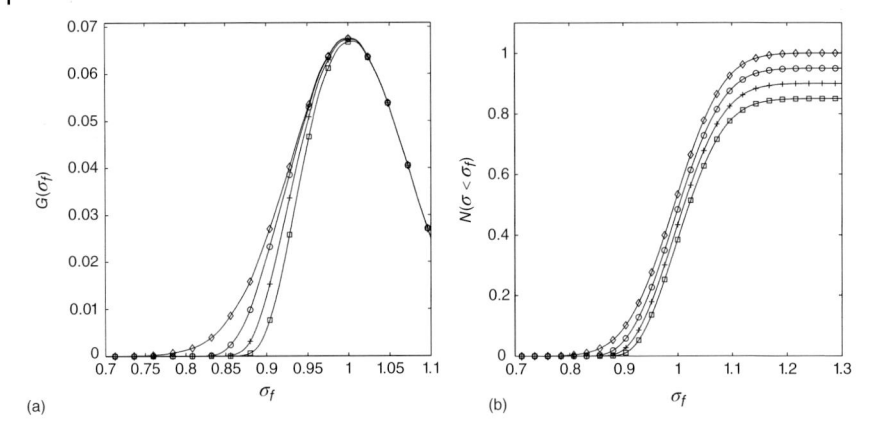

Figure 7.2 The evolution of the strength distributions for $\sigma_0 = 0.6$, $k_B T = 0.006$ and $k_B \Theta = 0.005$. (a) Evolution of the threshold distribution $G(\sigma_f)$ for $1 - U = 0$ (diamond), 0.05 (circle), 0.1 (plus), and 0.15 (square). (b) Evolution of the cumulative distribution $N(\sigma_0 < \sigma_f) = \int_0^{\sigma_f} G(x)dx$ for the same $1 - U$ values (and symbols). From Scorretti et al. (2001), with kind permission from EDP sciences.

weak fibers, which are the first to break. One can further assume that 90% of the unbroken fibers have thresholds greater than some $\sigma_{\min}$, which is, on the first approximation, a linear function of U: $\sigma_{\min} \approx [1 - \sqrt{2\pi k_B \Theta U}]$. Only the fibers with $\sigma_{\text{th}} > \sigma_{\min}$ plays an important role in supporting the load. Using these, one finds in the limit $U \approx 1$, $\sigma_0 > 0.3$, $2\sqrt{2k_B T} < (1 - \sigma_0)$, and $\sqrt{2k_B \pi \Theta} < (1 - \sigma_0)$, the failure time is

$$\tau_f \approx \tau_0 \exp\left[\frac{(1-\sigma_0)^2}{2k_B T_{\text{eff}}}\right], \tag{7.25}$$

where

$$k_B T_{\text{eff}} \approx k_B T \left(\frac{1}{1 - \frac{\sqrt{2\pi k_B \Theta}}{2(1-\sigma_0)}}\right), \tag{7.26}$$

and

$$\tau_0 = \frac{2\sqrt{2\pi k_B T}}{\left(\sigma_0 - \sqrt{2\pi k_B \Theta}\right)\left[1 + \exp\left[-\frac{\sqrt{2\pi k_B \Theta}(1-\sigma_0)}{k_B T}\right]\right]}. \tag{7.27}$$

As can be seen, the effective temperature is an increasing function of the disorder. Of note, this is different from the previous estimate of simple addition, which was calculated only for the time of first fiber breaking. These predictions of multiplicative effect of noise were verified using numerical simulations. In Figure 7.3(a), the variation of τ_f as a function of $1/k_B T$ for different values of $k_B \Theta$ is shown. Note that for the same value of T, τ_f decreases by orders of magnitude as Θ is increased. The validity of Eq. (7.25) can be checked by plotting τ_f/τ_0 as a function

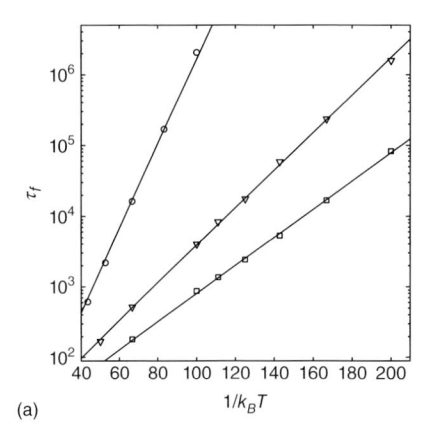

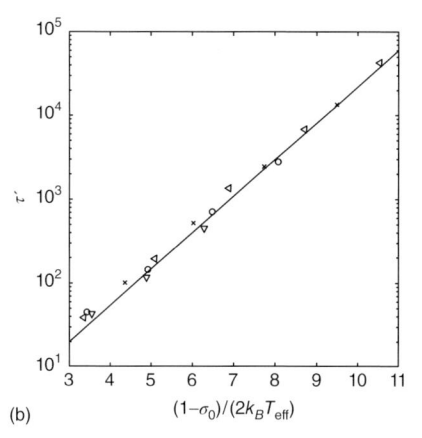

Figure 7.3 (a) Variation of failure time τ_f for a system loaded with $\sigma_0 = 0.45$ with noise parameter values $k_B\Theta = 0$ (circle), 0.02 (triangle), and 0.04 (square). The simulation points match very well with the analytical prediction (Eq. (7.25)); (b) Variation of scaled time $\tau' = \tau_f/\tau_0$ with $(1-\sigma_0)^2/(2k_B T_{\text{eff}})$. The continuous line is from Eq. (7.25) and the different points correspond to different noise parameter values, $k_B\Theta = 0.01$ (inverted triangle), 0.005 (circle), 0.0025 (cross), and 0.00125 (tilted triangle). From Scorretti et al. (2001), with kind permission from EDP sciences.

of $(1 - \sigma_0)^2/(2k_B T_{\text{eff}})$ (see Figure 7.3(b)). As can be seen, the simulation points collapse on a single curve, verifying the analytical estimate.

The fact that heterogeneity greatly enhances the effective temperature of the system explains the experimental fact that the temperature needed to have the measured lifetimes for some materials under creep test, is orders of magnitude higher than the actual temperature (Guarino et al., 1999a; Pauchard and Meunier, 1993). Also, for heterogeneous materials, the effect of actual change in the temperature is negligible on failure time, while it is important for homogeneous materials.

For fluctuating load, mimicking thermally activated failures, the waiting time distribution is also studied (Yoshioka et al., 2012). It shows two possible forms. At high load values, the process continuously accelerates toward a macroscopic failure. At high temperature and low load, there is first a slowing down and then eventual acceleration to failure. In slow rupture, the waiting time distribution follows a power-law with exponent value 2, while for accelerating phase, it is always 1, irrespective of the range of interaction. We discuss about the time-dependent relaxation of the strain rate in more detail in the next section.

One can generalize the above theory for time-dependent loading (Ciliberto et al., 2001). $1/\tau_f(W)$ can be interpreted as the damage density per unit time under a load W. For an arbitrary load $W(t)$, the total damage density at time t is

$$I(t) = \int_0^t \frac{1}{\tau_0} \exp\left[-\frac{(1 - F(t')/N)^2}{2k_B T_{\text{eff}}}\right] dt'. \tag{7.28}$$

7 Subcritical Failure of Heterogeneous Materials

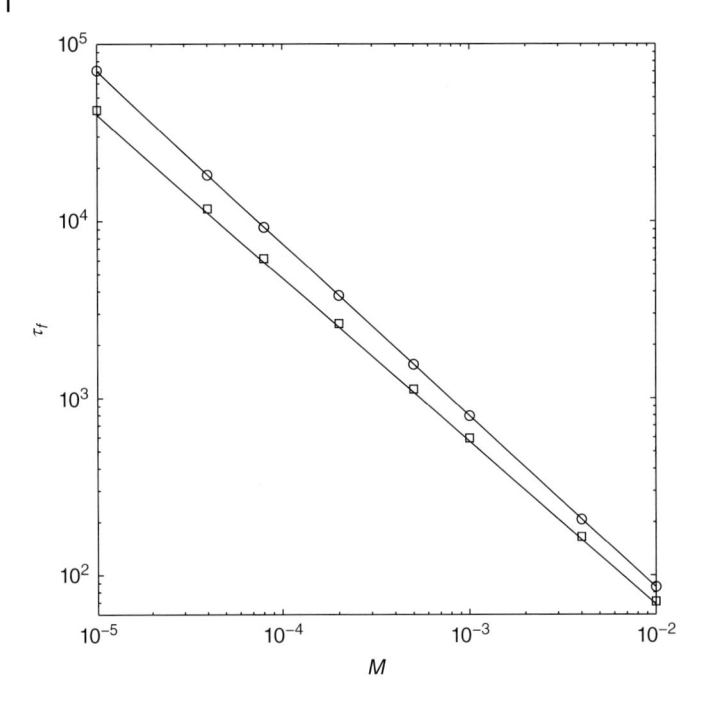

Figure 7.4 Failure time τ_f as a function of slope M of the linear loading. The continuous lines are from Eq. (7.29) and simulations are for $N = 1000$, $k_B\Theta = 0.005$ (circle), and 0.02 (square). From Ciliberto *et al.* (2001), with kind permission from Elsevier.

Therefore, the failure time can be obtained from $I(\tau_f) = 1$. For the particular case of linear loading $W(t) = Mt$, one can approximately get (for $M \ll \sqrt{2k_B T/\pi}$)

$$\tau_f = \frac{1 - \sqrt{2k_B T}\delta_M}{M(1 + \sqrt{2kBT/(2\delta_M)})}, \tag{7.29}$$

where

$$\delta_M = \ln\left(\sqrt{\frac{2k_B T}{\pi}}\frac{1}{M}\right). \tag{7.30}$$

This estimate agrees well with numerical results (see Figure 7.4). The estimates also agree for other time-varying functions, for example, sinusoidal variation.

The time-dependent loading gives the opportunity to verify Kaiser effect (Kaiser, 1953) (see Lavrov, 2003 for review) in this case (Ciliberto *et al.*, 2001). According to this effect, the acoustic emission from a stressed metal sample is zero, if the value of the stress is smaller than the previously applied maximum. Even though this effect was also found in rock (Kurita and Fujii, 1979), its existence was questioned for other heterogeneous materials (fiber glass, wood, etc.) (Guarino *et al.*, 1998). It was seen that in this model with temperature Kaiser effect is not present, since fibers can break to release energy for any loading. In the absence of any noise, however, Kaiser effect can be seen in global load sharing fiber bundle models.

7.1.2
Failure Due to Fatigue in Fiber Bundles

Subcritical failures in fiber bundles can also be looked at from the point of fluctuating failure probability. In the presence of finite external noise (temperature), a breaking of fiber can be thought of as overcoming a barrier, much like the subcritical Griffith's like crack propagation criterion in solids. The failure probability of a fiber, for which the failure threshold is σ_{th}^i, under a load per fiber value σ_0 and temperature T, can be taken as (Pradhan and Chakrabarti, 2003a)

$$P(\sigma_0, T) = \frac{\sigma_0}{\sigma_{th}^i} \exp\left[-\frac{1}{T}\left(\frac{\sigma_{th}^i}{\sigma_0} - 1\right)\right], \quad \text{for } 0 \leq \sigma_0 \leq \sigma_{th}^i$$
$$= 1, \quad \text{otherwise.} \tag{7.31}$$

As can be seen, all fibers now have a failure probability for any σ_0. Let us again consider a homogeneous system for which all fibers have equal failure strength denoted by σ_{th}. If the fraction of surviving fibers at time t is U_t, then

$$U_{t+1} = U_t \left[1 - P\left(\frac{\sigma_0}{U_t}, T\right)\right]. \tag{7.32}$$

In continuum limit, this can be written in a differential form as follows:

$$-\frac{dU_t}{dt} = \frac{\sigma_0}{\sigma_{th}} \exp\left[-\frac{1}{T}\left(\frac{\sigma_{th} U}{\sigma_0} - 1\right)\right], \tag{7.33}$$

which gives the failure time as

$$\tau = \int_0^\tau = \frac{\sigma_{th}}{\sigma_0} \exp\left(-\frac{1}{T}\right) \int_0^1 \exp\left[\frac{1}{T}\left(\frac{\sigma_{th}}{\sigma_0}\right) U\right] dU,$$
$$= T \exp\left(-\frac{1}{T}\right) \left[\exp\left(\frac{\sigma_{th}}{\sigma_0 T}\right) - 1\right], \tag{7.34}$$

for $\sigma_0 < \sigma_{th}$. For $\sigma_0 > \sigma_{th}$, however, $\tau = 0$ as all fibers will immediately fail. The aforementioned expression can further be simplified for small T and $\sigma_0 \to \sigma_{th}$, to give $\tau \approx T \exp\left[(\sigma_{th}/\sigma_0 - 1)/T\right]$. Therefore, the failure time approaches infinity as $T \to 0$.

One can also obtain the avalanche size distribution for this case. Note that here is no waiting time for failure, that is, some fibers will break every time. Therefore, the definition of avalanche in this case will be the number of fibers breaking per unit time. Solving for $U(t)$ from Eq. (7.33), also using Eq. (7.34), one gets

$$U(t) = \frac{\sigma_0 T}{\sigma_{th}} \ln\left[\frac{\tau - t}{T \exp(-1/T)} + 1\right]. \tag{7.35}$$

As mentioned earlier, one can think of dU/dt as the avalanche size Δ. Then

$$\Delta^{-1} \sim \frac{\tau - t}{T \exp(-1/T)} + 1 \sim \tau - t \tag{7.36}$$

for $T \to 0$. Since $\tau - t$ corresponds to the cumulative probability $\int_\Delta^\infty D(\Delta)d\Delta$ of avalanches beyond t, one gets

$$P(\Delta) \sim \Delta^{-\beta}, \quad \text{where} \quad \beta = 2. \tag{7.37}$$

Figure 7.6 compares the avalanche distribution with numerical simulations to get good agreement. One can also look at the situation where the failure thresholds have quenched randomness. The avalanche size distribution exponent value remains the same, and the failure time expression remains similar in form while fitted with such functions.

In the previous case, there is no waiting time for the failures, that is, for each scan of the system, one or more fibers fail. However, if the initial load σ_0 is sufficiently low, there will be a limit below which waiting time will start to appear, that is, there will be time steps for which no fiber fails. It is interesting to look at the phase boundary between the two phases where there is zero waiting time and where waiting time is finite (Pradhan et al., 2013). The initial load for which the waiting time first starts to appear is denoted by σ_w. Clearly, this value will depend upon the noise T. One can make an approximate estimate as follows: At $\sigma_0 = \sigma_w$, at least one fiber must fail initially. Then, the same load is redistributed among the rest of the fibers, which increases the failure probability further. Therefore, one just needs to ensure that (considering homogeneous system with failure threshold σ_{th})

$$N \exp\left[-\frac{1}{T}\left(\frac{\sigma_{th}}{\sigma_w} - 1\right)\right] \geq 1, \tag{7.38}$$

giving

$$\sigma_w \geq \frac{\sigma_{th}}{1 - T \ln(1/N)}. \tag{7.39}$$

This estimate is consistent with the fact that in the absence of any noise ($T = 0$), $\sigma_w = \sigma_{th}$. Furthermore, for the heterogeneous threshold distribution, it was conjectured that analogous to the homogeneous case, the form could be

$$\sigma_w \geq \frac{\sigma_f}{1 - T \ln(1/N)}, \tag{7.40}$$

where σ_f is the stress for which the system fails without noise. The variations of σ_w with T match approximately with numerical results (see Figure 7.5).

When the load $\sigma_0 < \sigma_w$, the dynamics continues through a punctuated equilibrium phase. Therefore, it is of interest to look at both the avalanche size distribution and the distribution of the waiting times by which these avalanches are separated. Note that the avalanche considered here is different from that of the previous case (Eq. (7.37)) where the avalanche size was simply the rate of breaking (in the absence of any waiting time). Here, an avalanche is the total number of fibers broken between two successive waiting times. Because of this change in definition, the exponent value of the avalanche size distribution also changes from the previous case. The present case still gives a power-law distribution with $P(\Delta) \sim \Delta^{-\beta}$, where $\beta = 2.5$ (see Figure 7.6). This form is universal with respect to

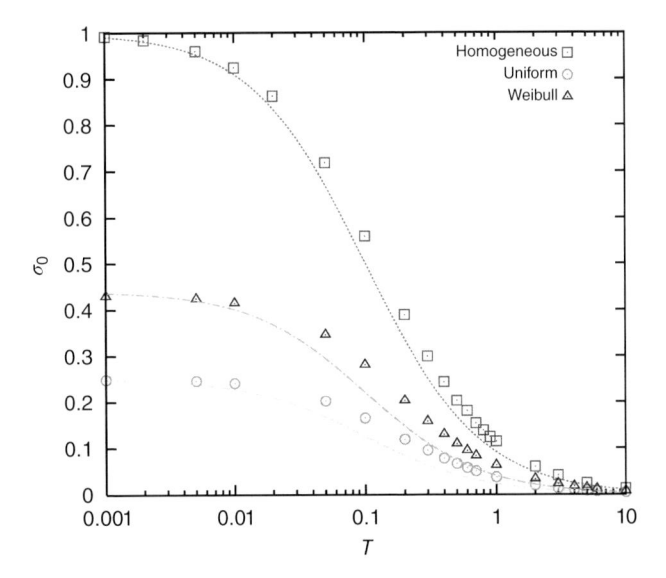

Figure 7.5 The phase boundary between failure with no waiting time (region above the curve) and failure with waiting time (region below the curve) for homogeneous system as well as uniform and Weibull threshold distributions. The continuous lines are the approximate predictions of Eqs. (7.39) and (7.40), and the points are numerical simulations. System size is $N = 20\,000$, averages taken for 100 samples. From Pradhan and Chakrabarti (2013), with kind permission from American Physical Society.

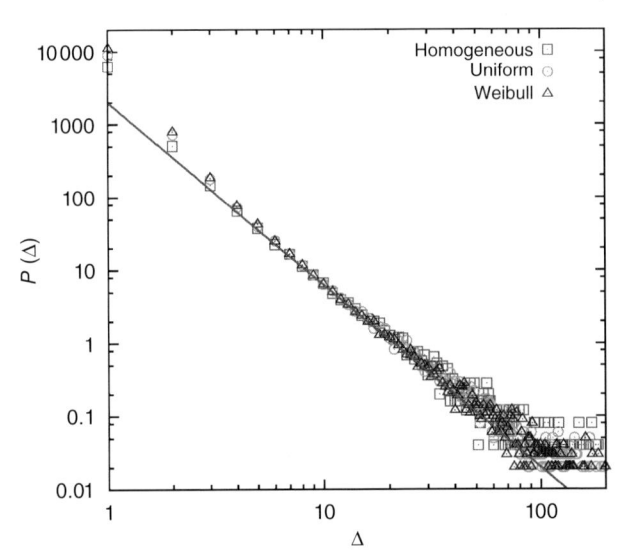

Figure 7.6 The avalanche size distributions for $\sigma_0 < \sigma_w$ in homogeneous system as well as those with uniform and Weibull threshold distributions. Avalanches are defined as the number of fibers breaking between two successive waiting times. The size distribution exponent value is -2.5. From Pradhan et al. (2013), with kind permission from American Physical Society.

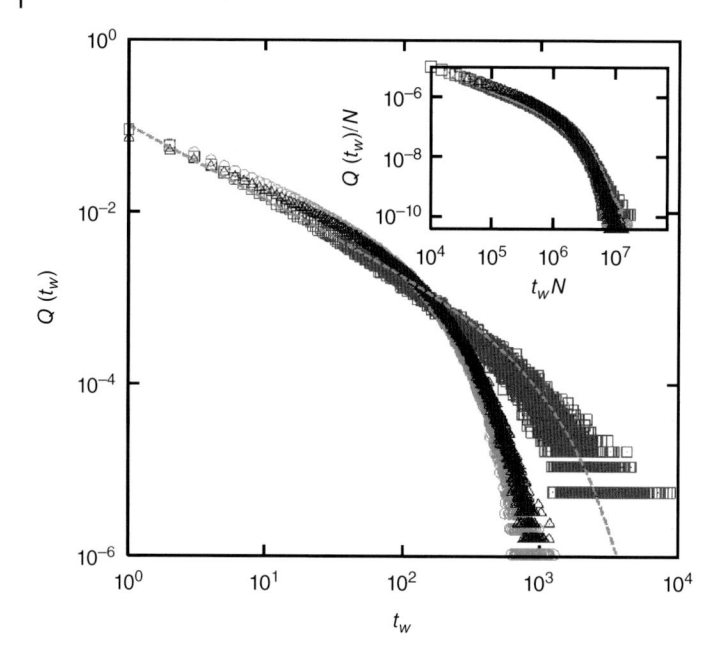

Figure 7.7 The simulation results for the waiting time distributions (squares for homogeneous bundle, circles for uniform distribution, and triangles for Weibull distribution). The dotted line is the Gamma function (Eq. (7.41)), with $\gamma = 0.15$ for homogeneous case and 0.26 for the other two cases. The inset shows the finite-size data collapse. From Pradhan et al. (2013), with kind permission from American Physical Society.

different threshold distributions (delta function, uniform, Weibull, etc.). The waiting time statistics is also important for this model. It is seen that the waiting time distribution can be fitted to a Gamma form (Corral, 2003, 2004, 2006b)

$$Q(t_w) \sim \frac{\exp(-t_w/a)}{t_w^{1-\gamma}}, \tag{7.41}$$

where t_w is the waiting time, defined as the idle time (number of scans with zero fiber breaking) between two successive avalanches, and a is a constant. For a homogeneous case, one gets $\gamma = 0.15$, while for the heterogeneous cases (say, uniform or Weibull distribution of thresholds) $\gamma = 0.26$ (see Figure 7.7). The plot $Q(t_w)/N$ versus $t_w N$ gives a goos data collapse (see inset of Figure 7.7).

7.1.3
Creep Rupture Propagation in Rheological Fiber Bundles

The creep failure in materials was also modeled by considering the relaxation dynamics of the strain developed explicitly. Particularly, the viscoelastic behavior in fiber bundles was modeled by (Hidalgo et al., 2002a) a Kelvin–Voigt element (see also Hao et al., 2012). The constitutive relation reads

$$\sigma_0 = \theta\dot{\epsilon} + Y\epsilon, \tag{7.42}$$

where θ is a damping factor and Y is the elastic constant. The time evolution of strain is, therefore

$$\epsilon(t) = \frac{\sigma_0}{Y}[1 - e^{-Yt/\theta}] + \epsilon_0 e^{-Yt/\theta}, \tag{7.43}$$

where ϵ_0 is the strain at $t = 0$. Clearly, at $t \to \infty$, Hooke's law is obeyed. In addition to this, a damage criterion was defined as follows: If the strain on a fiber exceeds a given value ϵ_d, the fiber breaks. Now, because of the asymptotic validity of Hooke's law, one expects the damage in the model to be same as that in the cases of usual stress-controlled failures. The difference, however, is that the failures are distributed in time and are not instantaneous. Now, if one considers the global load sharing mechanism and the cumulative distribution of the failure strain is written as $P(\epsilon)$, then the time evolution of the system under a steady load ϵ_0 can be written as

$$\frac{\sigma_0}{1 - P(\epsilon)} = \theta\dot{\epsilon} + Y\epsilon. \tag{7.44}$$

Two situations can arise, depending upon the value of σ_0. When σ_0 is low, the system reaches a steady state with finite fraction of fibers intact. However, beyond a critical value of σ_0, such steady state does not exist and complete failure occurs. The critical load, σ_f is the solution of the equation $\sigma_f = Y\epsilon_f[1 - P(\epsilon_f)]$, where ϵ can be determined by solving $d\sigma_0/d\epsilon_s|_{\epsilon_f} = 0$ (Sornette, 1989). Since $\sigma_0(\epsilon_s)$ is maximum (σ_f) at ϵ_f, one can write in the vicinity of ϵ_f

$$\sigma_0 \approx \sigma_f - C(\epsilon_f - \epsilon_s)^2, \tag{7.45}$$

where C is determined by the specific form of the distribution of failure thresholds. An important quantity to study is the failure time with load beyond the critical value. Particularly, when $\sigma_0 = \sigma_f + \Delta\sigma$, where $\Delta\sigma \ll \sigma_f$, one can solve Eq. (7.44) and using Eq. (7.45), one gets

$$\tau_f \sim \int \frac{1 - P(\epsilon)}{\Delta\sigma - A(\epsilon_c - \epsilon)^2}, \tag{7.46}$$

which is to be evaluated close to ϵ_c. One thus gets the relation

$$t_f \sim (\sigma_0 - \sigma_f)^{-1/2} \quad \text{for} \quad \sigma_o > \sigma_f. \tag{7.47}$$

This behavior was verified using numerical simulations with different failure threshold distributions (see Figure 7.8).

There have been many experiments to determine the failure properties in subcritical loading. In particular, the one with a cyclic loading was studied in Kun et al. (2007) with hot mix asphalt (HMA). Under cyclic loading with amplitude $\sigma_0 < \sigma_f$, the lifetime of the material was measured (number of load cycles before failure) as a function of σ_0. The deformation (ϵ) was also recorded. While the deformation and its derivative increase monotonically with N_{cycle} (see Figure 7.9), the failure

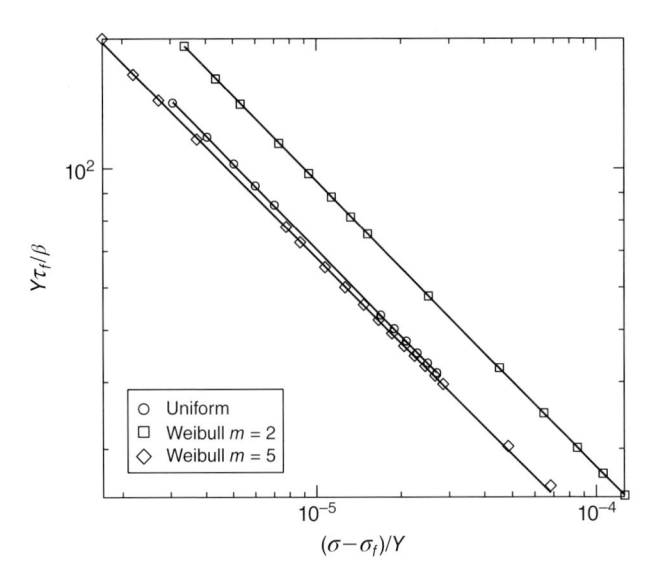

Figure 7.8 The failure time of the viscoelastic fiber bundle beyond critical loading for different failure threshold distributions (uniform and Weibull with two index values). The failure time is a power-law, as was shown in Eq. (7.47) From Hidalgo et al. (2002a), with kind permission from American Physical Society.

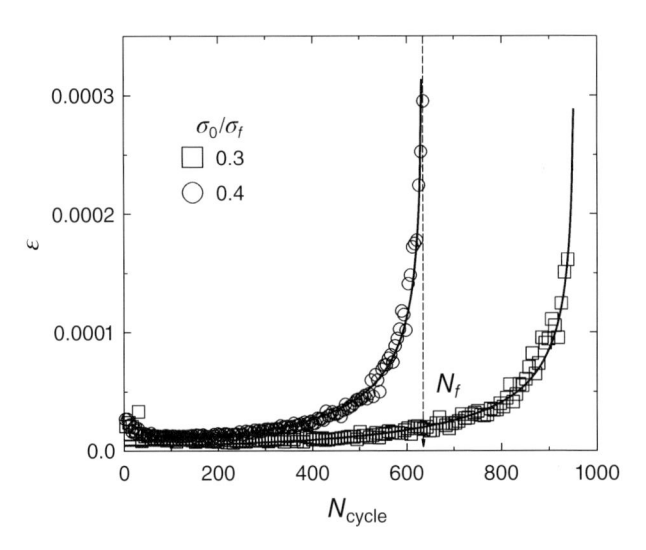

Figure 7.9 The deformation as a function of number of load cycles. The solid line is the fit from model. From Kun et al. (2007), with kind permission from IOP (UK).

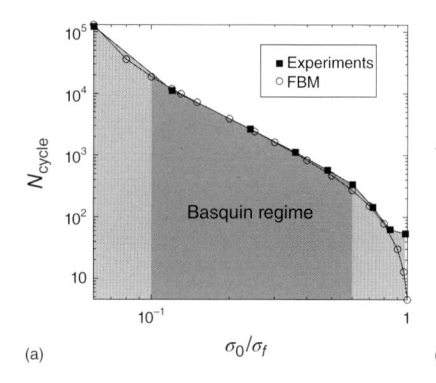

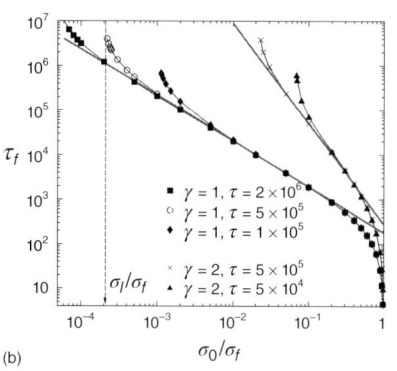

Figure 7.10 (a) Number of cycles to failure as function of σ_0. The experimental data fit very well with the model for $\gamma = 2$, $\tau = 15\,000$, and $a = 0.01$ using Weibull distribution for failure thresholds. (b) The lifetime t_f as a function of σ_0, the exponent value is same as the value of γ. From Kun et al. (2007), with kind permission from IOP (UK).

time (number of cycles) N_f decreases with σ_0. In fact, in the intermediate regime, the failure time decrease as a power-law (Basquin's (1910) law)

$$N_f \sim \left(\frac{\sigma_0}{\sigma_f}\right)^{-\alpha} \tag{7.48}$$

with $\alpha = 2.2 \pm 0.1$ (see Figure 7.10).

In order to model these failure properties, a modified fiber bundle model was proposed (Kun et al., 2007). The basic changes in the model are in two places. First, a fiber can break not only when its load (σ_i) exceeds some threshold σ_i^{th}, but also due to accumulation of damage. The amount of damage accumulated ($c_i(t)$) under a load $\sigma_i(t)$ in time t is assumed to be of the form

$$c_i(t) = a \int_0^t \sigma_i^\gamma(t')dt', \tag{7.49}$$

where both a and γ are positive. A fiber can tolerate an accumulated damage c_i^{th} beyond which it breaks. The thresholds c_i^{th} and σ_i^{th} are assumed to be independent. The second feature is the healing effect to capture such phenomena in microcracks. For this, a timescale τ was introduced, such that the damage accumulation is only effective within the finite range of timescale. The evolution equation can thus be written as

$$\sigma_0 = \left[1 - F\left(a \int_0^t e^{-(t-t')/\tau} \sigma^\gamma(t'dt')\right)\right][1 - P(\sigma(t))]\sigma(t), \tag{7.50}$$

where F and P are the cumulative distribution functions of failure threshold distributions of accumulated damage threshold and breaking thresholds of fibers, respectively. It is possible to solve for the failure time in the limit of $\tau \to \infty$. It shows the Basquin's law

$$\tau_f \sim (\sigma_0)^{-\gamma}, \tag{7.51}$$

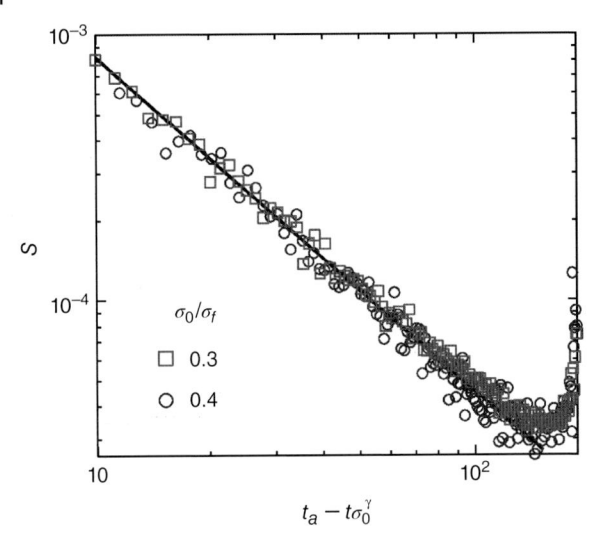

Figure 7.11 The experimental measure of the scaling function of the deformation time of asphalt specimens. From Kun et al. (2008), with kind permission from American Physical Society.

where the exponent γ, the damage accumulation exponent, is independent of the type of disorder. Also (Kun et al., 2008),

$$\epsilon(t) \sim (\tau_f - t)^{-1/(1+\gamma)} \tag{7.52}$$

The deformation of the system in fact follows a scaling of the form $\epsilon(t) = \sigma_0 S(t\sigma_0^\gamma)$, where the scaling function has the property $S(t\sigma_0^\gamma) \sim (t_a - t\sigma_0^\gamma)^{-1/(1+\gamma)}$, with $t_a = a(1+\gamma)$. The verification of this scaling law can be seen experimentally (see Figure 7.11).

As mentioned earlier, if the memory time τ is smaller than the lifetime of the system without healing, then a threshold load σ_l appears, below which the failure of the system is only partial. In fact, this brings about a continuous phase transition between complete and partial failures across the load σ_l. The relaxation time and failure time show power- law divergence with respect to difference from the load σ_l (see Figure 7.12) with distinct exponent values: $t_r \sim (\sigma_l - \sigma_0)^{-1/3}$ and $\tau_f \sim (\sigma_0 - \sigma_l)^{-2/3}$. One can also study the avalanche size distribution and its duration. Avalanche size distribution exponent follows the usual mean-field value 3/2, while the duration exponent value is 1.

7.1.3.1 Modification for Local Load Sharing Scheme

The different rheological models discussed so far are long-range type, in the sense that in case of failure of a fiber, the load carried by that fiber is to be shared equally among all the remaining fibers. However, it is interesting to see the effect of local stress concentrations in the system (Kun et al., 2003b). That can occur when a local load sharing rheological fiber bundle model is considered (Halász et al., 2012).

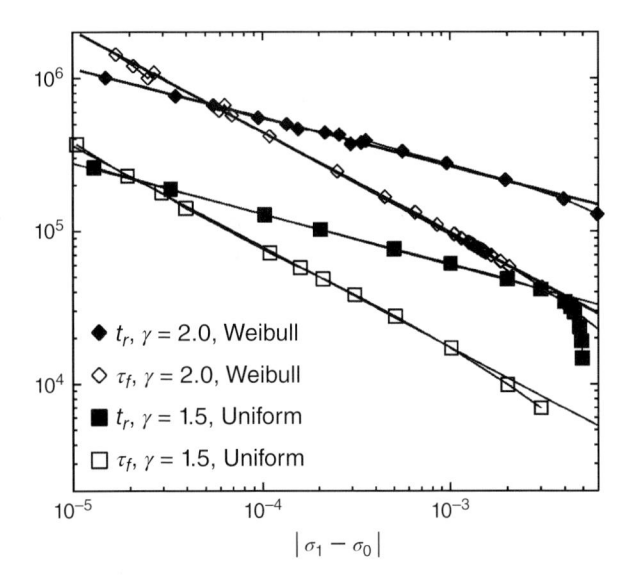

Figure 7.12 The dependence of relaxation time t_r and time to failure t_f on $|\sigma_1 - \sigma_0|$. The power-law exponents are $-1/3$ and $-2/3$. From Kun et al. (2008), with kind permission from American Physical Society.

The only difference with the previous studies is that when a fiber breaks, its load is to be shared equally by its nearest surviving neighbors. Now, a competition arises between the disorders in the failure thresholds and the stress concentrations due to localized load sharing. If the stress concentration is given very low importance ($\gamma \to 0$ in Eq. (7.49)), fibers break in accordance with their disorder thresholds, which are distributed randomly. Hence the damage is diffused (percolation-like). On the other hand, if the value of γ is sufficiently high, failure threshold disorders become irrelevant and a single crack growth leads to catastrophic failure.

An approximate estimate of the phase boundary between these two limits can be given as follows (Halász et al., 2012): Assuming that the damage thresholds are distributed uniformly in $C - W$ and $C + W$ and the stress failure thresholds are distributed uniformly in $[0:1]$, and a low σ_0 is applied such that it induces isolated breakings but no immediate avalanche. In that case, the loads on all fibers are σ_0 and those in the neighboring sites of the broken fibers are $5\sigma_0/4$. Now, under a load σ, the time (t_b) to break a single fiber having damage threshold c^{th} (note the $\sigma^{th} > \sigma_0$) is (from Eq. (7.49))

$$t_b(c^{th}, \sigma) = \frac{c^{th}}{a\sigma^\gamma}. \tag{7.53}$$

A single crack growth will take place when the next fiber to break will be the one with an already broken neighbor. In the worst case, it can have a threshold $C + W$, and its failure time should be shorter than the ones with intact neighbors and, in

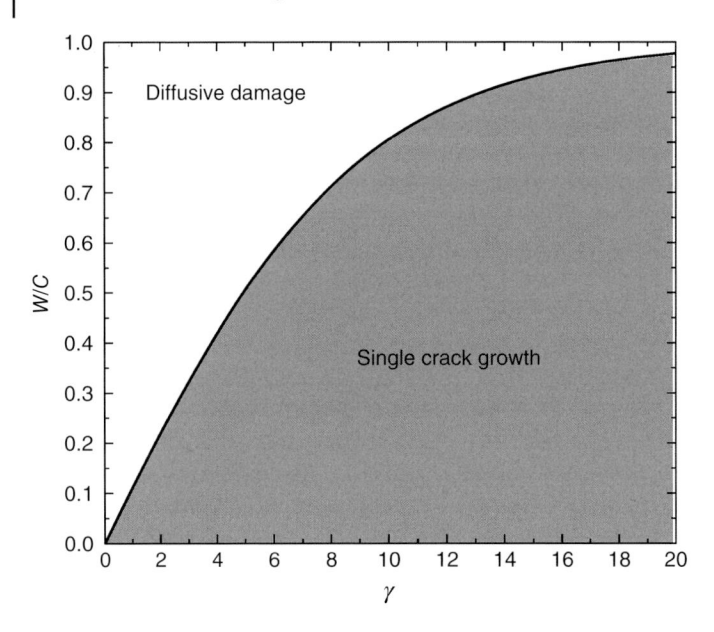

Figure 7.13 Phase diagram of the model separating percolation- and nucleation-type failures corresponding to Eq. (7.54). From Halász et al. (2012), with kind permission from American Physical Society.

the worst case, having failure threshold $C - W$. Hence, one gets

$$\frac{W}{C} < \frac{\left(\frac{5}{4}\right)^\gamma - 1}{\left(\frac{5}{4}\right)^\gamma + 1}. \tag{7.54}$$

Assuming equality in the aforementioned equation, a phase boundary between a percolation-like and a nucleation-like failure can be obtained (see Figure 7.13).

The macroscopic properties of the system remain well behaved. Particularly, time power-law, dependence of the time to failure on the applied load (Basquin's law), remains valid. It is also possible to define an avalanche statistics for the model. Near the peak load, it shows power-law distribution with exponent value 1.75, which is in contrast with usual local load sharing model statistics, where it is much sharper. The exponent value is independent of γ, but it depends on σ_0 (see Figure 7.14). The distribution of the duration, on the other hand, is sensitive to γ. It is also a power-law distribution, but the exponent depends on γ, while its cut-off is determined by σ_0 (see Figure 7.14). One can also measure the cluster size distribution. For diffusive damage region, it is similar to the cluster size distribution of percolation, while for higher γ values it becomes much sharper.

Another interesting aspect of this model was studied by Danku and Kun (2013b). They examined the spatial evolution of bursts. If one looks at the average number of failures in the intermediate steps of load redistribution within an avalanche,

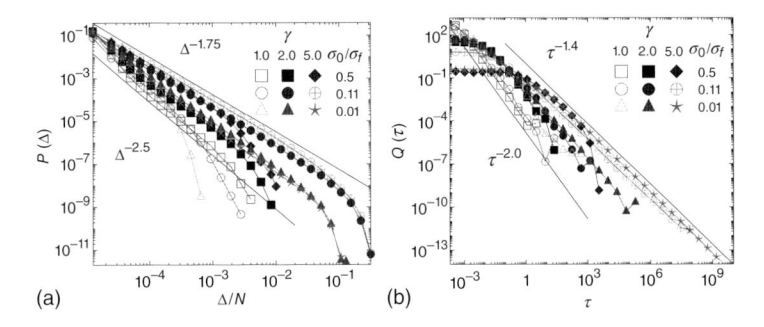

Figure 7.14 (a) The avalanche size distribution for different γ and σ_0 values. The exponent value is independent of γ but depends on σ_0. (b) The avalanche duration distribution. The exponent value here depends on γ, but σ_0 only sets the cut-off. From Halász et al. (2012), with kind permission from American Physical Society.

then it shows an asymmetry. It means that, within an avalanche, less fibers break at the beginning and then the number increases and reaches a peak after half the time of the avalanche is passed. This asymmetry (quantified by the skewness, i.e., the ratio of the third cumulant to the 3/2 power of the second cumulant) was then related to the fraction of broken fibers from the crack front (damage perimeter). For higher values of the asymmetry, this ratio is higher, signifying the correlation of spatial spreading with shape of the avalanche pulse. When the load redistribution is global, there is no notion of length, and the shape of the pulse is symmetric. It was also shown (Danku and Kun, 2013a) that these bursts are nonhomogeneous Poissonian processes with scale-free distribution of size and waiting time.

7.2
Dynamics of Strain Rate

Upon application of a subcritical load, the strain developed in the material shows interesting dynamics up to the point of rupture. Three regions in time are generally identified: the first one is the slow decrease of strain rate, which is often fitted to a power-law and is called the Andrade law (Andrade, 1910) for creep; the second regime is a quasi-constant region, where the strain rate changes very slowly; and finally, there is a power-law acceleration before the rupture event. There have been extensive investigations of these observations both experimentally and theoretically. In the following, we outline some of the attempts to understand these behaviors.

In Nechad et al. (2005b), creep rupture experiments were carried out using cross-ply glass/polyester composite materials and sheet modeling compound (SMC) composites with two combinations of cross angles. In total, 15 samples were taken, and the pressure and temperature were varied in the ranges 15–48 MPa and 60°–100°C, respectively.

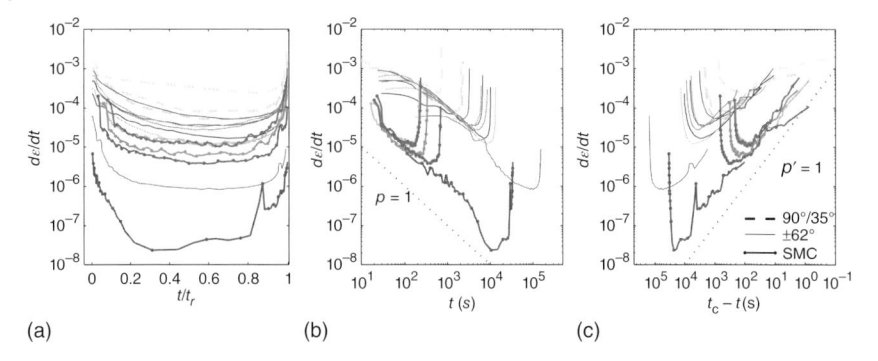

Figure 7.15 The dynamics of strain rate showing three regimes of creep. From Nechad et al. (2005b), with kind permission from American Physical Society.

As can be seen from Figure 7.15, the initial regime shows a decrease in the strain rate

$$\frac{d\epsilon}{dt} \sim t^{-p}, \tag{7.55}$$

which is the Andrade law for primary creep, with the exponent value $p \approx 1$. After this, the strain rate becomes almost a constant. Finally, before the rupture, the strain rate is accelerated, which fits to

$$\frac{d\epsilon}{dt} \sim (t_c - t)^{-p'}, \tag{7.56}$$

where $p' \approx 1$ for all the samples, and the fitted values of t_c are close to the actual failure times (τ_f) of the samples. A slightly different set of exponent values is seen for acoustic emission (AE) data of the samples (Nechad et al., 2005a); however, this is possibly due to the fact that AE catches all damages developing in the sample, while strain is a measure of damage and stress redistribution within the sample, and therefore is not the same physical process.

They also observed the proportionality between the time of the minimum of strain rate and failure time. The approximate relation is $t_m \approx 2\tau_f/3$, seen for all types of samples.

These observations are very general and verified in a wide variety of materials. For examples, in Amitrano and Helmstetter (2006), the creep ruptures in brittle rocks are reported; in Rosti et al. (2010), creep ruptures were studied in ordinary papers. It was seen that apart from the strain rate showing Andrade creep (exponent value close to 0.7), the standard deviation of the strain rate also decreases in a power-law with exponent value close to 0.55. Its significance is that as this exponent is smaller than the strain rate exponent, the fluctuation increases in relative terms and the dynamics becomes more intermittent with time. The proportionality between the strain rate minimum and rupture point was also found, but with a larger proportionality constant (0.83) than is usually seen in other experiments (Nechad et al., 2005a).

Apart from solids that are generally subjected to creep tests, in a recent experiment (Leocmach et al., 2014), protein gels were used to determine the rupture properties. Gels were formed by slow acidification of a sodium caseinate solution that behaves like a brittle material. Under a constant load, the strain rate dynamics shows the typical three regimes that are generally seen for brittle fracture. The exponent value for the primary creep (p) was reported to be slightly smaller than 1 (~ 0.85). The value of p' remains 1. The time to failure and time of minimum strain rate are still related linearly with $t_m = (0.56 \pm 0.04)\tau_f$. Finally, the entire regime of strain rate dynamics was fitted with a relation as follows

$$\frac{\dot{\epsilon}(t)}{\dot{\epsilon}_m} = \lambda \left(\frac{t}{\tau_f}\right)^{-\alpha} + \frac{\mu}{1 - t/\tau_f}, \tag{7.57}$$

with λ and μ being the fitting parameters, once the value of α is fixed from the primary creep. From the data collapse, it was inferred that the tertiary acceleration is a crossover from the primary creep regime to crack growth regime.

These observations in creep ruptures are seen in various other failure phenomena as well, for example, volcanic eruptions and landslides (see Amitrano and Helmstetter, 2006 and references therein). The seismic acceleration before an earthquake (a large failure event) was linked with the tertiary strain rate acceleration in creep (Main, 2000).

There have been numerous theoretical attempts to model the dynamics of strain rate and energy release in creep rupture. There are models based on empirical or semi-empirical constitutive laws that reproduce the two power-laws in creep (Lockner, 1998; Shcherbakov and Turcotte, 2003). In addition, there are discrete element-based models, such as the fiber bundle model, for describing creep rupture and the effect of heterogeneity. We outline some of those attempts here.

In the already-introduced model (Politi et al., 2002), numerically the Andrade creep was predicted. This was later confirmed analytically by Saichev and Sornette (2005). However, it was shown that the power-law has a logarithmic correction, that is, the failure rate behaves as $1/(t \ln \gamma t)$, where γ is a constant. This often indicates a smaller exponent than 1 for the Andrade creep.

In the viscoelastic fiber bundle models (Kun et al., 2003a; Hidalgo et al., 2002a) described earlier, Hidalgo et al. (2002a) showed tertiary regime of acceleration with an exponent 1/2. But, the primary Andrade creep is not seen in general, except for the case of critical load, even then the exponent value comes out to be 2, which is larger than usually seen in experiments (Nechad et al., 2005a). This model was reformulated and extended in Nechad et al. (2005a) in terms of representative elements (RE). Contrary to the definition of an equal load sharing fiber bundle model, this RE version is defined at a mesoscopic scale, where each RE consists of a set of fibers which may be in different orientation within the matrix, but effectively producing a mean-field effect. It assumes a power-law distribution for failure thresholds. Even that does not give the Andrade creep in the original form of the model. The desired behavior is seen when Eyring dashpot is used. The deformation equation is now governed by

$$\dot{\epsilon} = K \sinh(b\sigma_1), \tag{7.58}$$

where the stress in the dashpot σ_1 is given by $\sigma_1 = \sigma_0/(1 - P(\epsilon)) - Y\epsilon$. With the rupture threshold cumulative distribution assumed to be in the form $P(\epsilon) = 1 - \left(\frac{\epsilon_{01}}{\epsilon+\epsilon_{02}}\right)^\mu$, where ϵ_{01} and ϵ_{02} are constants with $\epsilon_{01} \le \epsilon_{02}$, the differential equation governing ϵ becomes

$$\dot{\epsilon} = K \sinh\left(\frac{b\sigma}{\epsilon_{01}^\mu}(\epsilon + \epsilon_{02})^\mu - bY\epsilon\right). \tag{7.59}$$

In the primary creep regime with $b\sigma_1 \gg 1$ and $\epsilon \ll \epsilon_{02}$, one gets $\dot{\epsilon} \sim 1/t$ giving the Andrade creep with $p = 1$. It is also independent of the choice of the threshold distribution.

The solution showing the tertiary regime is seen by neglecting ϵ_{02} compared to ϵ. Close to failure, when ϵ is large, the term $Y\epsilon$ is small compared to $(\sigma_0/\epsilon_{01}^\mu)(\epsilon + \epsilon_{02}^\mu)$ when $\mu > 1$. This leads to

$$\dot{\epsilon} \approx \frac{K}{2} \exp\left(\frac{b\sigma_0 \epsilon^\mu}{\epsilon_{01}^\mu}\right). \tag{7.60}$$

The approximate solution is

$$\epsilon(t) = A(-\ln(\tau_f - t))^{1/\mu}, \tag{7.61}$$

$$\dot{\epsilon} = \frac{A}{\mu}(-\ln(\tau_f - t))^{1/\mu - 1}\frac{1}{\tau_f - t}, \tag{7.62}$$

with $A = \epsilon_{01}(b\sigma_0)^{-1/\mu}$. The solutions are correct up to $\ln(\ln(\tau_f - t))$ correction terms for ϵ and up to $\ln(\tau_f - t)$ for $\dot{\epsilon}$, so that only the leading order power-law is reliable in the self-consistent solution. Therefore, this model with Eyring rheology shows both Andrade creep and divergence in tertiary regime, with exponent values 1. This also gives the proportionality of minimum of strain rate and failure time, but gives the proportionality constant to be $1/2$ rather than $\approx 2/3$ observed experimentally.

In a more recent work (Jagla, 2011), extending the analogy with seismic precursors, the fiber bundle model was modified to explain the dynamics of strain rate from a more microscopic point of view as opposed to the mesoscopic modeling approach outlined above that assumes a nonlinear rheology for its REs. This model has a set of fibers having different stiffness constants k_i, equal failure thresholds σ_{th} but are hanging from different coordinated u_i. Hence, effectively the stretching required to break one fiber is $\epsilon = \sigma_{th} + u_i$. Furthermore, the values of u_is relax with time such that the mechanical energy is minimized. The simplest way can be to consider the fibers such as harmonic chains (change in u_i proportional to force alone). However, that alone will not give the creep dynamics, it would rather

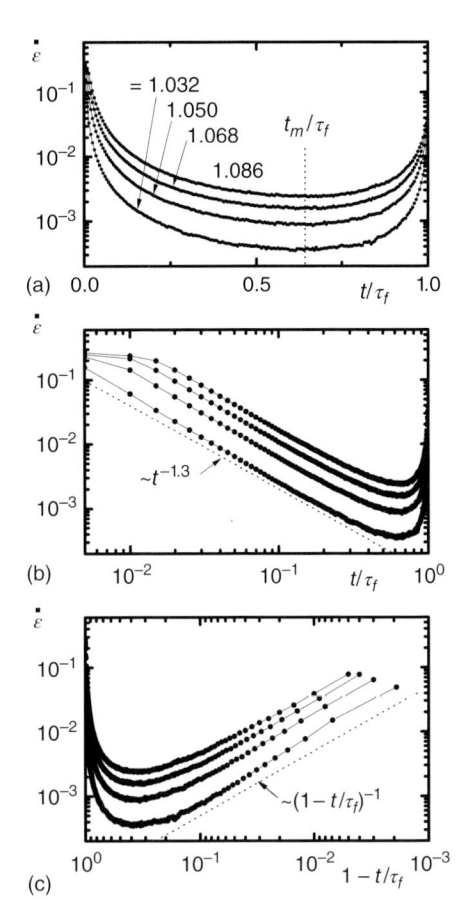

Figure 7.16 The temporal variation of strain rate for creep rupture in a modified fiber bundle model is shown for a system size 1000 × 1000. The three regimes of creep, (a) quasi-stationary regime, (b) Andrade creep regime, and (c) divergence before rupture, are seen. The creep rate minimum is more shifted toward the failure time, which is seen in the experiments. The Andrade law exponent value, however, is greater than what is seen in experiments. From Jagla (2011), with kind permission from American Physical Society.

give a viscous fluid. Therefore, the variation was assumed to be proportional to the fluctuation of force in the neighborhood

$$\frac{du_i}{dt} = \lambda \nabla^2 \frac{\delta E}{\delta u_i} = \lambda \sum_j (\sigma_i - \sigma_j) \tag{7.63}$$

where E is the mechanical energy. The index j runs on the intact fibers only.

The simulation results showing the temporal variation of strain rate for a force greater than the critical force for rupture, show three regimes (Figure 7.16). However, the exponent value in the Andrade creep regime is greater than what is usually seen in experiments. The tertiary creep acceleration exponent is, however, unity. The minimum of strain rate t_m is about 0.6 times the failure point τ_f, which is generally seen experimentally.

7.3
Summary

The failure of a solid before its critical threshold is reached in terms of applied stress is a vital issue. This is related to the stability of all the structures that we see around us.

Here, we have described the models and corresponding experimental observations of subcritical failures in brittle materials. Particularly, an estimate of the time to failure is crucial on application of some load below the failure load. We intended to extend the Griffith's theory approach of viewing the process of crack nucleation as an energy barrier problem, the external noise source being the temperature (see Section 7.1). However, it was also seen that the presence of heterogeneity in the solids has a significant effect in changing the effective value of the temperature. We summarized the versions of fiber bundle model that attempt to reproduce this effect at least qualitatively (see Eq. (7.26)). The relationship between time to failure and applied load has a power-law form (see Eq. (7.51)), which is an empirical observation known as the Basquin's law. This effect too can be captured by suitably modified fiber bundle models (see Figure 7.10). The dynamics of subcritical failure is also intermittent, showing scale-free size distribution of avalanche sizes and its durations. The behaviors are similar to what have been described in Chapter 6, however the exponent values are somewhat nonuniversal and depend on the applied stress (see Figure 7.14).

Another important observation in the dynamics of subcritical failure is the relaxation of strain rate. As has been described in Section 7.2, the dynamics has three regimes, namely the initial power-law decay of Andrade creep, the intermediate quasi-constant part, and the final power-law divergence leading to catastrophic failure (see Eq. (7.57)). These experiments are also explained by numerical modeling, although only qualitatively.

8
Dynamics of Fracture Front

In model-1 fractures, where the two parts of the solid are pulled apart to create fracture, the fracture front is the interface between the fractured part of the solid and the part which is still intact. The fracture propagates by the motion of the fracture front. The impression left by the front on the fracture surface gives rise to the roughness properties of the fracture surfaces discussed earlier. The dynamics of the front tells a lot about the about the fracture processes such as the roughness properties of the fractured surfaces and the intermittency in the front propagation detected in the experiments. In this chapter, we discuss the dynamics of fracture front.

8.1
Driven Fluctuating Line

Numerous theoretical attempts were made and are still being made to explain the roughness behavior of propagating fracture front. One branch of those attempts considers the fracture front as an elastic line propagating through a medium with quenched disorders and the roughness being the "footprint" left behind by the fracture front (Bouchaud et al., 1993). Motivated by the stress concentration in an elastic medium, these models consider a nonlocal elastic term for the fluctuating line and study the fluctuation of the line while it is being driven quasi-statically through the random medium.

The theory of fracture propagation in homogeneous materials was modified first by Gao and Rice (1989) to include the effect of random toughness in the motion of the two-dimensional fracture front. It was further simplified by considering the effect of first-order perturbation only, where the modification in the corresponding stress intensity factor for mode-I propagation depends only on in-plane coordinates. This effectively enables the study of in-plane projection of the fracture front, which is essentially a fluctuating line driven through a disordered medium. The first attempt to model propagating fracture front by elastic line was made by Schmittbuhl et al. (1995b). Here, an interfacial crack propagation was considered through an easy plane (along $z = 0$). The two solid parts on both sides have much higher toughnesses, and hence the crack is essentially always confined

8 Dynamics of Fracture Front

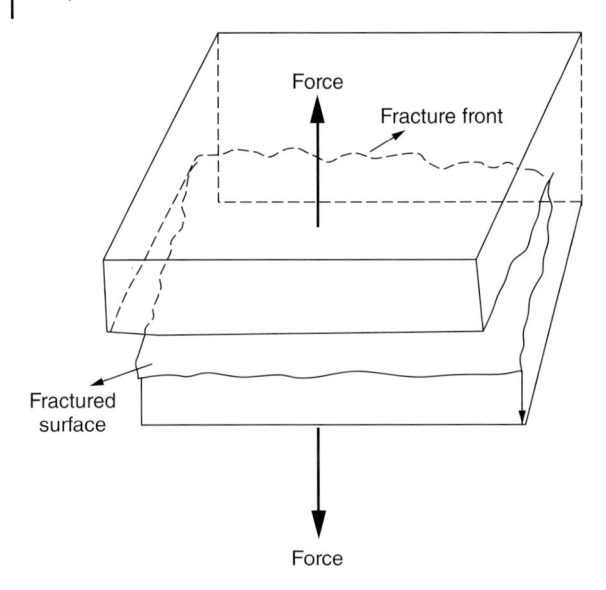

Figure 8.1 A schematic diagram showing the interfacial crack propagation. The crack front is modeled as a fluctuating elastic line.

in the $z = 0$ plane (see Figure 8.1 for a schematic description). The randomness of the medium was represented by a local toughness factor $K_c(x, y)$, which has a fluctuating component δK_c over an average K_c^0 of much higher amplitude than the fluctuation. The stress field at a point at a distance r from the fracture front in mode-I fracture has the diverging form $K/(2\pi\sqrt{r})$, considering the front to be straight. If the front is slightly distorted, up to first-order perturbation, $K(x)$ can be written as

$$\frac{K(x)}{K_0} = 1 + \frac{1}{2\pi} p.v. \int_{-\infty}^{+\infty} \frac{y(x' - y(x))}{(x' - x)^2} dx', \tag{8.1}$$

where K_0 is the stress intensity factor when the crack front is straight under the same loading condition and $p.v$ refers to the principal value. The quasi-static propagation of this model was then studied. The weakest point on the line was advanced, and a new value of toughness was assigned to it and so on. This model has been widely studied not only for fracture front propagation, but also for domain wall motion in magnets (Durin and Zapperi, 2000; Lemerle et al., 1998), contact lines of liquid menisci (Ertas and Kardar, 1994; Rolley et al., 1998; Le Doussal et al., 2009) and so on. (see Kardar, 1998 for a review).

The driven interface model shows a depinning transition when the driving force is high enough to sustain a continued steady velocity. This was first thought of as a critical phenomenon in Fisher (1985). As the growth of velocity is the signature of the transition, velocity is taken as the order parameter for the transition: $v \sim (f^{\text{ext}} - f_c^{\text{ext}})^\theta$. This transition behavior has been studied widely both theoretically (Narayan and Fisher, 1992, 1993; Nattermann et al., 1992; Rosso et al., 2003; Kolton et al., 2006) and experimentally (Måløy et al., 2006; Tallakstad et al., 2011; Ponson, 2009). The renormalization group calculations suggest $\theta = 0.78$ (to the

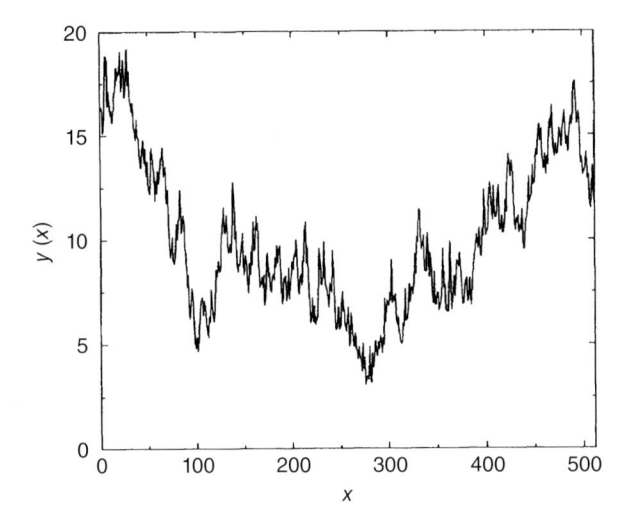

Figure 8.2 A snapshot of the crack front. From Schmittbuhl et al. (1995a), with kind permission from American Physical Society.

first order of smallness) (Ertas and Kardar, 1994) and $\theta = 0.59$ (to the second order of smallness) (Chauve et al., 2001). Experimental observation of this behavior for Botucatu sandstone gives $\theta = 0.80 \pm 0.15$ (Ponson, 2009), while numerical simulations give 0.625 ± 0.005 (Duemmer and Krauth, 2007) (see also Biswas and Chakrabarti, 2011).

The roughness properties of the front were measured when the front reached a steady state, after an initial time dependent on the system size where the roughness does not change with time anymore. The roughness exponent was measured using various methods. The power spectrum of the rough interface (see Figure 8.2) behaves as

$$P(f) \sim f^{-(2\zeta+1)} \tag{8.2}$$

where ζ is the roughness exponent. The simulation results presented in Schmittbuhl et al. (1995b) give an estimate $\zeta \approx 0.35 \pm 0.05$. It was also confirmed by measuring the standard deviation of the rough surface and the distribution of the distance between successive crossing the average of the interface by the interface. Finally, a scaling form for the width of the surface was proposed

$$W(\overline{h}, L) \sim L^\zeta F\left(\frac{\overline{h}}{L^{1/z'}}\right), \tag{8.3}$$

where $\overline{y}$ is the average position of the front and z' is the dynamical exponent. Figure 8.3 shows the scaling plot of the width using this scaling function.

Rosso and Krauth (2002) proposed a method for estimating the roughness exponent without actually simulating the dynamics of the entire time up to relaxation. First, they discretized the space and time variables, and therefore the configuration at a time t is given by $h^t = \{h^t_1, h^t_2, \ldots, h^t_L\}$ with $h_{L+i} = h_i$, where h^t_i is the

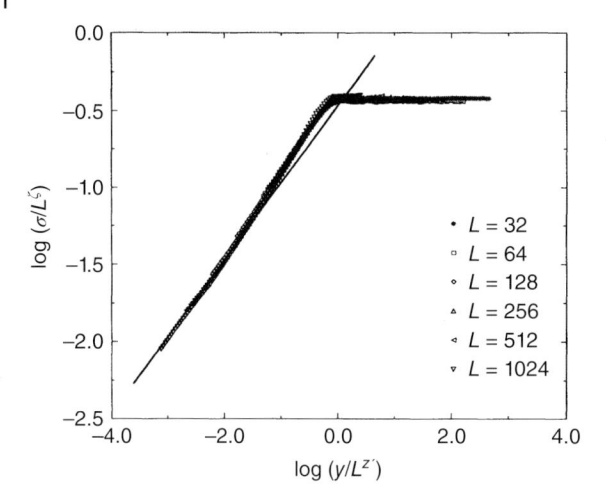

Figure 8.3 The data collapse for the RMS fluctuation of the front assuming the scaling form of Eq. (8.3). The exponent values used for collapse are $\zeta = 0.35$ and $z = 1.5$. From Schmittbuhl et al. (1995a), with kind permission from American Physical Society.

height of the ith element of the front (measured in y-direction). The long-range elastic force was calculated by summing over the periodic images

$$f^{el}[h_i] = \sum_{j=1}^{L-1} [2\psi'(j/L) - \pi^2/\sin^2(\pi j/L)] \frac{h_i^t - h_j^t}{L^2}, \tag{8.4}$$

where $\psi = d\Gamma(x)/dx$ and $\Gamma(x)$ is the gamma function. Then, the no-passing theorem was used which states that if two interfaces do not cross at some time, then they do not cross at any subsequent time. The height coordinate was then iterated to find the zeros of the velocity. The only parameter is the lower cut-off value of the velocity, which will be taken as zero. It was checked that the results do not depend on this cut-off. The width of the height profile was then calculated, which gives the roughness exponent rather accurately to be $\zeta = 0.388 \pm 0.002$ (see Figure 8.4).

The above mentioned theoretical modelings do not account for the fact that the roughness exponent essentially shows a crossover behavior from a high value ~ 0.6 at small length scales to a smaller value (~ 0.38) at large length scales (see Section of Chapter 5). There are some attempts to bring these two limits in a single model. Here, we discuss some of them.

Laurson et al. (2010) revisited the fluctuating elastic line problem with long-range kernel. A discretized version of the elastic string was considered. The local stress intensity factor has three terms: the elastic contribution ($K_i^{elastic}$, which is long range); the random pinning term (K_{i,h_i}^{random}, which represents an uncorrelated noise); and an external force (K_i^{ext}). The elastic force is long range with a power-law divergence near the crack tip, as mentioned earlier, and periodic boundary condition was taken. The total stress intensity factor on an element (K_i) of the discretized string is the sum of all these three terms. The simplified dynamics

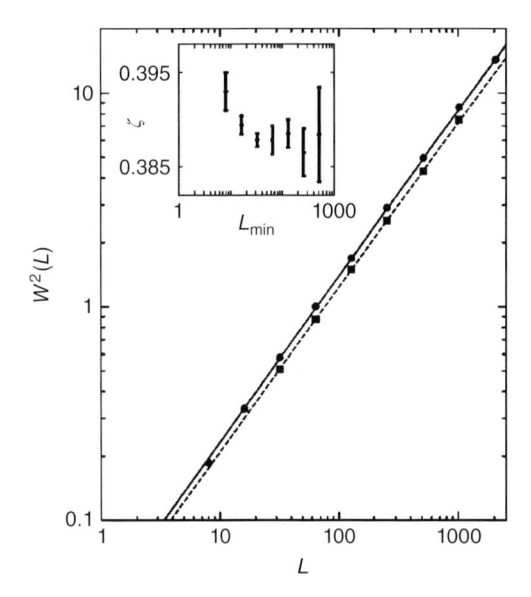

Figure 8.4 The mean-squared fluctuation of the long-range elastic front for different sizes (from $L = 8$ to $L = 2048$). Inset shows the roughness exponents for different system sizes. From Rosso and Krauth (2002), with kind permission from American Physical Society.

of the model is as follows: Wherever the local stress intensity factor is nonzero, the height of the string is increased by unity and a new random pinning force is assigned to it. Mathematically,

$$v_i(t) = h_i(t+1) - h_i(t) = \theta(K_i), \tag{8.5}$$

where $v_i(t)$ is the local velocity of element i at time t and θ is the heavy-side step function. This simplification ignores the fact that the local displacement should be proportional to the local stress intensity factor, however this does not change the scaling behavior. In order to maintain the quasi-static nature of the dynamics, as the crack front advances, a fraction of the external drive K_i^{ext} is withdrawn. The withdrawn part is proportional to the velocity. If one views this as an SOC system, this is equivalent to dissipation, which is otherwise not present in the model.

Two methods were used to study the avalanche clusters in this model: the avalanche map (AM) and the waiting time matrix (WTM), as discussed earlier. The avalanche map is a matrix $A(x,y)$, which has a unique value at all points which is swept by a single avalanche. The local clusters a^{AM} are then defined as the spatially connected sites with same value of the $A(x,y)$ matrix. Similarly, for the WTM, a local cluster is defined by the area of a spatially connected region, where the local velocity exceeds some threshold, which could be set to zero in this case of quasi-static driving.

The avalanche size distribution follows a power-law with a cut-off dependent on the proportionality constant k of the withdrawal of the external drive with velocity.

One can write

$$P(s) = s^{-\beta} F_c(s/c^{-1/y_c}). \tag{8.6}$$

The simulation results indicate $\beta = 1.25 \pm 0.05$ and $1/y_c = 0.725 \pm 0.08$, independent of the strength of the elastic interaction. The size of the spatially connected clusters in an avalanche also follows a power-law distribution but with different exponents:

$$P(a) = a^{-\beta_a} \mathcal{G}_c(a/c^{-1/y_c}), \tag{8.7}$$

with the numerical estimates $\beta_a = 1.52 \pm 0.05$ (measured by the avalanche map method) and $\beta_a = 1.53 \pm 0.05$ (determined by the waiting time matrix method).

A relationship between global avalanche size and local cluster size is obtained as follows: On average, it is expected that the probability of number of clusters increasing and decreasing should be equal (since these two processes balance each other). In an uncorrelated event, one would expect that the number of disconnected clusters (N) will follow a random walk of s steps with a reflecting boundary at $N = 1$. One would expect $\langle N(s) \rangle \sim s^\alpha$ with $\alpha = 1/2$. Thus, the local cluster size scales as $a \sim s/\langle N(s) \rangle$, that is, $s \sim a^2$. This implies (considering $P(a)da = P(s)ds$)

$$\beta_a = 2\beta - 1. \tag{8.8}$$

This is valid for this model ($\beta_a \approx 1.5$, $\beta \approx 1.25$). Note that for the scaling to be valid, the avalanche size must be smaller than the cut-off scale. In the presence of a correlation, when $\langle N(s) \rangle \sim s^\alpha$ in general, one would have $\beta_a = (\beta - \alpha)/(1 - \alpha)$.

It was observed experimentally earlier that the roughness exponent shows a crossover behavior from a larger value in the small length scale to a smaller value in the large length scale (see Chapter 5). In this model, it was claimed that the smaller value of the roughness exponent is expected to be valid only in the length scale, larger than the Larkin length $L_c \sim \Gamma_0^2$, where Γ_0 is the elastic strength. Indeed, while scaled by a factor Γ_0, a crossover is observed in the aspect ratio of the clusters from a value $\zeta = 0.55 \pm 0.05$ to a smaller value 0.39 ± 0.03. Similar behavior is also observed in the case of the RMS fluctuation of the from show in Figure 8.5. This is verified by studying the power spectrum as well. This shows that the crossover lengthscale is proportional to the Larkin length for the model.

To directly compare this model with experiment, the WTM clusters with a velocity cut-off were measured. The scaling form is still same

$$P(a) = a^{-\beta_a} \mathcal{G}_v(a/v_{th}^{-1/y_v}) \tag{8.9}$$

with $\beta_a = 1.53 \pm 0.05$ and $1/y_v = 1.8 \pm 0.01$. In Figure 8.6 the data from simulation and experiments were compared to get good agreement.

8.1.1
Variation of Universality Class

Even though the elastic line depinning in the context of mode-I fracture is generally considered to have the $1/r^2$-type elastic force, a more general form

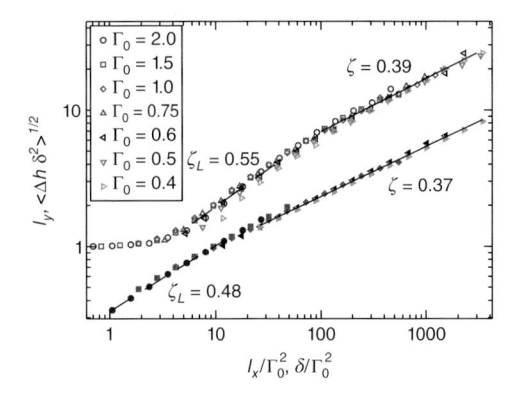

Figure 8.5 The scaling of the aspect ratio of the local clusters of an avalanche and the RMS fluctuation of the line. The axis is scaled by the Larkin length and it shows that the crossover from a larger to a smaller roughness exponent occurs for a length scale proportional to the Larkin length. From Laurson et al. (2010), with kind permission from American Physical Society.

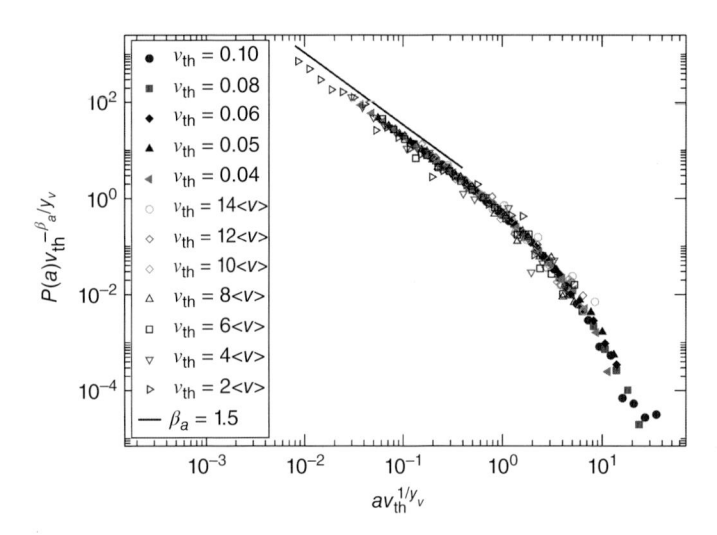

Figure 8.6 The scaled size distribution of the WTM clusters for various values of the velocity threshold. Direct comparison with experiments with Plexiglas shows good agreement. From Laurson et al. (2010), with kind permission from American Physical Society.

such as $1/r^\gamma$ can also be studied (Tanguy et al., 1998). For large values of γ, the behavior tends to the nearest neighbor Edwards–Wilkinson class and for $\gamma \to 0$, the mean-field limit is obtained. A continuous variation of roughness exponent with γ between $1 \leq \gamma \leq 3$ was numerically obtained ($\zeta = (3 - 2\gamma)/3$). In a more recent study (Laurson et al., 2013), the evolution of the shape of avalanche was studied for different values of γ. The average avalanche size is known to follow (Sethna et al., 2001; Papanikolaou et al., 2011) the scaling form

$$\langle s(\tau) \rangle = \int_0^\tau \langle v(t|\tau) \rangle dt \propto \tau^x, \qquad (8.10)$$

where v is the front velocity. Therefore, in general, one assumes the form

$$\langle v(t|\tau) \rangle = \tau^{x-1} F\left(\frac{t}{\tau}\right). \qquad (8.11)$$

Assuming that the early-time ($t/\tau < b \ll 1$) growth of the avalanche is of the form $t^{\delta'}$ and requiring the above-mentioned scaling to be valid at all times one gets $\langle v(b\tau|\tau) \rangle \propto (b\tau)^{\delta'} \propto \tau^{x-1}$, giving $\delta' = x - 1$. Finally, to arrive at the expression for $\langle v(t|\tau) \rangle$, one has to multiply by $(1 - t/\tau)^{x-1}$ to get the symmetric deceleration part of the avalanche toward the end. However, the symmetric nature of the avalanche shape is not generally valid (Baldassarri et al., 2003). In fact, the time reversal symmetry is broken for the case of smaller interaction range, although it is less apparent for higher range of interaction. In general, to account for the small asymmetry observed, Laurson et al. (2013) multiplied a factor $1 - a(t/\tau - 1/2)$ with the expression $\langle v(t|\tau) \rangle$. This choice is, however, not unique. The full expression now reads

$$\langle v(t|\tau) \rangle \propto \tau^{x-1} \left[\frac{t}{\tau}\left(1 - \frac{t}{\tau}\right)\right]^{x-1} \left[1 - a\left(\frac{t}{\tau} - \frac{1}{2}\right)\right]. \qquad (8.12)$$

Now the simulation was done for the discrete interface model, each element of which experiences the force

$$F_i = \Gamma_0 \sum_{j \neq i} \frac{h_j - h_i}{|x_j - x_i|^\gamma} + \eta(x_i, h_i) + f^{\text{ext}}, \qquad (8.13)$$

where η is the uncorrelated quenched disorder and f^{ext} is the external force. Three regimes are seen for x values by varying γ; $x = 2.0 \pm 0.01$ for $\gamma \leq 1$, 1.79 ± 0.01 for $\gamma = 2$, and 1.56 ± 0.01 for $\gamma \geq 3$. Now, the experimental data from planar crack front propagation give $x = 1.67 \pm 0.15$ (see Figure 8.7), making $\gamma = 2$, the closest candidate for the elastic force term. Thus, it gives an experimental justification for using the $1/r^2$-type elastic force term in modeling crack front propagation and also provides a measure of evolution of universality with the range of elastic interaction.

8.1.2
Depinning with Constant Volume

In general, these propagation models can be numerically integrated using an equation of the form

$$\begin{aligned} h_i(t+1) &= h_i(t) + 1 \quad &\text{if} \quad G_i(t) > 0 \\ &= h_i(t) \quad &\text{otherwise} \end{aligned} \qquad (8.14)$$

where $h_i(t)$ is the height of the ith element of the discretized chain (fracture front line) at time t. The form of $G_i(t)$ depends on the particular model. Consider the simplest form

$$G_i(t) = k(h_{i-1}(t) + h_{i+1}(t) - 2h_i(t)) + \eta_i(h_i(t)) + f_i(t), \qquad (8.15)$$

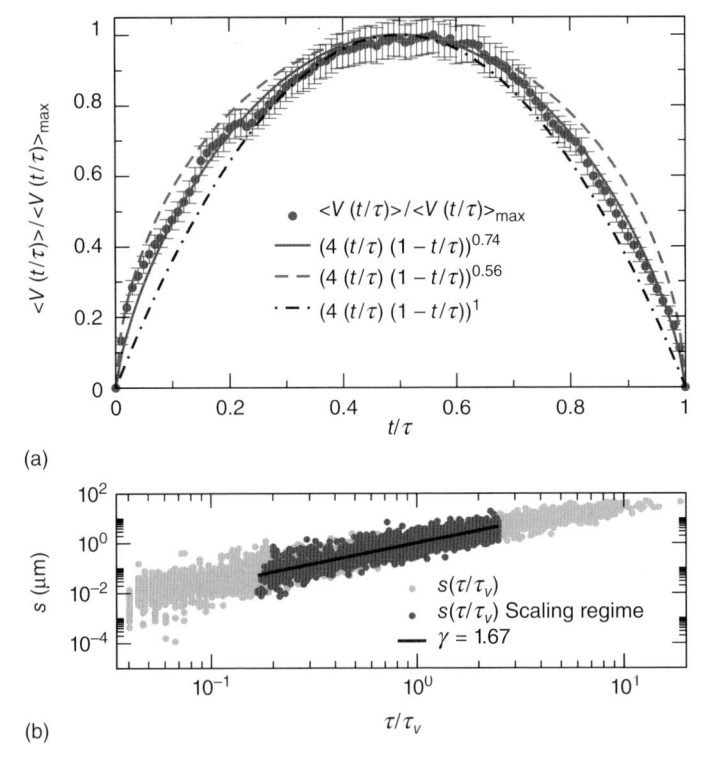

Figure 8.7 The shape of the average avalanche in planar crack propagation, (a) shape and fit corresponding to $x = 1.74 \pm 0.08$, (b) scaling of avalanche with $x = 1.67 \pm 0.15$. From Laurson et al. (2013).

where η_i is the quenched noise and $f_i(t) = f_0$ is the site- and time-independent external force in the EW model. The other part is, as usual, the local curvature at position i and is due to the elastic nature of the interface. This form can be different for different models.

In all previous studies, the external force was taken as a constant at every site (also independent of time). Let us discuss the following picture: A fluid of constant volume is injected, and the base of the system is breaking due to its pressure (see Figure 8.8). Now, clearly, the parts where fracture has propagated more will have higher force. Therefore, the force at any point i will now be ρh_i. Throughout the time, the volume was kept conserved. Therefore, the external force has a long-range correlation in the sense that an advancement of fluid at any part will deplete the water level and therefore the external force globally.

The velocity of the front was measured with time for different values of the initial height (see Figure 8.9). For low values of initial height, the velocity decreases to zero, that is, this is a pinned state. However, when the initial height was increased, the dependence of velocity with time becomes non-monotonic. This is due to the fact that although the interface is mostly pinned, at some points, the front has propagated more and due to the concentration of force at those

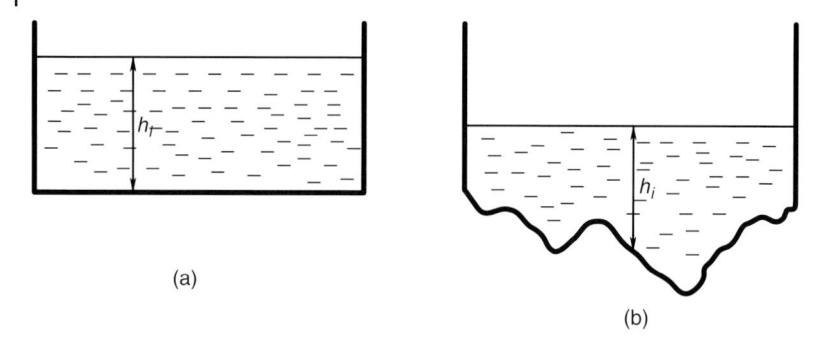

Figure 8.8 Fracture due to fluid of constant volume. The forces on the more advanced points are higher.

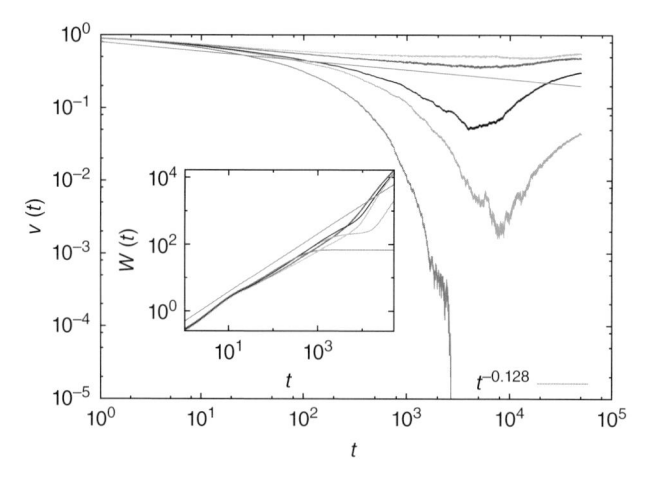

Figure 8.9 The velocity of the interface with time; nonmonotonic behavior is seen when fingering takes place. The best-fit power-law decay matches with the EW result.

points, a fingering-like behavior takes place. At still higher values of the initial height, at some points, a power-law decay was obtained and the exponent is same as that of the EW model. Finally, when the initial height is very high, the interface is depinned.

8.1.3
Infinite-range Elastic Force with Local Fluctuations

A common aspect of the interface depinning models is that the elastic force is either local (EW or KPZ) or decaying as inverse square (or some other power-laws) with distance, where dominance of the local configuration is clear.

Here, we discuss a model for front propagation where the elastic force is fully nonlocal, that is, the net amount of stretching of the string is felt with undiminished amplitude in all parts of the string (Biswas and Chakrabarti, 2011), however this is not a mean-field model.

Figure 8.10 Photograph of birds flying in a line forming shapes similar to the ones seen in infinite-range models (see Figure 8.11). From OpenStax College. Drag Forces, OpenStax-CNX Web site. http://cnx.org/content/m42080/1.8/, Feb 19, 2014.

In general, these propagation models can be numerically integrated using an equation of the form Eq. (8.14). The form of $G_i(t)$ depends on the particular model. For this infinite-range model, this term is complete nonlocal, and has the following form

$$G_i(t) = \frac{1}{L}\sum_j \left[\sqrt{(h_j(t) - h_{j+1}(t))^2 + 1} - 1\right]$$

$$sgn(h_{i-1}(t) + h_{i+1}(t) - 2h_i(t)) + \eta_i \overline{(h_i(t))} + f^{\text{ext}} \quad (8.16)$$

Let us denote this as model I. A further simplification was also proposed, where the nonlocal extension term was replaced by a constant, that is,

$$G_i(t) = C sgn(h_{i-1}(t) + h_{i+1}(t) - 2h_i(t)) + \eta_i \overline{(h_i(t))} + f^{\text{ext}}, \quad (8.17)$$

where C is a constant. Let us denote this by model II. This model for "depinning" transition can somewhat mimic a flock of birds flying in a line (see Figure 8.10). The elastic force denotes the tendency of a bird to fly along with its two nearest neighbors, the pinning force may be considered as some measure of "(un)fitness" of a bird (that is why it is taken as independent of height or in this case, distance traveled), and the external force is the urge or necessity of flying (lack of food, presence of enemy, etc.). Of course, $v \to 0$ limit does not apply for a flock of flying birds. But, here we can measure the velocity from the rest frame of the slowest moving bird. Then, the "depinning" transition would indicate whether the birds fly "together" or get scattered in the long time limit. If the average velocity of the flock is zero with respect to the slowest moving bird, then they fly "together" or the flock is "pinned." If, however, the average velocity is finite in the rest frame of the slowest bird, then in the long time, the flock will be "scattered" or is said to be "depinned."

The nature of disorder used has similar properties as before. It has a random value between $[0 : -1]$ with probability $1 - p$ and zero otherwise. In the following, we present results for $p = 0$ case, but the exponent values do not change for finite p values. The evolution rule is analogous to the equation of motion presented above for an elastic chain. But, the major difference is that, in this case, it is not an "equation of motion" but only a rule chosen for the movement of

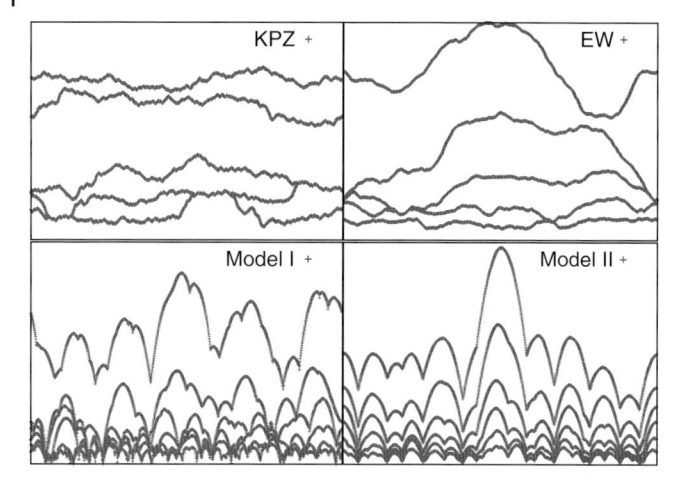

Figure 8.11 The snapshots of fracture fronts, shown at different times, in EW, KPZ, and in two infinite-range models discussed in the text. From Biswas and Chakrabarti (2011).

the birds. Therefore, the choice of Δt is arbitrary and we fix it to unity in this case.

In Figure 8.11, the advancements of the fracture front at different times following EW, KPZ, and the two infinite-range models are shown. The models show depinning transitions when the external force is more than a critical value. As usual, the velocity is taken as the order parameter ($v \sim (f^{\text{ext}} - f_c^{\text{ext}})^\theta$), where the values of the order parameter exponent are 0.83 ± 0.01 and 1.00 ± 0.01.

8.2
Fracture Front Propagation in Fiber Bundle Models

The elastic line modeling of fracture front in general takes the effect of the solid through the pinning centers and in terms of long-range elastic force between the elements of the fracture front line. The fiber bundle model has been used to model mode-I crack propagation recently. This approach is rather different. The randomness, as usual, enters through the failure thresholds of the fibers. Furthermore, the effect of breaking, that is the stress redistribution takes care of the elastic property of the solid. The fracture front as such is not considered, it arises here as an emergent property of the model. In the following, we describe those attempts briefly.

8.2.1
Interfacial Crack Growth in Fiber Bundle Model

The large difference between the theoretical prediction of roughness exponent from fluctuating line model and experiments lead to many efforts in this area

(Schittbuhl et al., 2003) (see also Alava and Zapperi, 2004; Schmittbuhl et al., 2004). The assumptions made in the theory of fluctuating line depinning model and its calculations are: (i) fracture fronts are essentially small perturbations over a long-range elastic string driven through a disordered medium, (ii) the fracture front is a single-valued function, thus ruling out islands and overhangs, and (iii) the fracture front has no damage zone ahead of it. While none of these are necessarily true, all of these may not contribute in the roughness exponent value. Nevertheless, the idea of modeling fracture front as a front advancing due to coalescence of fracture zones created ahead of it, rather than as an elastic line moving due to competition of elastic and pinning forces, was due to lifting the above-mentioned assumptions.

Experimentally, there are two typical values of roughness exponent obtained for different materials depending on the length scale involved in the experiment. These are close to 0.35 and 0.60 respectively for large and small length scales. It has been suggested (Gjerden et al., 2012) (see also Stormo et al., 2012; Gjerden et al., 2013) that these two mechanisms are due to two different physical processes, namely the coalescence (of small damages ahead of fracture front) in small length scales (gradient percolation) and fluctuating elastic line in large length scales. An attempt was made to obtain these two limits in a single model, under the general framework of the fiber bundle models of failure.

The idea of the coalescence model was to consider an elastic half plane which is connected to an infinitely rigid plate by discrete interface made up of an array of linear elastic springs with breaking thresholds taken randomly from a uniform distribution between zero and one.

In this model, the fibers are arranged in a square lattice and each fiber acts on an area $a \times a$. As in the general FBM, the fibers are attached to two elastic plates. However in this case, although the top plate is considered to be infinitely rigid, a finite elasticity E is considered for the bottom plate. A fiber has to carry a load in accordance with its local (u_i) and global (D) displacements

$$f_i^{\text{ext}} = -k_i(u_i - D), \tag{8.18}$$

where k_i is the elastic constant of the fiber. Regarding the load transfer mechanism, which determines the behavior of the system, the load is transferred via Green's function

$$u_i = \sum_j G_{ij} f_i^{\text{ext}},$$

$$G_{ij} = \frac{1-v^2}{\pi E a^2} \int \int_{-a/2}^{a/2} \frac{dx'\,dy'}{|r_i(x,y) - r_j(x',y')|} \tag{8.19}$$

where v is the Poisson's ratio.

Since the goal is to study the propagation of a fracture front, the loading scheme has to be such that the fracture propagates from one side and there is a well-defined front. One way to ensure that is to take a linear gradient in the values of the failure thresholds of the fibers. In this way, one limiting case will be the gradient percolation.

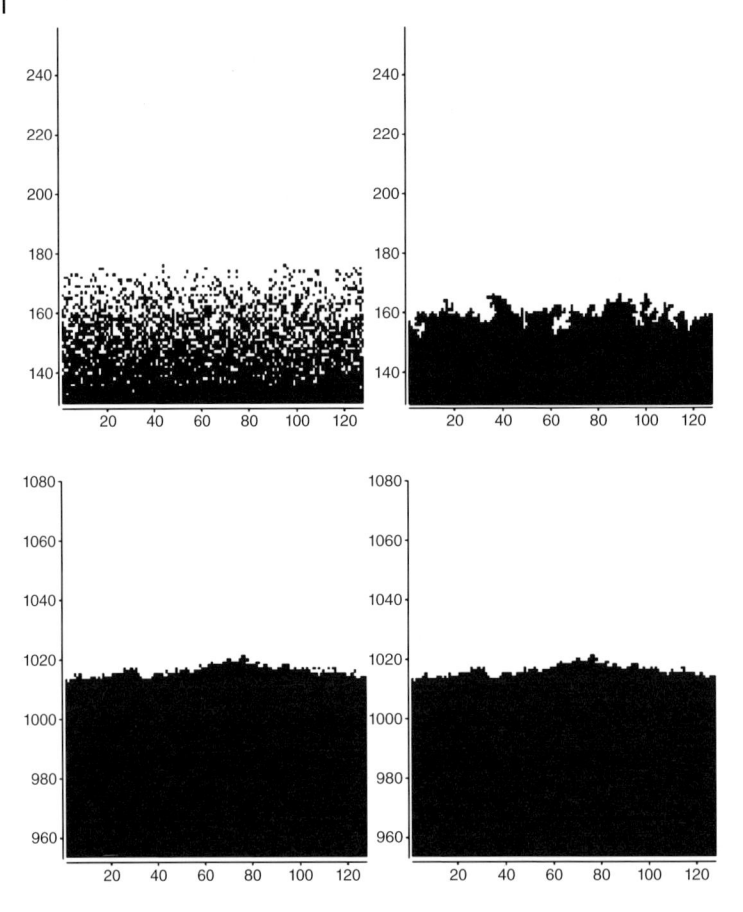

Figure 8.12 Snapshots of the system with high elastic modulus (upper figures) and low elastic modulus (lower figures). The left panel is the direct snapshot, while the right panel depicts only the largest connected cluster. A visible change in the roughness properties is seen. From Gjerden et al. (2012).

By changing the Young's modulus from a high to low value, the model shows a transition from the situation where the propagation of the front is coalescence-controlled to the situation where it behaves like fluctuating elastic line. In Figure 8.12, snapshots of the system are shown. For higher elastic modulus, the largest connected cluster is chosen and its roughness properties are analyzed by removing the overhangs. This is done by doing an averaged wavelet coefficient (AWC) analysis. Here, a wavelet transformation is done on the single-valued front $h(x)$ and then the resulting wavelet coefficient $w(a,b)$ is averaged for each length scale a at position b. For self-affine surface such as this, the averaged coefficient scales as

$$W(a) \sim a^{\zeta+1/2}, \tag{8.20}$$

8.2 Fracture Front Propagation in Fiber Bundle Models

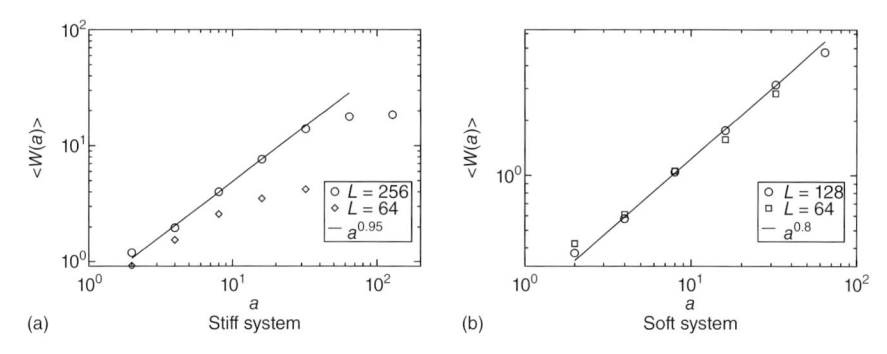

Figure 8.13 The averaged wavelet coefficients for the two systems. From Gjerden et al. (2012).

where $W(a) = \langle |w(a,b)| \rangle_b$. For the two extreme cases, two exponents (see Figure 8.13) are obtained: $\zeta_1 = 0.45$ for the stiff system and $\zeta_2 = 0.3$ for the soft one. However, the stiff system exponent is somewhat smaller than the one reported in the case of coalescence models (0.6) (Schittbuhl et al., 2003) and also the soft system exponent is smaller than the one reported for numerical studies of fluctuating lines (0.39) (Rosso and Krauth, 2002). However, this model attempts to link the two extreme cases of fracture front propagation in a single model.

8.2.2
Crack Front Propagation in Fiber Bundle Models

Modeling the intermittent dynamics and avalanche statistics of fracture front moving through disordered solids has been done using several approaches, as discussed earlier. Here, we mention the SOC dynamics of a force-controlled fiber bundle model that gives realistic results in terms of the avalanche statistics and fracture front morphology (Pontuale et al., 2013).

The model considers some modification over the standard formulation of the fiber bundle model. In particular, the model aims at describing the slow propagation of the front by a self-organized dynamics. In that case, the system needs to be slowly driven to bring the subcritical system to criticality and a dissipation mechanism needs to be present to bring the supercritical system to criticality. Apart from that to ensure continued motion of front, a regenerative mechanism needs to be present. In this model, a quasi-static driving mechanism is used and when a fiber breaks, a fraction of its load is dissipated and a new fiber is regenerated in front of it to ensure continued motion.

The model is considered in one dimension. Fibers are arranged in a line and each fiber is assigned a breaking threshold. Stress is then applied slowly, until the weakest fiber breaks. This fiber is then replaced by a new one with a new failure threshold drawn from the same distribution. A fraction δ of the released stress is dissipated and the remaining $(1 - \delta)$ fraction is redistributed. For the redistribution, a distance-dependent rule was adopted. This is because a crack front in

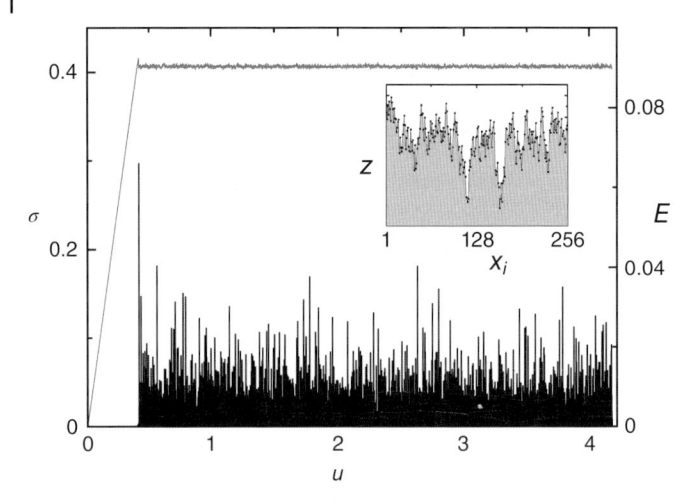

Figure 8.14 Avalanches of energy release (lower curve) and instantaneous bundle stress (upper curve). Inset shows a snapshot of the crack front. From Pontuale et al. (2013), with kind permission from EPL.

elastic medium generates a inverse square stress field in the medium. Similarly, if a fiber at position x_i breaks under a stress σ_i, then the stress of a fiber situated at position x_j is increased by an amount $\Delta\sigma_j$ given by

$$\Delta\sigma_j = C\frac{\sigma_i}{(x_i - x_j)^2} \quad \text{for} \quad j \neq i \quad \text{and} \quad \delta\sigma_i = 0, \tag{8.21}$$

where periodic boundary condition was employed and C is a normalization constant to ensure $\sum_j \Delta\sigma_j = (1-\delta)\sigma_i$.

Figure 8.14 shows a time series of the energy release $E = \sum_i \sigma_i^2/2$ and also the snapshot of the configuration of the fibers, that is, the fracture front. The time series shows that after some transient, the system reaches a stationary state where the average stress saturates and further perturbations in the system can yield responses of all sizes in various quantities, which is a signature of SOC. The different quantities that show scale-free behaviors are the avalanche area A, defined as the total number of failed fibers in an avalanche, the duration τ of an avalanche, and also the size of the disconnected clusters C of an avalanche. These size distributions do not depend on the dissipation, but obey the following scaling form

$$P(\Delta) = \Delta^{-\beta_\Delta} f_\Delta(\Delta/\Delta_0) \tag{8.22}$$

where the cut-off value Δ_0 is determined by the dissipation and obeys $\Delta_0 \approx y^{-y_\Delta}$. The exponent values obtained from data collapse are the following: for size of avalanches (total area) A, $\beta_A = 1.2$, for duration of the avalanche τ, $\beta_\tau = 1.5$, and for the size of the disconnected clusters C, $\beta_C = 1.5$ (see Figure 8.15). These values agree with those found in experiments (see Bonamy, 2009 for a review).

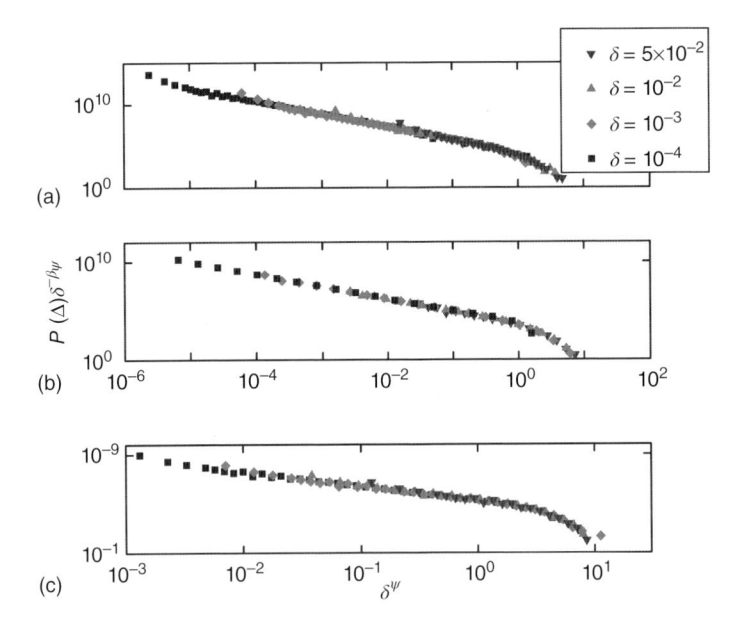

Figure 8.15 The scaled size distributions of (a) disconnected clusters in an avalanche C, (b) total area or the number of failed fibers in an avalanche, and (c) duration of an avalanche. From Pontuale et al. (2013), with kind permission from EPL.

To obtain the roughness of the fractured surface, a height variable h_i was defined at each site. It is assumed that when a fiber breaks, a local advancement of the crack front takes place in the direction perpendicular to the direction of the length, which is proportional to the breaking stress of that fiber. Hence $h_i(t) = \sum_{k=1}^{t} \sigma_{k_i}$. One can then compute its spatial fluctuation in the stationary state

$$W_r^2 = \langle (h_{i+r} - h_i) \rangle \tag{8.23}$$

which should scale as $r^{2\zeta}$. In this case, the exponent value is $\zeta = 0.35$, again the cut-off being controlled by the dissipation factor. In the absence of dissipation, the system shows a transition between pinned state, where the front does not propagate, and a propagating phase, where the front propagates with a finite velocity.

8.2.3
Self-organized Dynamics in Fiber Bundle Model

As we have discussed earlier, while in the global load sharing version, the load at which the system completely fails scales with system size linearly, and for the local load sharing case, the increase is only sublinear (in fact, $N/\log(N)$; N being system size (Pradhan et al., 2010)). The implication being the critical load per fiber σ_c at which the system fails becomes finite for the global load sharing case and goes to zero in the large system size limit for the local load sharing case. Therefore, the

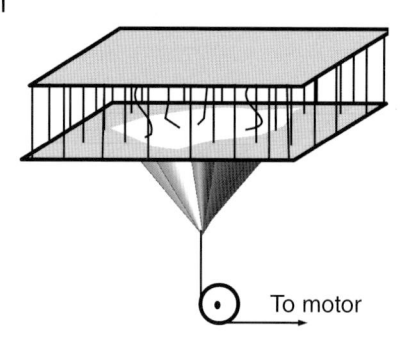

Figure 8.16 A schematic diagram of our model. Load is slowly increased at a single point on the lower platform. The patch of broken fibers is indicated by the white portion in the lower platform. From Biswas and Chakrabarti (2013b), with kind permission from American Physical Society.

observations such as divergence of relaxation time and proper scale-free distribution of avalanches, if any, are not seen in the local load sharing case; unlike in the global load sharing version, where these can be analyzed in detail (Pradhan *et al.*, 2010).

However, the situation can be quite different if one makes the initial applied load localized (at an arbitrarily chosen central site (Biswas and Chakrabarti, 2013b); see Figure 8.16) in a local load sharing fiber bundle model in greater than one dimension (for one dimension, the damage interface cannot increase). Let this load be increased at a slow but constant rate. Initially, no load is present on any fiber except for one at the central site. As the applied load increases beyond the failure threshold of this central fiber, it breaks and the load carried by it is redistributed among its nearest neighbors and so on. Here we study two versions of the model. Model I: In general, whenever a fiber breaks, the load carried by that fiber is redistributed equally among its nearest surviving neighbor(s), hence the load of the broken fiber remains concentrated around the damage. In this way, the fibers that are newly exposed to the load, say, after an avalanche, have a relatively low load compared to the ones that are accumulating load shares from the earlier failures and are still surviving. As we see later, this helps in maintaining a compact structure of the cluster or patch of the broken fibers. This local force redistribution is justified from the point of view that the newly exposed fibers are presumably further away from the point of loading and therefore have to carry a smaller fraction of the load at the original central site.

The fibers on the perimeter of the failed or damaged region, which together are carrying the entire load, increase in number with time. Hence, the load per fiber decreases during an ongoing avalanche, which is assumed to be a much faster process compared to the external load increase. However, as the load on the bundle increases at a constant (but slow) rate, the load per fiber along the boundary will tend to increase (see Figure 8.17). Eventually, a dynamically stable state will occur when the load per fiber fluctuates around a stable value and the system has reached a self-organized state. In this dynamical state, failure of fibers in the process of

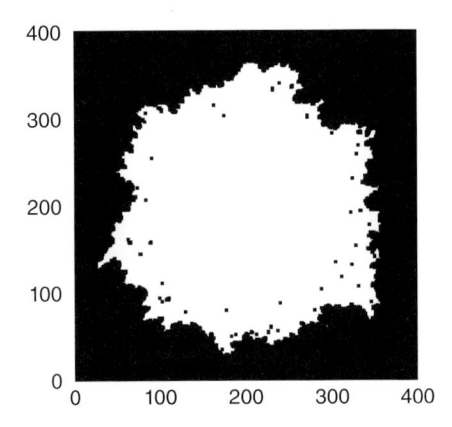

Figure 8.17 A snapshot of the patch of broken fibers when the load redistribution is among the nearest surviving neighbors (Model I). The fiber failure thresholds are uniformly distributed between [0:1]. From Biswas and Chakrabarti (2013b), with kind permission from American Physical Society.

avalanches is seen to have a scale-free size distribution, suggesting the state to be a self-organized critical one.

A simpler version of this model, Model II, was studied in which the load of a broken fiber was equally shared by all the surviving fibers that have at least one broken neighbor, that is, it was equally distributed along the boundary of the broken or damaged patch. Due to the fact that local fluctuations are ignored in the process of load redistribution, this version is analytically tractable using a mean-field-like approach and the numerical results compare well.

In the following, we first discuss the numerical results for the nearest neighbor load sharing model (Model I) and then go over to the simpler version of uniform load sharing along boundary (Model II), giving the analytical estimates for the later and comparing them with numerical results.

As mentioned earlier, usually one does not find a critical load that remains finite in increasing the system size for local load sharing fiber bundle models. This is the result of extreme value of statistics, in which the strength of the sample is determined solely by its weakest point. In that case, as the load, no matter how small, is applied system wide, there will always be a large enough weak patch, which will be broken and due to concentration of the load on the boundary and the fact that more load is nucleated if the fracture progresses, keep the patch growing, leading to a system-wide failure without any further input from outside (see Section 4.3).

However, the situation will be very different when the initial load, too, is applied locally and not on all elements of the system. Initially, the load was applied on one fiber, say the one at the middle. Then, its value was increased continuously at a very slow rate. When this fiber breaks, the load carried by it is redistributed equally among its four neighbors. Since the load is always increased (only on the fibers that are already carrying a nonzero load), this breaking and redistribution dynamics continues. In general, whenever a fiber breaks, the load carried by that fiber is equally shared by the surviving fiber(s) nearest to it. When the islands break (all four neighbors already broken), its load is redistributed among its nearest surviving neighbor(s) (searched along the perpendicular axes; other searches are slow and do not change the critical behaviors), no matter how far they are located. Note that in this way of redistribution, the fibers that are newly exposed to the load have a rather low load per fiber compared to the ones that are gathering and withstanding the loads of broken fibers from a few steps earlier. This ensures that the compact structure of the patch of the broken fibers is maintained, since a rather weak but new fiber is more likely to survive than a stronger but old one, thereby eliminating the possibilities of fingering like effects from the patch boundary. Also note that, following the usual definition of avalanche, the external load was not increased when a fiber failed or more fibers failed due to load redistribution. In other words, the process of load redistribution happens in a much faster rate than load increase, which is quasi-static.

As we see, the application of load at a point and subsequent localized redistribution rules bring the system into a self-organized critical state. One of the signatures of self-organization is the stationarity in macroscopic quantities. In this case, the most important quantity is the average value of the load per fiber. It was found that this quantity stabilized after some initial transients. In fact, the entire distribution function of the load per fiber, and not only its first moment, becomes stable. Similarly, the stability in the distribution function of the failure threshold values of the surviving fibers was also found, each of which carry a nonzero load. In Figure 8.18 the saturation of the average load per fiber value is shown. The inset shows the stationary distribution of the load per fiber values and failure threshold values. While these results are for the initially uniform (in [0:1]) threshold distribution, the same phenomenon occurs for other threshold distributions, such as Weibull, Gaussian, and triangular. The average stationary values change for different distributions.

Note that usually the two basic ingredients of a self-organized critical system are external drive and dissipation (Bak *et al.*, 1987; Manna, 1991). In the present case, although there is an external drive, there is no explicit dissipation. But the effect of dissipation enters the system from the fact that the effective system size (number of surviving fibers carrying nonzero load) is an increasing function of time.

As indicated earlier, a quasi-static load increase protocol was followed. In effect, this means increasing the load per fiber uniformly on all those surviving fibers that already carry a nonzero load, until the weakest (having the smallest difference between load and failure threshold) fiber breaks. Then the system was allowed to adjust by breaking the fibers and redistribution of the loads, and come to a state when no further fiber breaks. All the fibers broken in between, constitute one avalanche. This process was repeated in every time step. The size distribution of avalanches measured in this way is plotted in Figure 8.19. This shows a power-law distribution with exponent value close to 1.35 ± 0.03. This exponent value remains unchanged when other types of threshold distributions (not shown here) are used. The duration of an avalanche (number of times the load was redistributed in an avalanche) was also measured. This is also a power-law decay with exponent value close to 1.53 ± 0.02 (see Figure 8.20). It is interesting to note that avalanche size

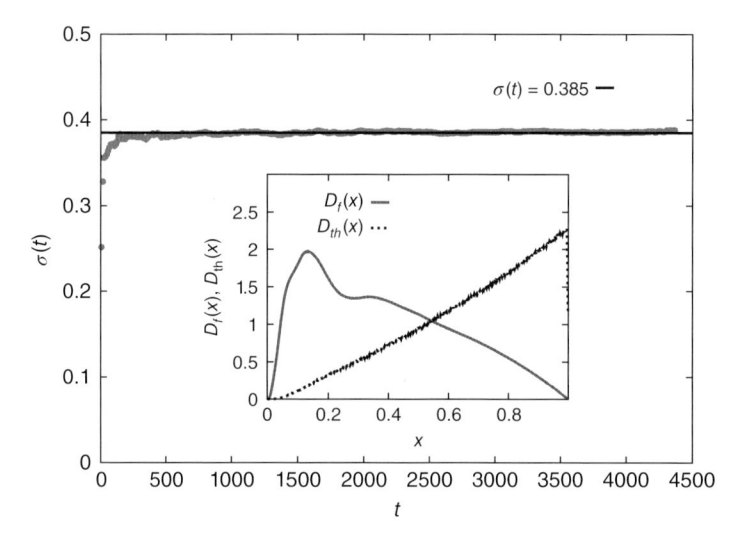

Figure 8.18 The time evolution of the average load per fiber on the broken patch boundary (the breaking thresholds are distributed uniformly in [0:1]) for the nearest surviving neighbor load redistribution (Model I). The saturation is close to 0.385. The inset shows the stationary distribution of the load per fiber value ($D_f(x)$) and the failure threshold values ($D_{th}(x)$) of the surviving fibers on the boundary. From Biswas and Chakrabarti (2013b), with kind permission from American Physical Society.

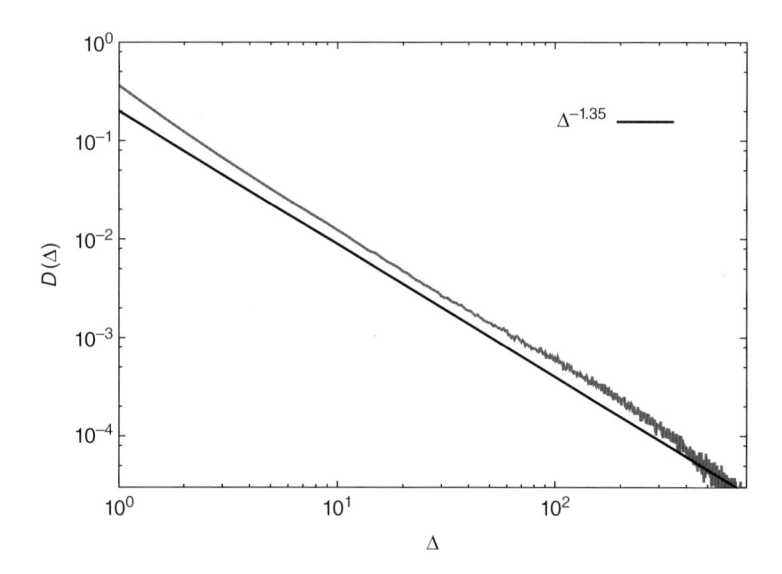

Figure 8.19 The distribution of the avalanche sizes for Model I (load redistribution is among the nearest surviving neighbors). The distribution is a power-law with exponent value 1.35 ± 0.03. From Biswas and Chakrabarti (2013b), with kind permission from American Physical Society.

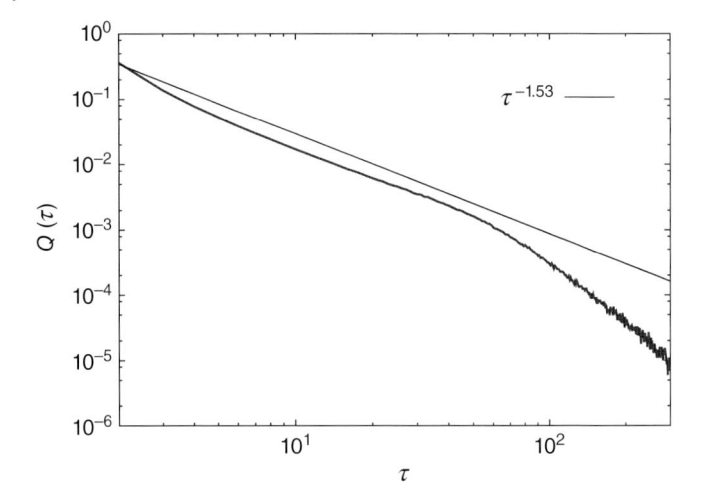

Figure 8.20 The distribution of the duration of an avalanche for Model I (the nearest surviving neighbor load sharing case). This shows a power-law decay with exponent value 1.53 ± 0.02. From Biswas and Chakrabarti (2013b), with kind permission from American Physical Society.

and duration exponent values are rather close to the stochastic sandpile or Manna universality class (Manna, 1991; Lübeck, 2004).

A simpler version of the model discussed so far is the one where the load carried by a broken fiber is equally redistributed among all the fibers that have at least one broken neighbor. This is another way of saying that the load is redistributed uniformly along the entire boundary, once a fiber breaks (see Figure 8.21).

Since the size of the boundary is large after the transients, the behavior of the model in the present form is tractable via a mean field-like approach. In the following, we first discuss the stationary distribution functions of load per fiber value and failure thresholds for different initial failure threshold distributions; comparing the estimates with numerical results. Then, we follow the same procedure for avalanche size distribution.

Same as the earlier version of the model, the load distribution and the threshold distribution along the patch boundary reach a stationary state after some transient. In this case, however, one can also make an analytical estimate, along with the numerical simulations, to get a good agreement.

Consider the failure thresholds to be distributed uniformly between the limits $[a:1]$. First, let us take the case when $a = 0$. If one measures the average load per fiber with time (one time step being one avalanche), one finds that after a transient, it saturates to a given value (see Figure 8.22). For this case, the value is close to $1/3$. In analytically estimating it, assume that dynamically all values of force (between $[0:1]$, that is, the range of the threshold distribution) are generated with equal probability. Once this is assumed, we have the probability of having a load on a fiber between x and $x + dx$ as $P(x)dx$. Now, the probability that the fiber has a threshold greater than x is $2(1 - x)$. Therefore, the probability that we will find a

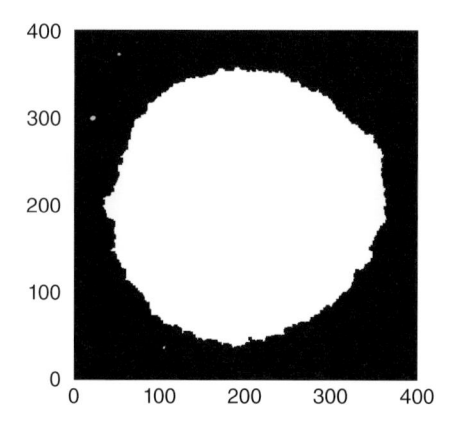

Figure 8.21 A snapshot of the patch of broken fibers when the load redistribution is uniform along the broken patch boundary, that is, Model II. The failure thresholds are uniformly distributed between [0:1]. From Biswas and Chakrabarti (2013b), with kind permission from American Physical Society.

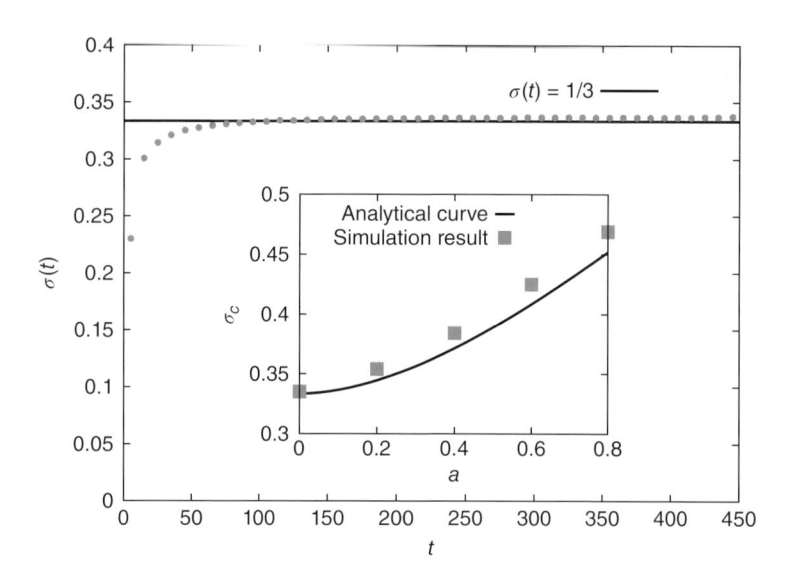

Figure 8.22 The time evolution of the average load per fiber (when the threshold is uniformly distributed between [0:1]) for Model II (i.e., uniform load redistribution along the patch boundary). The solid line is the analytical estimate (1/3) and the simulation result is close to that value. The inset shows the comparison of the simulation results and analytical estimates of the saturation load per fiber value, when the threshold is distributed uniformly between [a:1]. From Biswas and Chakrabarti (2013b), with kind permission from American Physical Society.

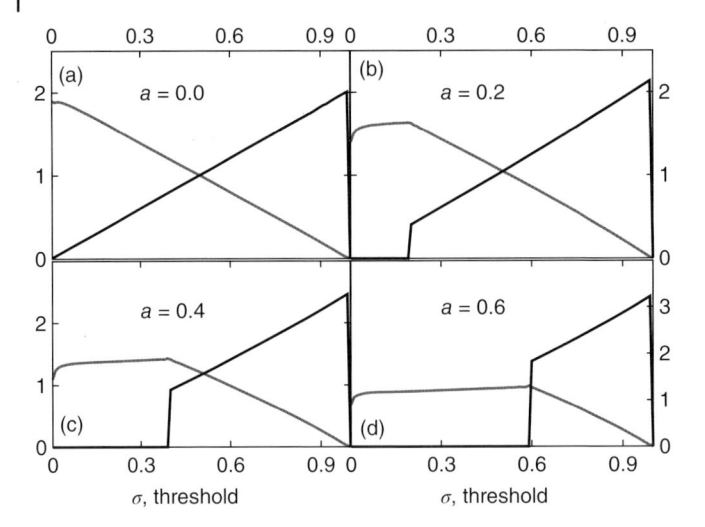

Figure 8.23 Simulation results for the probability density functions of the load per fiber (curves decreasing with load) and failure thresholds of the fibers (curves increasing with threshold), both only along the boundary of the patch of broken fibers, for different values of the lower cut-offs in the otherwise uniform threshold distributions for Model II (uniform load redistribution along the patch boundary). From Biswas and Chakrabarti (2013b), with kind permission from American Physical Society.

fiber carrying a load between x and $x + dx$ is $2(1-x)P(x)dx$. With $P(x) = 1$ in this case (as $a = 0$), we have the normalized probability density function for the force distribution as $D_f(x) = 2(1-x)$. This is very close to what was obtained numerically (Figure 8.23). Of course, the average $\int_0^1 2(1-x)x\,dx = 1/3$, which is again close to what is seen numerically (Figure 8.22). In the same way, one can calculate the probability density function for the threshold, which is $D_{\text{th}}(x) = 2x$ and is also seen numerically (Figure 8.23).

Now consider the case when a is finite, that is, the threshold distribution function has a lower cut-off. Now, since all fibers have threshold higher than a and it was assumed before that force is uniformly distributed, the probability density function for force will be uniform between $[0:a]$ and then decreasing linearly as before (can be obtained along the same line as before). From normalization condition, the height of the uniform part is $\frac{2}{a+1}$. The average force comes out to be

$$\langle \sigma_c \rangle = \frac{a^2}{1+a} + \frac{2}{1-a^2}\left[\frac{1}{6} - \frac{a^2}{2} + \frac{a^3}{3}\right], \tag{8.24}$$

where the angular brackets denote spatial average. This prediction of this calculation was compared with the values obtained numerically in Figure 8.22.

In order to show universality of the above-mentioned phenomenon with respect to different threshold distributions, the forms of the threshold distributions were varied (Gaussian, triangular, etc.). Here, we discuss the case of Weibull distribution, which is more abundant in real situations.

A general form of the Weibull distribution is $\alpha'\beta' x^{\alpha'-1} e^{-\beta' x^{\alpha'}}$, where α' and β' are two parameters. Let us consider the particular case when $\alpha' = 2$ and $\beta' = 1$. As before, the probability that the threshold is greater than x is proportional to $\int_x^\infty x' e^{-x'^2} dx' \sim e^{-x^2}$. Now, since the probability density function for force is uniform, the probability density function of force on the survived fibers will simply be

$$D_f(x) = \frac{2}{\sqrt{\pi}} e^{-x^2}. \tag{8.25}$$

Similarly, the probability that the force is lower than x is proportional to x; using the form for threshold distribution it is straightforward to get the probability density function for threshold distribution of the survived fibers, which is

$$D_{\text{th}}(x) = \frac{4}{\sqrt{\pi}} x^2 e^{-x^2}. \tag{8.26}$$

Both of these functions are in good agreement with numerical simulations (see Figure 8.24). Also, the saturation value of load per fiber can be calculated as

$$\int_0^\infty x D_f(x) dx = \frac{2}{\sqrt{\pi}} \int_0^\infty x e^{-x^2} dx = \frac{1}{\sqrt{\pi}}, \tag{8.27}$$

which is also in good agreement with simulation value (see inset Figure 8.24).

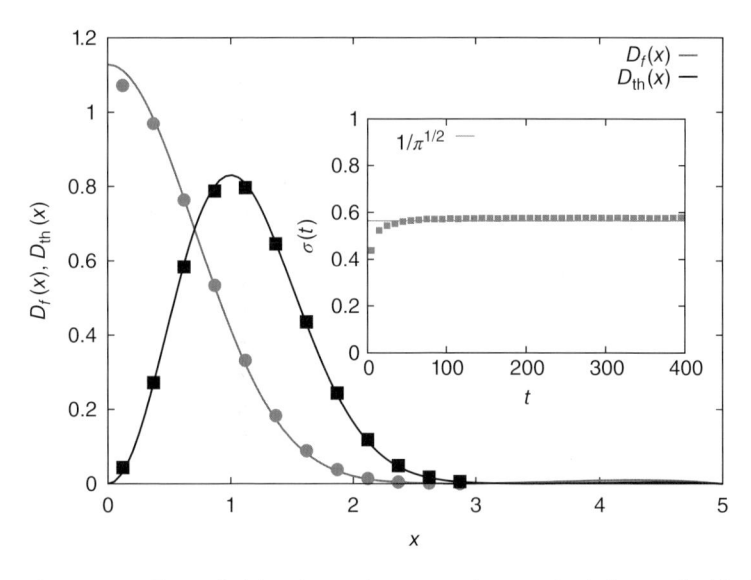

Figure 8.24 The probability density functions of load per fiber ($D_f(x)$) and failure thresholds ($D_{th}(x)$) of the fibers in the boundary of the broken patch for Weibull threshold distribution with $\alpha' = 2$ and $\beta' = 1$ for Model II. The solid lines are analytical predictions (Eqs. (8.25) and (8.26)) and the points are simulation results. The inset shows the time variation of the average load per fiber value, which saturates close to the analytical estimate $1/\sqrt{\pi}$. From Biswas and Chakrabarti (2013b), with kind permission from American Physical Society.

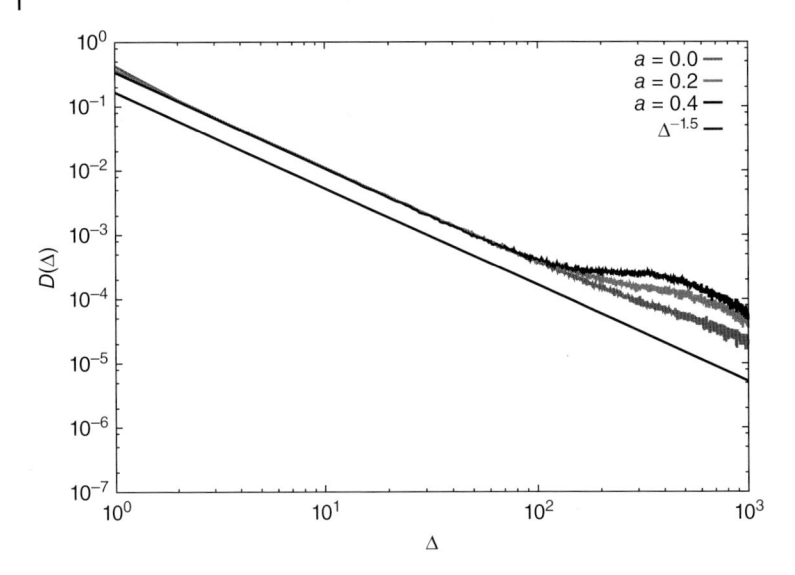

Figure 8.25 The distribution of the avalanche sizes is plotted for zero and finite lower cut-offs for Model II. The distribution function is a power-law with exponent value 1.50 ± 0.01, which is also our estimate from scaling arguments. From Biswas and Chakrabarti (2013b), with kind permission from American Physical Society.

The size distribution of avalanches is measured as before. Now, the size distribution exponent was found to be close to 3/2 (see Figure 8.25), which is in agreement with the mean-field predictions of avalanche size distributions in SOC models (Lübeck, 2004). Note that similar exponent value was obtained as before when one measures avalanches only near the critical point in global load sharing models (Pradhan et al., 2006) and also in other versions of self-organized models with fiber regeneration (Moreno et al., 1999).

The duration distribution is a power-law as before, with exponent value close to 2.00 ± 0.01, which is again in agreement with mean-field predictions of SOC models (see Figure 8.26).

Now, to estimate the value of the avalanche size exponent, first assume that the average load per fiber on the perimeter of the damaged region has a distribution, which is Gaussian around its mean: $P(\sigma) \sim e^{-(\sigma-\sigma_c)^2/\delta\sigma}$. Hence, on a dimensional analysis, mean-squared fluctuation $\delta\sigma \sim (\sigma - \sigma_c)^2$. Also, the avalanche size Δ scales as $(\delta\sigma)^{-1}$ since it may be viewed as the number of broken fibers after a load increase of $\delta\sigma$. This gives

$$(\sigma - \sigma_c) \sim \Delta^{-1/2}. \tag{8.28}$$

The probability of an avalanche being of the size between Δ and $\Delta + d\Delta$ is $D(\Delta)d\Delta$. Now, the deviation from the critical point scales (Pradhan et al., 2002) with the cumulative of all avalanches up to that point; giving $(\sigma - \sigma_c) \sim \int_\Delta^\infty D(\Delta)d\Delta$. If one

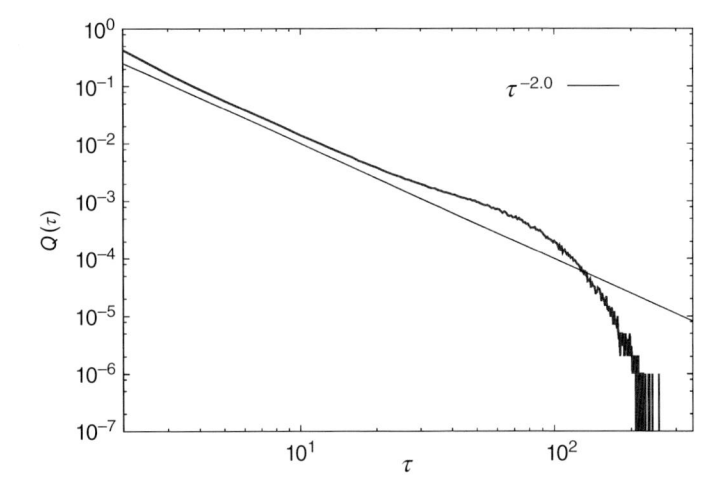

Figure 8.26 The distribution of the duration of an avalanche is plotted for Model II. This shows a power-law decay with exponent value 2.00 ± 0.01. From Biswas and Chakrabarti (2013b), with kind permission from American Physical Society.

takes $D(\Delta) \sim \Delta^{-\beta}$, then

$$(\sigma - \sigma_c) \sim \Delta^{1-\beta}. \tag{8.29}$$

Comparing Eqs. (8.28) and (8.29), one has $\beta = 3/2$. Therefore, the probability density function for the avalanche size becomes $D(\Delta) \sim \Delta^{-3/2}$, which fits well with simulation results (Figure 8.25).

Note that these results were also checked for triangular lattices, where the number of nearest neighbors is six. As one can see, the analytical estimates are independent of lattice topology, hence predict same behavior for triangular lattice as well, which is what also obtained numerically.

8.3
Hydraulic Fracture

Often it is important to determine the fracture pattern and breaking strength of a material under hydraulic pressure. Such situations particularly arise in the case of borehole diggings for oil reservoirs and the like. An incompressible fluid, such as water, is generally injected creating a high pressure. The rock breaks under that pressure creating easy paths for oil extraction. The difficulty in studying hydraulic fracture in spite of its high importance is the lack of accessibility. The quantities measured mainly are the acoustic emissions and fluctuation of pressure of the injected fluid.

A typical behavior of pressure with time can be seen in Figure 8.27 (Galindo Torres and Muñoz Castaño, 2007). The pressure linearly increases when the system

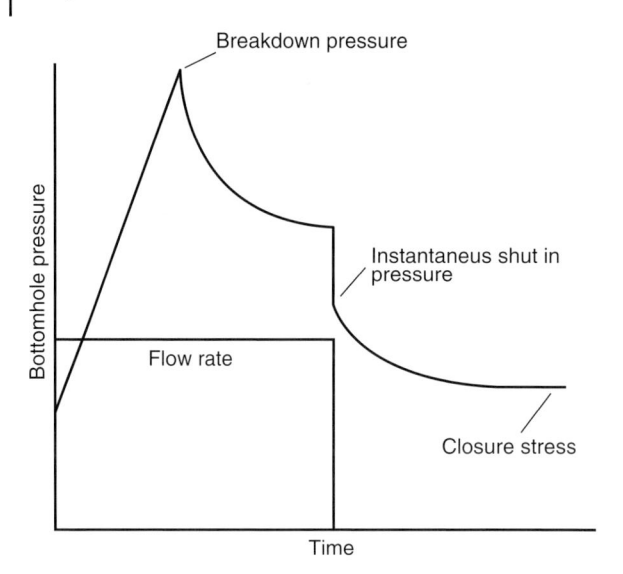

Figure 8.27 The typical variation of pressure with time (flow rate) in hydraulic fractures. From Galindo Torres and Muñoz Castaño (2007), with kind permission from American Physical Society.

can sustain it without breaking. After a given point, the rocks break triggering a sudden drop in pressure. Beyond this point, the pressure fluctuates, indicating propagation of fracture in the rocks. When the fluid injection is stopped (called the instantaneous shut in pressure point), fracture further propagates due to the current pressure, and the tectonic pressure tends to balance it. An equilibrium is reached when these two forces balance.

Because of its importance, hydraulic fracture has been modeled in various ways. Here we mainly discuss some microscopic models of hydraulic fracture for heterogeneous materials and their relationships with experimental observations.

In Tzschichholz and Herrmann (1995), a beam lattice model in square lattice was used to model hydraulic fracture. Since the fracture propagates in two dimensions, two-dimensional models are used. In this model, the coordinates and rotation angles are the degrees of freedom. The beams have a material-specific threshold of stress, beyond which it irreversibly breaks. In order to simulate the effects of heterogeneous materials, this threshold distribution is taken from a power-law with negative exponent of magnitude smaller than 1. Numerical simulations indicate interesting behavior in the static and dynamic properties of this model. To begin with, the radius of gyration of the broken patch grows with the number of broken bonds in a power-law greater than one, signifying its fractal nature. Particularly, $N \propto N^{d_f}$ with $d_f \approx 1.4$ which finally crosses over to a value 1.25 at a volume dependent on the distribution of disorder (higher the disorder, larger is the crossover volume).

The dynamical behavior shows the typical intermittent propagation with power-law size distribution of bursts (with lifetime exponent about 0.54) separated by waiting time also decaying in a power-law (inverse time; similar to earthquake statistics).

There are many other studies, for example, with triangular spring lattice models where the interplay of pressure diffusion and elastic deformation and fracturing of solid materials are considered (Flekkøy and Malthe-Sørenssen, 2002). As before, the springs break beyond a certain point of stretching. However, the triangular lattice produces an isotropic elastic behavior as opposed to the square lattices.

Galindo Torres and Muñoz Castaño (2007) modeled the rock using an array of Voronoi polygons. To construct a Voronoi polygon in two dimensions, first, a set of randomly distributed points (Voronoi points) is taken. Then, to each point, one can associate a Voronoi polygon as the points in the plane that are closer to that Voronoi point than any other. The adjacent polygons have cohesive forces between them. The fractal dimension of the broken patch was studied and it was found to be about 1.22, which is much smaller than the previous estimates with square lattices. The effect of loss of fluid was also studied, and the results are comparable with continuum mechanics predictions. In addition to these two-dimensional models, three-dimensional models of hydraulic fracture are also studied (Wangen, 2013).

8.4
Summary

When a crack is opened (in mode I) through an easy plane, it has hardly any out of plane fluctuation. The crack front is almost fully confined in two dimensions. The modeling approaches for those kinds of fracture are discussed here. A first-order correction to the linear elastic theory is used to model the fracture front as an elastic line driven through random pinning (Section 8.1). The values of the exponent for roughness and avalanche statistics depend on the nature of the elastic force used.

In addition to the elastic line depinning, the interfacial fracture propagation has been modeled with fiber bundles. The local load sharing scheme of fiber bundle has in general no critical dynamics associated with it, however the damage area can be localized either by introducing a gradient in the failure thresholds (see Section 8.2.1) or by making the applied load localized (Sections 8.2.2 and 8.2.3). The interface generated in this method gives a self-organized critical dynamics with avalanche statistics and roughness exponents close to what are observed in experiments (discussed in Chapter 5).

9
Dislocation Dynamics and Ductile Fracture

9.1
Nonlinearity in Materials

Nonlinearity in the stress–strain response curve is a signature of ductile materials. It causes deformation of the sample as a response to the applied stress. Most ductile metals show nonlinear plastic behavior at normal temperature, pressure, and atmospheric conditions. It is now realized that the presence of defects within a material and the mobility and rearrangements of the defects under applied stress are primarily responsible for the occurrence of the nonlinear stress–strain response, plastic flow, and deformation of the material. In this chapter, we will discuss how defects such as dislocations facilitate deformation by slip in crystal planes, the effect of motions of dislocations, and the factors that affect these motions.

9.2
Deformation by Slip

The usual method of plastic deformation in metals is by sliding of blocks of the crystals over one another along definite crystallographic planes, called slip planes. It is usually the plane of largest atomic density, and the slip direction is mostly the closest-packed direction within the slip plane. The planes of greatest atomic density are also the most widely spaced planes in crystal structure. The resistance to slip is generally less for these planes than for any other set of planes. Slip occurs when the shear stress exceeds a critical value. The atoms of one block close to the slip plane move by an integral number of atomic distances along the slip plane, and a step is produced in the polished surface which shows up as a line when viewed by a microscope from the top. The slip process is much facilitated in the presence of dislocations and their motions, and the critical shear stress needed for the slip is reduced drastically.

Statistical Physics of Fracture, Breakdown, and Earthquake: Effects of Disorder and Heterogeneity, First Edition.
Soumyajyoti Biswas, Purusattam Ray, and Bikas K. Chakrabarti.
© 2015 Wiley-VCH Verlag GmbH & Co. KGaA. Published 2015 by Wiley-VCH Verlag GmbH & Co. KGaA.

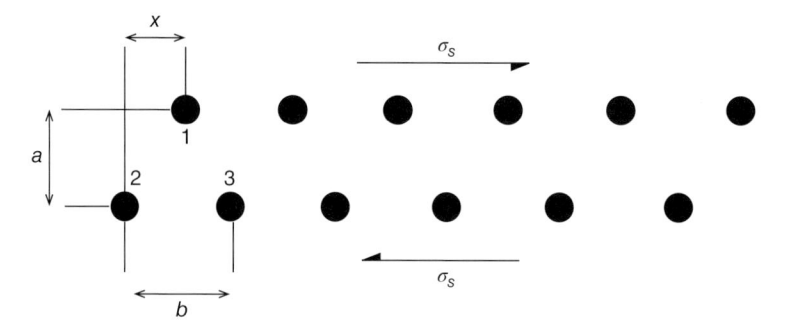

Figure 9.1 Shear displacement of one plane of atoms over another atomic plane. Following Cottrell (1953).

9.2.1
Critical Stress to Create Slip in Perfect Lattice

If slip is assumed to occur by the translation of one plane of atoms over another, one can estimate the shear stress required for such a movement in a perfect lattice. Consider two planes of atoms subjected to a homogeneous shear stress (see Figure 9.1). The shear is assumed to act in a slip plane along the slip direction. Let us assume that the atoms along the slip direction are separated by a distance b, and the gap between adjacent lattice planes is a. Let us further assume that the shear has caused a relative displacement x in the slip direction between the pair of adjacent lattice planes.

The shearing stress is zero when the two planes are in coincidence, and it is also zero when the two planes have moved by a distance b, so that the point 1 in the top plane is over point 2 or point 3 in the bottom plane. The shearing stress is also zero when the atoms of the top plane are midway between points 2 and 3 of the bottom plane, since this is a symmetry position. Between these positions, each atom is attracted toward the nearest atom of the other row, so that the shearing stress can be thought of as a periodic function of the displacement.

As a first approximation, the relationship between shear stress and displacement can be expressed by a periodic sine function

$$\sigma_s = \sigma_m \sin \frac{2\pi x}{b} \tag{9.1}$$

where σ_m is the amplitude of the sine wave and b is the period. For small displacement, Hooke's law should apply in the elastic limit, and

$$\sigma_s = G\gamma = \frac{Gx}{a} \approx \sigma_m \frac{2\pi x}{b} \tag{9.2}$$

This provides an expression for the maximum shear stress at which slip should occur,

$$\sigma_m = \frac{G}{2\pi} \frac{b}{a} \sim \frac{G}{2\pi} \tag{9.3}$$

if we approximate $b \sim a$.

The shear moduli for metals lie in the range of 20–150 GPa. Eq. 9.3 predicts that the theoretical shear stress should be in the range 3–30 GPa. However, the actual values of shear stress required to produce plastic deformation in single crystals of ductile metals are in the range of 0.5–10 MPa—orders of magnitude lower than the calculated stress. Even if more refined calculations are used to correct the sine wave assumption, the value of σ_m cannot be equal to the observed shear stress.

Since the theoretical shear strength of metal crystals is at least 100 times greater than the observed shear strength, it must be concluded that a mechanism other than bodily shearing of planes of atoms is responsible for slip. Existence of defects, such as dislocations, in crystals explains this discrepancy between theoretical and experimental values.

9.3
Slip by Dislocation Motion

The motion of dislocation through a lattice requires a stress far smaller than the theoretical shear stress we estimated. In a perfect lattice, all the atoms are in minimum energy positions. When the crystal is in equilibrium with an applied shear stress, the atoms are displaced from their equilibrium lattice positions and a restoring force equal to the shearing force develops within the crystal so that the net force on an atom becomes zero. In the presence of a dislocation, the atoms far away from the dislocation are still in minimum energy positions but at the dislocation, only a small movement of the atoms is required to go from one minimum energy state to the other as the shear stress is applied. The result is the motion of dislocation and an overall slip of the crystal planes.

Referring to Figure 9.2, the extra half plane which was initially at the lattice position 3, with application of a small force comes at 4. Since the atoms around the dislocations are symmetrically placed on opposite sides of the half plane, equal

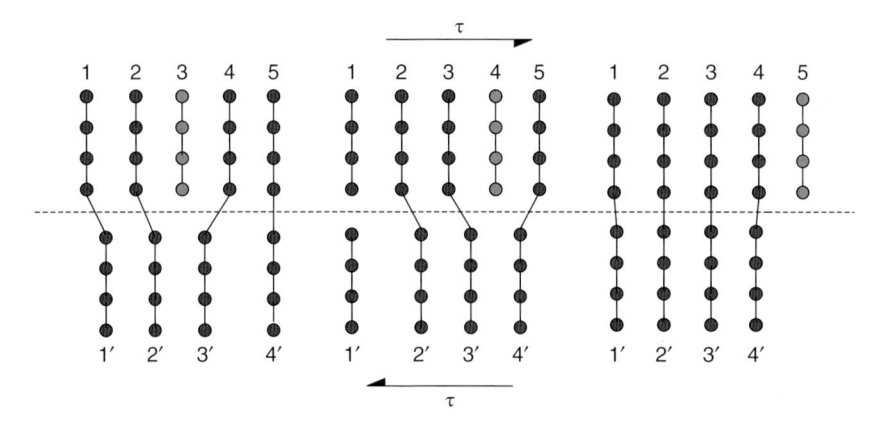

Figure 9.2 Atom movements near dislocation in slip and movement of an edge dislocation. Following Dieter (1928).

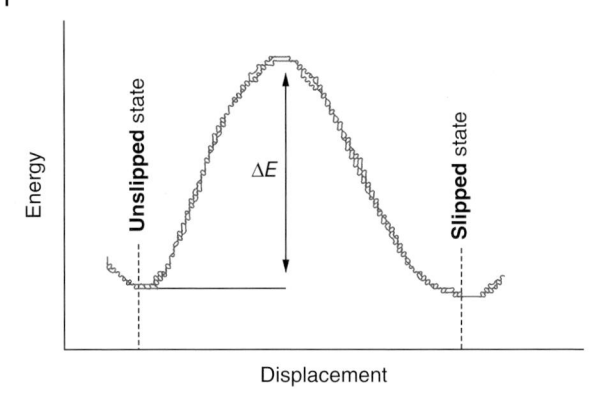

Figure 9.3 Energy change from unslipped to slipped state. Following Cottrell (1953).

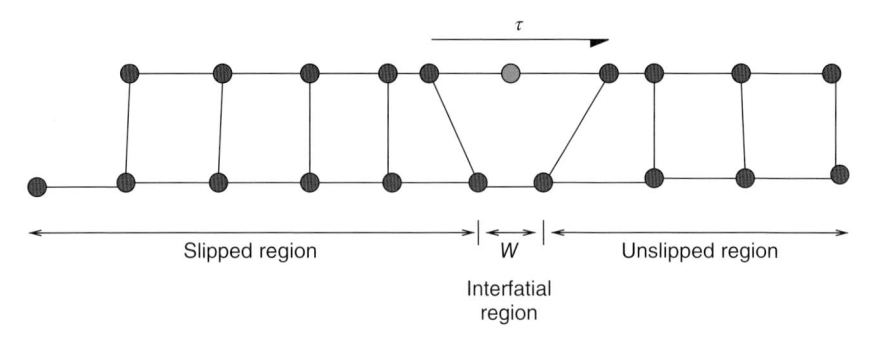

Figure 9.4 Stages in growth of slipped region due to dislocation. Following Cottrell (1953).

and opposite forces oppose and assist the motion. Thus, in a first approximation, there is no net force on the dislocation, and stress required to move the dislocation is almost zero. The continuation of the process creates a motion of the dislocation toward right and when the extra half plane of atoms reaches a free surface, it results in a slip step of one Burgers vector. That is why the movement of the dislocation produces a step, or slip band, at the free surface.

Cottrell (1953) proposed an instructive way of looking at the slip by dislocation motion. Considering that the plastic deformation occurs by the transition of the entire plane from an unslipped to a slipped state (Figure 9.3), the process involves an energy barrier ΔE which the system has to overcome to go from unslipped to slipped state. Instead, one can assume that all the atoms along the slip plane will not move simultaneously to go from unslipped to slipped state. To minimize the energetics of the process, the slip will grow at the expense of unslipped region by the advancement of an interfacial region (Figure 9.4). The interfacial region is the dislocation. To minimize energy for the transition, we expect the interface thickness w to be narrow. The distance w is the width of the dislocation which is given by the Burgers vector associated with the dislocation.

9.4 Plastic Strain due to Dislocation Motion

We can calculate the shear strain that results from the motion of a dislocation in a crystal. Consider a crystal of volume $V = hld$ containing straight-edge dislocations, where h, l, and d are height, length, and width of the sample. Under an applied shear stress acting on their slip plane in the direction of their burgers vector b, as shown in the Figure 9.5, the dislocations will glide in the way as described above. If the shear acts clockwise, the positive dislocations (with b) shift to right and the negative ones to left. Therefore, the top surface of the crystal is displaced plastically relative to the bottom surface as demonstrated in the figure. If a dislocation moves completely along the slip plane to a distance d, it contributes with a distance b to the total displacement D. The dislocation produces a slip offset of length b, so that the shear strain (Figure 9.5) is $\gamma = b/h$. The total macroscopic plastic strain is the sum of all the small strains due to a very large number of individual dislocations.

Now consider the case where a dislocation has moved part way through the crystal along the slip plane (Fig 9.5). Since b is very small compared with L or h, the displacement δ_i for a dislocation at an intermediate position between $x_i = 0$ and $x_i = L$ would be proportional to the fractional displacement x_i/L.

$$\delta_i = \frac{x_i b}{L}$$

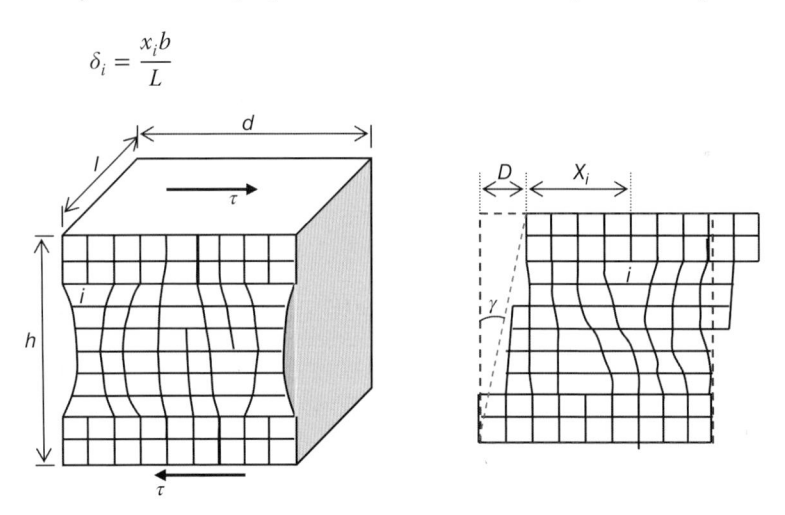

Figure 9.5 Shear strain associated with passage of single dislocation and motion of dislocation through a crystal. Following Dieter (1928).

The total displacement of the top atomic plane of the crystal relative to the bottom for many dislocation motions on slip planes is

$$\Delta = \sum \delta_i = \frac{b}{L} \sum_i^N x_i$$

where N is the total number of dislocations that have moved in the volume of the crystal. The macroscopic shear strain produced due to the slips is

$$\gamma = \frac{\Delta}{h} = \frac{b}{hL} \sum_i^N x_i = \frac{bN\bar{x}}{hL} = b\rho\bar{x}$$

in terms of the average slip $\bar{x} = \frac{\sum_i^N x_i}{N}$ of the dislocations and $\rho = N/hL$, the total length of the dislocation line per unit volume.

The shear strain rate can be written as

$$\dot{\gamma} = \frac{d\gamma}{dt} = b\rho\frac{dx}{dt} = b\rho\bar{v} \qquad (9.4)$$

From Eq. 9.4, we see that if we want to describe macroscopic plastic deformation in terms of dislocation behavior, we need to know

1) the crystal structure in order to evaluate b
2) the number of mobile dislocations, ρ
3) the average dislocation velocity, $\bar{v}$

within which the quantities ρ and $\bar{v}$ depend on stress, time, temperature, and prior thermodynamical history.

9.5
When Does a Dislocation Move?

Dislocation mobility is an extremely important factor in determining the plasticity of a crystal. The plasticity as well as ductility increases with increasing mobility of the dislocations. There are some parameters that control this mobility. These parameters are related to the nature of the dislocations, crystal structure, and external effects. In the following, we discuss these parameters and their effect on the dislocation mobility.

9.5.1
Dislocation Width

If we consider the dislocation as a transition region between slipped and unslipped areas of a slip plane, the width w of a dislocation is a measure of the sharpness of the transition. When the transition occurs within an interval of, say, one or two atomic spacings, the dislocation is narrow; when it occurs over several atomic spacings, the dislocation is wide (Figure 9.4). This width dependence of dislocation motion can be nicely explained by Peierls–Nabarro theory (see, e.g., Joós and

Duesbery, 1997). According to this theory, the critical stress required to move a dislocation is given by

$$\sigma_p \sim e^{-w}$$

where w is the dislocation width. The force required to move a wide dislocation is less than the force required for a narrow dislocation. Then, for a sufficiently large stress, the mobility of wide dislocation will be much more than narrow dislocation. Therefore, a crystal with large dislocation width will exhibit more plasticity and ductility.

9.5.2
Dependence on Grain Boundaries in Crystals

The structure of the crystal affects the motion of the dislocation to a great extent. Especially, existence of grains in crystals influences mobility of the dislocation.

The existence of several defects breaks the translational and orientational symmetries of a crystal. The crystal may contain many grains. Grains are the regions where atoms are oriented identically but the atomic planes are oriented randomly from each grain to the other. The interfaces between the grains are called grain boundaries (Figure 9.6).

When a dislocation moves, it has to cross the grain boundaries to go from one grain to the other. Since the orientation of each grain is different, the dislocation has to change its direction of motion while crossing the boundaries. This acts as a obstacle to the motion of the dislocation and reduces its mobility. For materials with large number of grains, the mobility of dislocations is restricted and such materials act more like brittle materials. Existence of lesser grains increases ductility in materials.

9.5.3
Role of Temperature

We have already discussed that dislocation is the transition region from unslipped to slipped plane. According to Figure 9.3, there is an energy barrier ΔE between

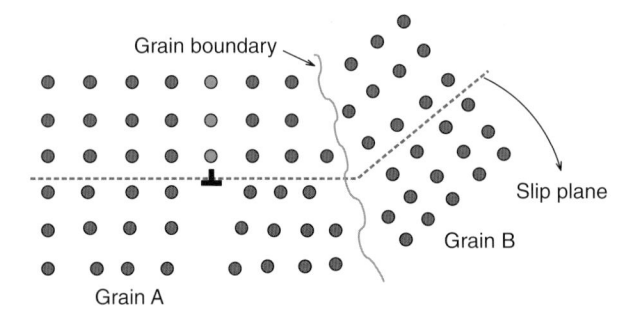

Figure 9.6 Randomly oriented atomic planes and grain boundaries.

the slipped and unslipped regions. This barrier can be overcome by means of thermal energy. The dependence of dislocation velocity on temperature is given by Arrhenius equation:

$$v \sim \exp(-\Delta H/RT) \tag{9.5}$$

where ΔH is analogous to activation energy required for dislocation motion, R is the gas constant, and T is the temperature. Therefore, with increasing temperature, the dislocations become more and more mobile. A material which at low temperature does not show deformation, can introduce both slip and deformation and hence ductility at higher temperature.

9.5.4
Effect of Applied Stress

The energy needed to overcome the barrier in Figure 9.3 can be given by means of external stress. The dislocation mobility increases with increasing stress acting on the material. Generally, two different types of dependence are taken

$$v \sim e^{-1/\tau} \tag{9.6}$$

$$v \sim \tau^n \tag{9.7}$$

where τ is the applied stress on the object. Similar to temperature, increase in applied stress also introduces higher mobility of the dislocations.

9.6
Ductile–Brittle Transition

During the study of dislocation motion, we found that the mobility of the dislocations is not an inherent property of a material, rather it depends on various parameters such as temperature, pressure, and applied stress. A material with a particular crystal structure may exhibit ductile or brittle properties depending on the aforementioned parameters. For example, with increasing or decreasing temperature or applied stress, most materials show transition between states, which exhibit predominantly ductile or brittle properties. This transition of a material from ductile to brittle or vice versa is known as ductile–brittle transition. Experiments on rocks have shown that the brittle rocks show ductile properties at higher temperatures and pressures.

9.6.1
Role of Confining Pressure

Under sufficiently large pressure ($\sim$MPa), brittle materials are experimentally observed to exhibit ductile properties. The brittle–ductile transition with increasing pressure can be demonstrated in experiments on marble at room temperature.

With increasing confining pressure, there are some important features one observes in the stress–strain curves of materials: before macroscopic failure, the strain increases very markedly when the confining pressure exceeds a certain threshold value. This change from the occurrence of macroscopic fracture at strains of less than a few percent to a capacity for undergoing distributed strains of larger magnitudes is taken as defining the brittle–ductile transition. There is an increasing tendency for the stress–strain curve to continue rising up to large strains and with greater slope, that is, there is a greater extent and degree of strain-hardening at higher pressures.

With increasing applied pressure, the stress acted upon a material increases. We have already discussed that one of the important factor in a material that determines the brittle or ductile features is dislocation density. The higher the dislocation density, the more brittle the fracture will be in the material. The idea behind this theory is that plastic deformation comes from the movement of dislocations. As dislocation density increases in a material due to stresses above the material's yield point, it becomes increasingly difficult for the dislocations to move because they pile into each other. Therefore, a material that already has a high dislocation density deforms and fractures in a brittle manner.

9.6.2
Role of Temperature

Increase in temperature has, in general, a very important role in promoting ductility in materials. At sufficiently high temperatures, ductility can be achieved in compression tests at confining pressures of considerably less than 1000 MPa. However, an increase in temperature alone at atmospheric pressure is normally ineffective in attaining ductility. An experimental observation with dunite shows that even under 3000 MPa confining pressure, this rock is still brittle at room temperature (Shimada, 2000), but it becomes ductile at temperatures above about 1200 K.

As temperature increases, the atoms in the material vibrate with greater frequency and amplitude. This increased vibration allows the atoms under stress to slip to new places in the material (i.e., break bonds and form new ones with other atoms in the material). This slippage of atoms is seen on the outside of the material as plastic deformation, a common feature of ductile fracture. As the temperature is lowered, atomic vibration decreases, and the atoms do not want to slip to new locations in the material. Therefore, when the stress on the material becomes high, the atomic bonds break without any slip. This decrease in slippage causes little plastic deformation before fracture and we get a brittle-type fracture. The explanation provided above also holds good with Arrhenius equation Eq. (9.5), which shows the increase of ductility in materials with increasing temperature (Figure 9.7).

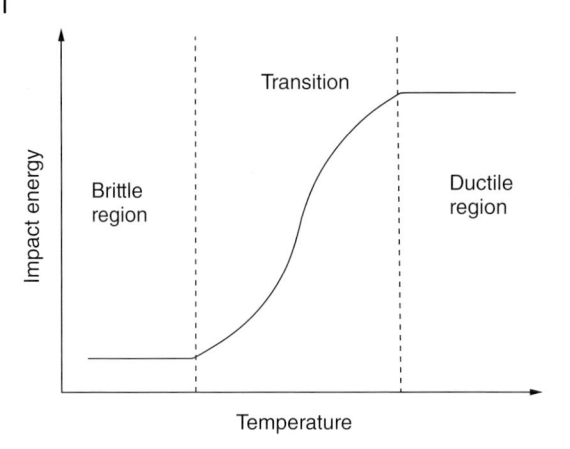

Figure 9.7 Ductile–Brittle transition with variation in temperature.

9.7
Theoretical Work on Ductile–Brittle Transition

Khanta, Pope, and Vitek, in their KDV model (Khantha et al., 1995), tried to explain the origin of plasticity and the ductile–brittle transition directly from the statistical mechanics description of dislocations and their motions. In this model, like the unbinding of vortices in the XY-model or unbinding of dislocations in melting transitions, the onset of ductile behavior corresponds to a cooperative dissociation of many dislocation dipoles driven primarily by thermal fluctuations, and assisted by the applied stress. The mutual interactions between any two dislocations are taken from the Kosterlitz–Thouless concept of thermally induced dislocation screening. The ductile–brittle transition temperatures for several materials are predicted. The theory does not explain the strain rate dependency of the transition temperature which has been considered subsequently in the models of Hirsch and Roberts (1996).

A mechanism of cooperative generation of dislocation pairs near the crack tip of a loaded crystal above a critical temperature (Khantha et al., 1997) explains some unique features of the ductile–brittle transition: (i) the sharp increase of the fracture toughness in a narrow temperature interval around the ductile–brittle transition temperature and (ii) the strain rate dependency of ductile–brittle transition temperature.

Monte Carlo simulations show that dissociation of dislocation dipoles occur at temperatures well below the Kosterlitz–Thouless transition temperature in the presence of applied loads. Simulation results show that the critical temperature is progressively reduced from the Kosterlitz–Thouless limit (at zero stress) as the applied stress or electric field is increased. This is in good agreement with the theoretical predictions (Ling, 1999; Minnhagen et al., 1995; Ambegaokar et al., 1980).

Picallo et al. (2010) introduced a lattice model to study the plasticity and ductile–brittle transition in disordered materials. The model is a generalization of the random fuse model, where every bond in the lattice is considered to be a fuse which can carry current up to a certain strength (randomly selected from a distribution) and becomes an insulator when the current through the bond reaches the threshold. In the generalized model in Picallo *et al.* (2010), whenever the current in a fuse reaches its threshold, an oppositely directed current is passed through the fuse, the magnitude of the current being proportional to the threshold. The healing cycle is repeated number of times whenever the net current in a fuse reaches its threshold. The elastic analog of this model is that whenever an element reaches its threshold, a permanent deformation is imposed on the element. In this model, ductile fracture is obtained once the strain is completely localized and the fracture surfaces correspond to the minimum energy surfaces. The model exhibits a smooth transition from brittleness to ductility. It is to be noted that in all these models, nonlinearity is introduced with each basic bond element. Langer and Lobkovsky (1999) have studied a one-dimensional problem of decohesion from a substrate of a membrane that obeys viscoplastic constitutive equations. Plasticity is a fully dynamic phenomenon in this model. This gives rise to a dynamic ductile–brittle transition.

Brittle to ductile and quasi-brittle transitions have been studied in fiber bundle model. In the mean-field version of the model, the transition appears as a critical point with a power-law divergence of the relaxation time and scale-free avalanche size distribution of the ruptured fiber. This has been described in Chapter 6 in the context of fiber bundle model.

10
Electrical Breakdown Analogy of Fracture

As a safety measure, fuses are fitted in household electrical circuits, such that in case of an increase in the voltage drop, the fuses burn out. The fuse can be thought of as an element that conducts up to a certain range of the applied voltage and becomes nonconducting beyond that range. The process of the fuse becoming nonconducting is called fuse failure. On the other hand, a dielectric breakdown is just the opposite phenomenon, that is, a dielectric element is nonconducting up to a certain range of the applied voltage and becomes a conductor beyond that voltage. These two phenomena belong to the general class of failure processes when a physical attribute fails when the value of a perturbing force crosses a threshold value (Samanta et al., 2009). The most common example of such failures is the mechanical failure or fracture as we have discussed so far, and the above two examples are its electrical analog that we discuss in this chapter.

The electrical failure phenomenon has many similarities with its mechanical counterpart and is somewhat simpler to deal with. In the electrical case, the field is a scalar as opposite to the tensorial stress field in solids. Apart from that, the presence of disorders in the naturally occurring solids and that in the electrical networks are similar. In both the cases, the stress field suffers strong modification around such defects and beyond a critical limit the defect starts propagating. The propagation may stop or it may lead to global failure depending upon certain energy considerations.

The disorder in the electrical network can be modeled by the presence of randomly placed (p fraction) insulators in an otherwise conducting material or vice versa. In the simplest form, the system can be viewed as a regular lattice (with substitutional defects) and one looks at the failure in the global scale.

In dealing with a large lattice (the failure properties of which we are interested in), one can apply percolation theory (Stauffer and Aharony, 1994). For the disorder (say, conducting element in a nonconducting lattice) concentration p below the percolation threshold p_c, the system does not conduct globally, but beyond the percolation threshold, at least one connecting path spans the system, making it conducting.

As mentioned earlier, beyond the percolation threshold, the system is conducting. But, as the voltage is increased, some of the elements may fail (become nonconducting) and eventually beyond a certain voltage, the entire system may

Statistical Physics of Fracture, Breakdown, and Earthquake: Effects of Disorder and Heterogeneity, First Edition.
Soumyajyoti Biswas, Purusattam Ray, and Bikas K. Chakrabarti.
© 2015 Wiley-VCH Verlag GmbH & Co. KGaA. Published 2015 by Wiley-VCH Verlag GmbH & Co. KGaA.

become nonconducting. The current just prior to the global failure is called failure current $T_f(p)$. Understandably, it is a function of the disorder concentration p and for $p = 1$ it is simply (when scaled by the system size) the threshold current for each conducting element. On the other hand, for $p < p_c$, initially the sample is nonconducting and with an increase in the applied voltage (and consequent local failure events that makes those bonds conducting), the sample becomes conducting for $V \geq V_b(p)$, where $V_b(p)$ is the dielectric breakdown voltage. Similar to the case before, for $p = 0$, the breakdown voltage $V_b(0)$ (scaled by the system size) is the breakdown voltage of the nonconducting bonds. But as $\to p_c$, both the breakdown voltage and current go to zero (see Figure 10.1 for a phase diagram).

In this chapter, we discuss both the fuse network and dielectric breakdown problems. Its quantum analog is briefly described in Appendix D.

10.1
Disordered Fuse Network

The field lines inside a pure conductor are parallel to each other, if the sample is pure and is placed between a voltage difference. However, the presence of disorders (nonconducting elements) modifies the field lines, since the nonconducting path cannot support them. Consequently, there is a field concentration around the defect, which is similar to a case of stress concentration around a crack. The current density in the neighborhood of the defects becomes i_e from i (the current density far from the defect). The enhanced density can be written as

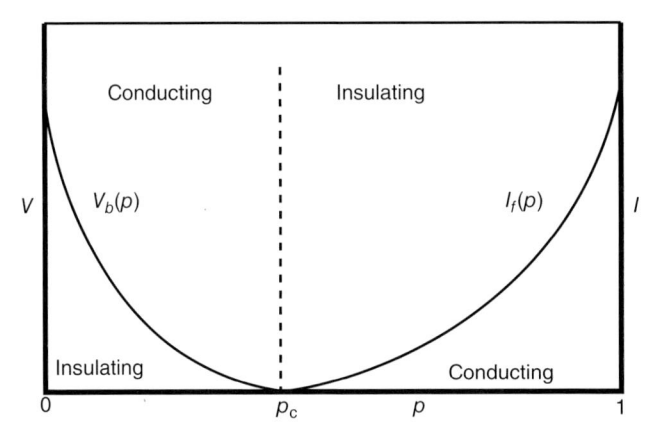

$$i_e = i(1 + k) \tag{10.1}$$

Figure 10.1 Phase diagram of a sample that has p fraction of conductors and $(1 - p)$ fraction of insulators placed at random. On the left-hand side, the sample is insulating for an applied voltage V which is less than the breakdown voltage $V_b(p)$ and conducting otherwise. On the right-hand side, the system is conducting for a current I less than the fuse current $I_f(p)$ and insulating otherwise.

with k as the enhancement parameter governed by the geometry of the defect. For example, for an elliptic defect with semi-major axis l and semi-minor axis b, one has $k = l/b$ (Chakrabarti and Benguigui, 1997). If the electrode surface area is A, the total current I is

$$I = Ai = \frac{Ai_e}{1+k}. \tag{10.2}$$

Failure takes place when i_e first reaches i_0, the fatigue limit of the sample. The failure current is given by

$$I_f = \frac{Ai_0}{1+k}. \tag{10.3}$$

Obviously, for large values of the enhancement factor k, the failure current is small. In the case of an elliptic defect, if $l \gg b$, the failure current can be substantially reduced. Hence, sharper and larger defects make the enhancement factor larger and consequently, the system becomes more vulnerable. This may lead to a situation where the failure of the first element leads to the failure of the sample (the defect continues to grow in size), which is called a brittle failure.

A single regular-shaped defect, as discussed so far, is a much simpler condition than the usual multiple randomly shaped defects present in real samples. For that case, we use percolation theory (Ray and Chakrabarti, 1985a, 1985b; Duxbury et al., 1986).

Let us start with a regular square lattice in two dimensions or three dimensions (cubic). If the system is made up of fully conducting elements and one single bond is removed parallel to the current flow direction, the failure current is

$$I_f = \frac{\pi}{4} L i_0 \tag{10.4}$$

in 2d, if Li_0 is the failure current of the pure lattice. The enhancement factor is $4/\pi$. A more realistic condition may be obtained by removing $1 - p$ fraction of bonds randomly. The most vulnerable defect and the enhancement factor are no longer possible to be calculated simply. We focus on two limits: (i) where the disorder concentration is low $p \to 1$ and (ii) the percolation critical point $p \to p_c$, beyond which the system looses the conductivity altogether.

10.1.1
Dilute Limit ($p \to 1$)

The very few defects present in this limit can be taken as independent. The *most probable dangerous defect* (the weakest point which causes the largest concentration of the current density) is a line of n defects perpendicular to the current direction in 2d and a disc for 3d (Duxbury et al., 1986, 1987). The current in the immediate neighborhood of the largest defect is

$$i_e = i(1 + k_2 n) \text{ (in 2d)}, \quad i_e = i(1 + k_3 n^{\frac{1}{2}}) \text{ (in 3d)}. \tag{10.5}$$

where i is the current density far from the defects. The $\sqrt{n}$ is present, since the current is deviated by n defects and spreads over the perimeter of the disc which is proportional to $\sqrt{n}$. The probability of a defect of size n (successively removed bond) in a lattice of total element L^d is

$$P(n) \sim (1-p)^n L^d. \tag{10.6}$$

The condition $P(n) \approx 1$ gives the most probable dangerous defect. Hence

$$n_c = -\frac{2}{\ln(1-p)} \ln L \text{ (in 2d)}, \quad n_c = -\frac{3}{\ln(1-p)} \ln L \text{ (in 3d)}. \tag{10.7}$$

From Eq. (10.5) and Eq. (10.7), we get

$$i_e = i\left[1 + k_2\left(-\frac{2\ln L}{\ln(1-p)}\right)\right] \text{ (in 2d)},$$

$$= i\left[1 + k_3\left(-\frac{3\ln L}{\ln(1-p)}\right)^{\frac{1}{2}}\right] \text{ (in 3d)}. \tag{10.8}$$

When $i_e = i_0$, one gets the expression for the failure current

$$I_f = \frac{i_0 L}{1 + 2k_2[\frac{\ln L}{\ln(1-p)}]} \text{ (in 2d)},$$

$$I_f = \frac{i_0 L^2}{1 + \sqrt{3}k_3[\frac{\ln L}{\ln(1-p)}]^{\frac{1}{2}}} \text{ (in 3d)}. \tag{10.9}$$

Obviously, when $p \to 1$, $I_f \to i_0 L^{d-1}$, which is the value of the current in the pure lattice. Also, the $I_f(p)$ versus p curve (see Figure 10.1) has an infinite slope at $p = 1$, since the presence of even a single defect can facilitate failure through cascades of local failures. The interesting thing is the system size dependence. For large L and not so large $\ln(1-p)$, the failure current density $i_f = I_f/L^{d-1}$ is of the form $1/\ln L$ in 2d and $1/(\ln L)^{1/2}$ in 3d.

10.1.2
Critical Behavior ($p \to p_c$)

When $p \to p_c$, where p_c is the percolation threshold, the highly disordered system resembles a node-link-blob picture (Stanley, 1977; Skal and Shklovskii, 1975; de Gennes, 1976) (see Figure 10.2). There is a correlation length ξ_p associated with the criticality, which means that the conducting clusters are self-similar up to that length scale. This means that the large cluster can be divided into blocks of size ξ_p, each containing a backbone that conducts the current and other dangling bonds that do not take part in the conduction process. The backbone consists of multiply connected bonds (blobs) and singly connected bonds (links or red bonds). The singly connected bonds carry the entire current and thus determine the current distribution in the whole sample. At the percolation point (p_c), ξ_p spans the whole system and the failure current consequently tends to zero.

The quantitative estimates of the failure current I_f can be made by considering this blob-node-link picture. If V is the total applied voltage, then it must be

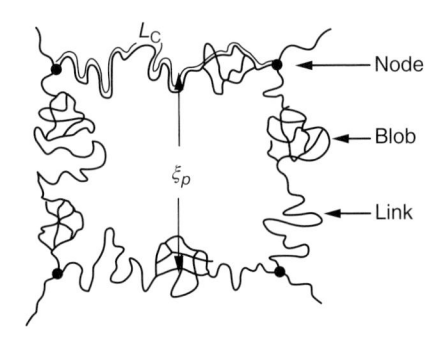

Figure 10.2 A portion of the node-link-blob superlattice model is shown near the percolation point p_c. The distance between two nodes of the lattice is of the order of ξ_p, while chemical length of the tortuous link of the superlattice is L_c.

distributed among L/ξ_p cells in series, each having a voltage drop (on average) $V_L \sim \xi_p V$. Now, there are L/ξ_p number of links along the length of the sample and ξ_p^{-1} such links. Therefore, if the link resistance is R_L, the sample resistance is given by $R \sim \xi_p^{d-2} R_L$. Therefore, the mean link current is of the order of $i_L \sim \xi_p^{d-1} V/R$. When i_L touches the failure current of a link, the value of the applied voltage is the failure voltage V_f. Assuming

$$V_f \sim (p - p_c)^{-y_1}. \tag{10.10}$$

and using the percolation theory $\xi_p \sim (p - p_c)^{-\nu}$ and the form $R \sim (p - p_c)^{-y_2}$, one has the scaling relation $y_1 = y_2 - (d-1)\nu$ (Samanta et al., 2009). Also from the relation $I_f = V_f/R$, one gets

$$I_f \sim (p - p_c)^{(d-1)\nu}. \tag{10.11}$$

The values of the correlation length exponent ν and conductivity exponent y_2 are both 1.33 in 2d and 0.88 and 2.0 in 3d, respectively. Hence, the failure voltage V_f, unlike the failure current, attains a finite value for 2d and diverges with an exponent about 0.2 in 3d as the percolation point is approached. The failure current vanishes in both dimensions.

10.1.3
Influence of the Sample Size

The failure current depends on the system size, since the most dangerous defect depends on the system size. Here, the most dangerous defect is the one having a length ξ_p along the direction parallel to the applied voltage and l_{max} perpendicular to it. The probability of the existence of a defect of size l is $P = g(l)(L/\xi_p)^d$, where $g(l)$ is the probability density function of defect size l. From the percolation theory, we have (Stauffer and Aharony, 1994; Sahimi, 1994)

$$P \sim \exp\left(-\frac{l}{\xi_p}\right)\left(\frac{L}{\xi_p}\right)^d. \tag{10.12}$$

Now, $l \approx l_{max}$ when $P \approx 1$ and consequently

$$l_{max} \sim \xi_p \ln L. \tag{10.13}$$

The current that flows in the immediate neighborhood of the defect is proportional to $(l_{max})^{d-1} I$ and one obtains

$$I_f \sim \frac{(p-p_c)^{(d-1)\nu}}{(\ln L/\xi_p)^{(d-1)}}. \tag{10.14}$$

This is the finite size correction over Eq. (10.11). This also illuminates the competition between the extreme and percolation statistics in the failure of disordered solids. In the regime where $\ln L > \xi_p$ (i.e., the Lifshitz length scale is greater than the percolation correlation length), the extreme statistics dominates.

In Li (1987), the system size dependence of I_f was proposed in the form $(\ln L)^{-\psi_f}$. The approximate range of ψ_f is given by

$$\frac{1}{2(d-1)} < \psi_f < 1. \tag{10.15}$$

Combining the dilute and critical limits (Eq. (10.9) and Eq. (10.11))

$$I_f = I_0 \frac{\left[\frac{(p-p_c)}{(1-p_c)}\right]^{\phi_f}}{1 + K\left[-\frac{\ln(L/\xi_p)}{\ln(1-p)}\right]^{\psi_f}}. \tag{10.16}$$

The values of the exponents and the constant K depend on the dimension and the type of percolation, and are listed in Table 10.1. The Eqs. (10.9) and (10.14) imply that the combined form is valid only for 2d. The three obvious aspects of the expression are

1) When $p = 1$, $I_f = I_0$, expectedly.
2) Near $p \approx 1$, $p - p_c$ is almost constant and one gets back Eq. (10.9).
3) Near the critical point (p_c), the denominator of Eq. (10.16) is almost unity and one recovers Eq. (10.11), with $\phi_f = (d-1)\nu$.

10.1.4
Distribution of the Failure Current

10.1.4.1 Dilute Limit ($p \to 1$)
The failure current I_f, being an extreme quantity, shows fluctuation between different samples and does not converge to a mean value, as it is not self-averaging.

Table 10.1 Theoretical estimates for the fuse failure exponent ϕ_f.

Dimension	Lattice percolation	Continuum percolation
2	$\nu (= 4/3)$	$\nu + 1 (= 7/3)$
3	$2\nu (\simeq 1.76)$	$2(\nu + 1)(\simeq 3.76)$

The failure current distribution in a sample of size L follows the form (see Chakrabarti and Benguigui, 1997, see also discussions in Section 4.2)

$$F_L(I) = 1 - \exp\left[-A_d L^d \exp\left\{-dA\ln L\left(\frac{\frac{I_0}{I}-1}{\frac{I_0}{I_f}-1}\right)^{d-1}\right\}\right]. \tag{10.17}$$

The failure current is obtained from the derivative of this cumulative distribution with respect to current. The derivative is maximum at the most probable failure current. It can be shown that I_f becomes the most probable failure current in the large system size limit. The cumulative distribution is meaningful for I up to I_0. It should become unity for $I = I_0$ and $I \to \infty$, but it dependent on size and defect concentration through I_f (see Eq. 10.9).

This form of the cumulative distribution is known as the Gumbel distribution (Gumbel, 1958). Another popularly used form in the engineering is the so-called Weibull distribution

$$F_L(I) = 1 - \exp\left[-rL^d\left(\frac{I}{I_f}\right)^m\right], \tag{10.18}$$

where m is a constant for large values (say above 5) of which, I_f becomes the most probable current.

10.1.4.2 At Critical Region ($p \to p_c$)

Near the critical point, the above-mentioned cumulative distribution function takes the form

$$F_L(I) = 1 - \exp\left[-A'_d L^d \exp\left(-\frac{k'(p-p_c)^\nu}{I^{\frac{1}{d-1}}}\right)\right], \tag{10.19}$$

where A'_d and k' are two constants.

10.1.5
Continuum Model

The ideas of the lattice percolation model discussed so far can be extended to the continuum case as well. A solid with defects can be viewed as a continuous field having some inclusions (representing the defects). For three dimensions, the defects are spherical holes, and in two dimensions they are circular. These holes may have overlaps (*Swiss-cheese model*), or there can be a conducting region passing between two nonoverlapping holes. The narrowest part of these channels (of cross section δ) determines the failure properties similar to the discrete case discussed earlier. Under certain (reasonable) assumptions, it is possible to obtain (Halperin et al., 1985) the failure properties of these systems in terms of the percolation correlation length ξ_p.

The dilute limit of this continuum case is similar to the discrete limit. Therefore, the results discussed earlier hold true here as well. But, as the critical point is approached, a large cluster of mean length ξ_p, but with different values of cross-sectional width δ, is formed. The representation of the backbone is in the form of a superlattice (de Gennes, 1976; Skal and Shklovskii, 1975) of tortuous singly connected blobs and links that cross at nodes in a separation

$$\xi_p \sim |p - p_c|^{-\nu}. \tag{10.20}$$

The chemical length L_c between any two nodes is given by

$$L_c \sim |p - p_c|^{-\theta}. \tag{10.21}$$

For bonds in the percolating backbone that are singly connected, $\theta = 1$ in all dimensions except $d = 1$ (Halperin et al., 1985; Stauffer and Aharony, 1994; Coniglio, 1982). There are $(L/\xi_p)^{d-1}$ parallel links between the two electrodes. Therefore, the current though one link is given by $i_L = (\xi_p/L)^{d-1} I$, giving the current density through a link of width δ to be

$$i \sim \frac{\xi_p^{d-1} I}{\delta^{d-1}}. \tag{10.22}$$

Naturally, the maximum density is obtained for the minimum width $\delta_{\min}$, which goes as L_c^{-1} (Halperin et al., 1985; Baudet et al., 1985; Benguigui, 1986). If i_0 is the threshold density of current at which the sample fails, then

$$i_0 \sim \xi_p^{d-1} L_c^{d-1} I_f \tag{10.23}$$

giving

$$I_f \sim (p - p_c)^{\nu+1} \text{ (in 2d)}, \quad I_f \sim (p - p_c)^{2(\nu+1)} \text{ (in 3d)}. \tag{10.24}$$

Hence, the exponent values are higher for the failure currents in the continuum case than the discrete case.

Using $R \sim |p - p_c|^{-\tilde{y}_2}$ and Eqs. (10.20) and (10.21), the failure voltage becomes $V_f \sim |p - p_c|^{\tilde{y}_1}$ with $\tilde{y}_1 = (d-1)(\nu+1) - \tilde{f}_2$ (Chakrabarti and Benguigui, 1997). $\tilde{y}_2$ is the conductivity exponent with values of about 1.3 and 2.5 in 2d and 3d, respectively. Hence, $\tilde{y}_1$ becomes about 1 and 1.3 in $2d$ and $3d$, respectively. Therefore, unlike the discrete case, where the failure voltage converges to a nonzero value in 2d and vanishes in 3d near the critical point, in the continuum limit it always vanishes at p_c.

10.1.6
Electromigration

An application of the electrical breakdown problem discussed so far can be found in the electromigration processes. For miniaturization of circuits in electronic gadgets, very thin conductors are used as connecting elements. Consequently, a large current density may be achieved. Now, because of the finite momentum of electrons, they can displace metallic ions from the equilibrium positions. Thus, depending upon the particulars of the material, a net material transport can occur (Tu, 2003) through grain boundary, surface or lattice diffusions. These transports and consequent deposition effects due to large current densities (more than 10^4 A/cm^2) are called electromigration. It causes void formation in cathode and extrusion in anode, hence increasing the risks of open and short circuits, respectively. The avoidance of such failures is an important problem.

Let us take the situation where $1 - p$ fraction of connections are removed from a 2d random resistor network. Consider a random walker that starts its journey from one of the electrode-free sides and goes to the other, jumping over the bonds irrespective of them being occupied or vacant. The constraint on the walker is that it is self-avoiding, that is, it does not visit same site twice. Now, for a given configuration of resistors, if n_0 and n_1 denote the number of occupied and unoccupied bonds, respectively, then the minimum value of n_0 is the shortest path (see Chayes et al., 1986; Stinchcombe et al., 1986 for theoretical studies). The mean shortest path is denoted by $\langle n_0 \rangle$, which is averaged over many configurations of resistors. The failure due to electromigration can be seen if the random walker burns all the resistors it crosses in its path. Note that the failure criterion for the resistors in this case is not a threshold value of current, but a threshold value (Q_0) in the charge accumulation since the time of application of a constant current I_0 on the network (for details, see Bradley and Wu, 1994; Wu and Bradley, 1994). Let t_1 be the time of failure of a resistor, then it must satisfy

$$\int_0^{t_1} I(t)dt = Q_0. \tag{10.25}$$

As mentioned earlier, the shortest path for a given configuration is the one in which the least number of resistors are to be burned. For a sample having an isolated defect of size n, the failure time τ_f is given by

$$\tau_f = (L - n)Q_0/I_0, \tag{10.26}$$

where LQ_0/I_0 is the failure time of the pure network. In this limit, the average failure time $\langle \tau_f \rangle$ is related to the longest probable defect of size n_c. This means that the burned resistors are also the ones carrying the highest currents.

The failure time τ_f decreases as the defect concentration is increased and vanishes at the percolation threshold p_c; since, beyond this point the shortest path is zero (sample becomes conducting from the start). The failure current I_f, the shortest path n_0, the number of broken bonds N_f, and the mean failure time τ_f all go to zero at p_c with the correlation length exponent ν.

10.2
Numerical Simulations of Random Fuse Network

One simple way to mimic the qualitative features (scale-invariant avalanche size distribution, roughness of the fractured surface, size effect, etc.) is to study the random fuse model (RFM). It was first proposed (de Arcangelis et al., 1985a,b) in its two-dimensional version. The model was defined in a square lattice with randomly placed (with probability p) conductors and insulators (probability $1 - p$).

As before, the applied voltage is increased slowly and V_f^1 is the voltage for the first failure, V_f^2 is that for the second failure, and so on, until the final failure occurs for V_f^{fin}. The first failure voltage V_f^1 initially decreases with decrease of p, attains a minimum value for $p \sim 0.7$ and then increases again for larger p values. But, V_f^{fin} increases monotonically and behaves almost identically as V_f^1 for $|p - p_c| < 0.08$. Both diverges near p_c with exponent 0.48. It seems that the divergence in V_f^1 for 2d is a pseudo-divergence.

Similar types of simulations were performed by Duxbury et al. (1987) for the range of p between 1 and $p_c = 0.5$. It was found that the p dependence of the failure current I_f fits well with the formula given in Eq. (10.16) with $\phi_f = 1$, where the theoretical prediction was 1.33. The disagreement may be due to finite size effect.

The finite value of V_f^1 at p_c was also found by studying the failure current I_f and conductance near the percolation threshold. The linear dependence of L/I_f on $\ln L$ following Eq. (10.9) was also verified for several p values (see Figure 10.3). The slope $\psi = 1$ was determined for the plots and its value increased as the percolation threshold was approached. In place of Eq. (10.17), the following form of the cumulative distribution function for the failure voltages was preferred

$$F_L(V) = 1 - \exp\left[-AL^2 \exp\left(-\frac{KL}{V}\right)\right]. \tag{10.27}$$

de Arcangelis et al. (1985b) and Duxbury et al. (1987) also determined the number of failed bonds N_f up to the global failure point and found it to be going to zero algebraically with an exponent value close to the correlation length exponent.

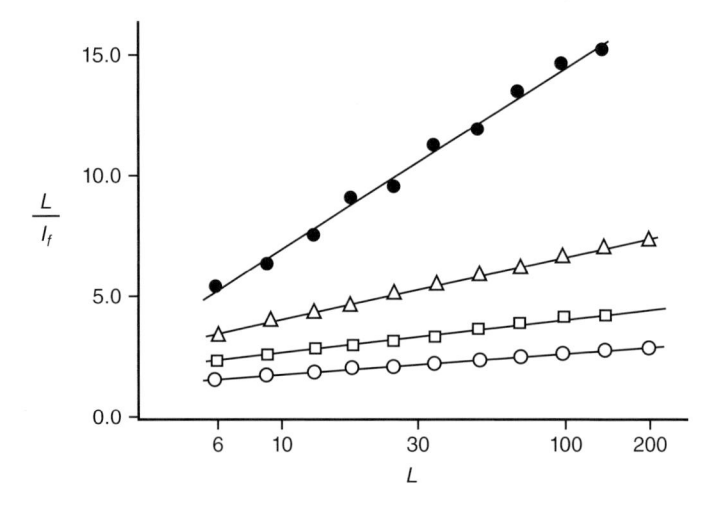

Figure 10.3 The plot of L/I_f versus $\ln L$ expressing their linear dependence (from Duxbury et al. (1987), with kind permission from American Physical Society). The curves shown from top to bottom correspond to initial impurity probability $p = 0.6, 0.7, 0.8,$ and 0.9, respectively.

10.2.1
Disorders in Failure Thresholds

Another way of studying the effect of disorder in random fuse model apart from the disorders in occupation probabilities is to use a distribution for the failure threshold voltages (or currents) through the resistors in an initially fully occupied lattice. One might argue that even for the percolation model, the fully occupied lattice is the coarse-grained limit of occupation probability higher than the percolation threshold. In Kahng et al. (1988), the authors considered an initially fully occupied lattice of random resistors, the failure threshold voltages of which are uniformly distributed within the limit $[1 - w/2 : 1 + w/2]$, where $w \leq 2$. The failure properties of the model sensitively depend on w and system size. For very small values of w, the system fails trivially due to growth of a linear crack. For higher values of w, the voltage is to be increased a few times for the failure to occur. But the failure damage is nucleated around the few weakest bonds and hence can be called a brittle failure. For still wider distribution of thresholds, that is, even higher values of w, the failure enters a region where a substantial fraction of the bonds fails due to increase in voltage (see Figure 10.4). Particularly, the voltage-driven bond breaking is of the order of L in a $L \times L$ lattice. This is then called a ductile failure regime. The crossover point w_c depends on the system size. In the thermodynamic limit, however, the failure is always brittle type. The general result proposed here was that $w_c \to 2$ as $L \to \infty$. The particular form in the unstable crack approximation gives $w_c(L) \sim 2(1 - \mathcal{O}(1/L))$ as $L \to \infty$, representing an upper bound for the quantity.

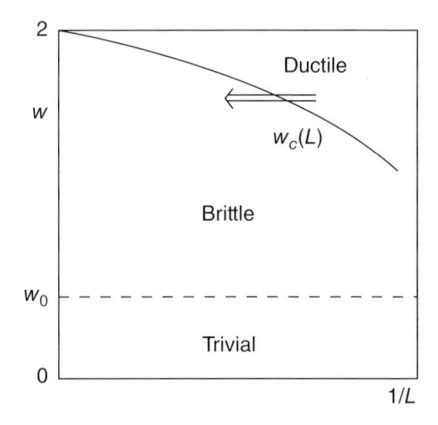

Figure 10.4 The phase diagram of the fuse model with continuously distributed disorder is shown. For the case when the disorder is very narrow $w < w_0$, the system trivially fails due to a single straight crack. For higher values of disorder, a voltage drive is required for failure, however it is still in the brittle regime for $w < w_c(L)$. For still higher values of w, the failure is ductile type, that is, further increase in the applied voltage is important than the initial failure. It is seen that the failure is always brittle type in the thermodynamic limit. From Kahng et al. (1988), with kind permission from American Physical Society.

The average voltage drop per unit length required to break the network varies as $\langle v_b \rangle \sim v_- + \mathcal{O}(1/L^2)$ as $L \to \infty$ in the brittle regime and $1/(\ln L)^y$ in the ductile regime, with $y \leq 0.8$. In addition to these, the distribution function of $\langle v_b \rangle$ was also studied. Particularly, one can measure the cumulative probability $F(v_b)$ for failure of the network for an average voltage v_b or less. In the brittle region, $1 - F(v_b)$ decreases exponentially in v_b. In the ductile region, however, the best fit is a double exponential (Gumble) form

$$F(v_b) = 1 - \exp\left[-a \exp\left(-bL^2 v_b^{-m}\right)\right] \tag{10.28}$$

where $L = 50$, $a = 28.5$, $b = 6.36$, and $m = 2.8$.

In Nukala and Simunovic (2004), the size dependence of the failure strength was found to be log normal from large-scale numerical simulations. In addition, the peak load of the system was found to scale as $I_{\text{peak}} = C_0 + C_1 L^\alpha$ with $\alpha \approx 0.96$ for a two-dimensional lattice. For three dimensions (Zapperi and Nukala, 2006), the failure strength distribution was again log normal and peak load scales as $I_{\text{peak}} \sim L^{1.95}$.

10.2.2
Avalanche Size Distribution

The catastrophic failure in the RFM is preceded by avalanches in which some of the bonds break and the current is redistributed. The current redistribution is again assumed to be much faster than any driving process. The distribution of these avalanches follows a power-law scaling. This is reminiscent of the acoustic emissions in fracture (see Section 6.1). The avalanche statistics in RFM is widely studied over the last few decades.

The mean-field limit of RFM coincides with the global load sharing fiber bundle model (FBM). The avalanche statistics is known exactly to have a power-law distribution $P(\Delta) \sim \Delta^{-\beta}$ with $\beta = 5/2$ (Hemmer and Hansen, 1992). Now, if one considers (Hansen and Hemmer, 1994) a series of parallel resistors in one dimension placed between two bars of conductors, the model is precisely the FBM with global load sharing, and similar statistics is expected to be valid. However, interesting behaviors can be observed when higher dimensional versions are studied. In Hansen and Hemmer (1994), the authors reported the avalanche size distribution (integrated over all currents) in a square lattice to be a power-law with exponent value close to 2.7. This is close to the mean-field limit mentioned earlier. Subsequent studies (Zapperi et al., 1997a) also show similar behavior in the limit of strong disorder in two dimensions. The mean-field prediction for the average size of avalanches will be $\langle \Delta \rangle \sim (I_c - I)^{-\gamma}$, which is proportional to the susceptibility, where $\gamma = 1/2$.

Later on, however, numerical simulations with higher accuracies (Zapperi et al., 2005b) found departures from the mean-field behaviors. Here, the failure properties of two-dimensional random fuse model were checked for different lattice geometries (square, diamond, and triangular). The general scaling form, as usual,

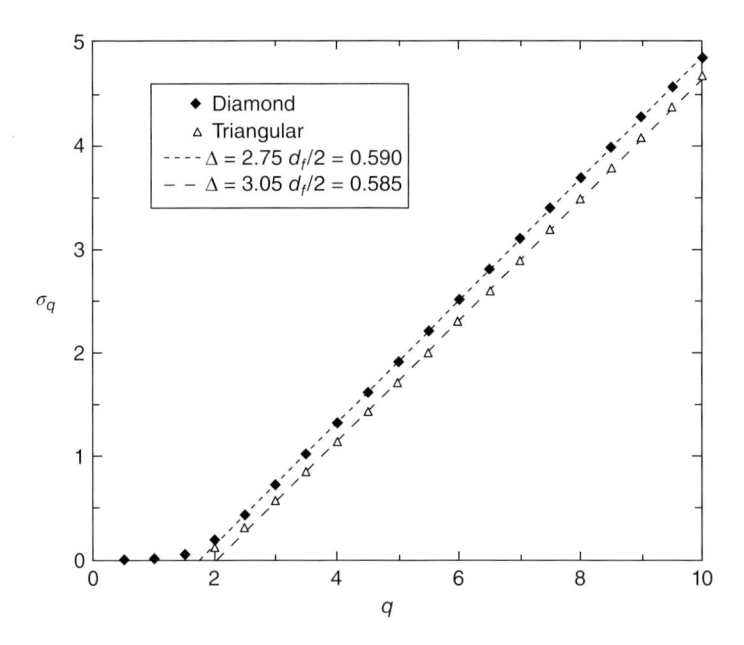

Figure 10.5 The variation of σ_q, the scaling exponent of the qth moment of the avalanche size distribution for diamond and triangular lattices. The lines do not coincide, indicating different values for avalanche size distribution exponent β for the two types of lattices. From Zapperi et al. (2005b), with kind permission from American Physical Society.

was taken as

$$P(\Delta, N) = \Delta^{-\beta} F(\Delta/N^{d_f/2}), \tag{10.29}$$

where N is the system size. Then, the qth moment of this distribution was computed. The qth moment $M_q \equiv \langle \Delta_q \rangle$ is expected to scale as N^{σ_q}, where $\sigma_q = 0$ for $q < \beta - 1$ and $\sigma_q = d_f(q + 1 - \beta)/2$ for $q > \beta - 1$. Figure 10.5 shows the variation of σ_q with q for triangular and diamond lattices. Surprisingly, even though the curves are parallel (indicating same value of d_f for the two lattices) they do not coincide (indicating different values for β). Indeed $\beta \approx 2.75$ for diamond lattices and $\beta \approx 3.05$ for triangular lattices. For both the lattices $d_f \approx 1.1$, indicating a rather weak system size dependence. The finite size data collapse for the distributions using these values of exponents yields satisfactory results (see Figure 10.6).

Avalanche statistics was also studied in three-dimensional fuse networks. The threshold distribution was considered to be both strong ($w = 2$) and weak ($w = 1$) (for threshold distribution uniform in $[1 - w/2 : 1 + w/2]$) (Räisänen et al., 1998). The finite size scaling of the breaking voltage in the strong disorder limit shows a behavior like $V_b \sim L/(\ln L)^y$ with $y \approx 0.3$, which is similar form of the two-dimensional model but with $y \approx 0.8$. The scaling is further supported in the number of broken bonds that scales as $N_b \sim L^3/(\ln L)^{0.3}$. The fact that the scalings are not simply proportional to volume indicates a departure from the mean-field behavior. Hence, the stress concentrations around the microcracks become an

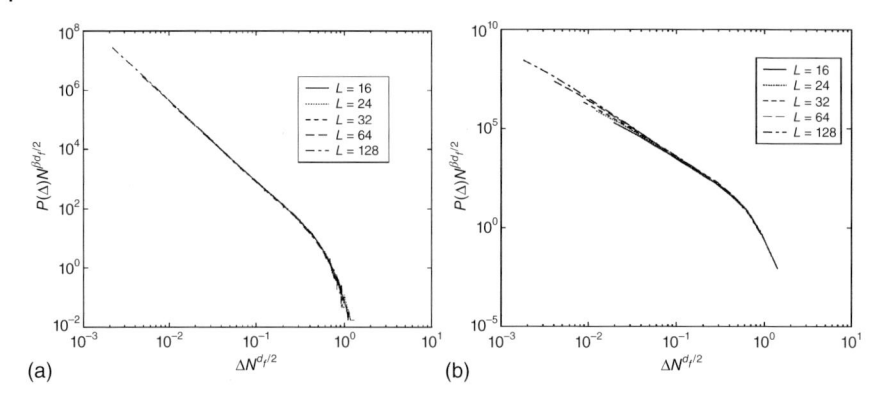

Figure 10.6 The data collapse for the avalanche size distribution with $\beta = 2.75$ and $d_f = 1.18$ for diamond lattice (a) and $\beta = 3.05$ and $d_f = 1.17$ for triangular lattice (b). From Zapperi et al. (2005b), with kind permission from American Physical Society.

important question. Räisänen et al. (1998) reported that the size distribution of avalanches for this three-dimensional system depends on the distribution of disorders. For the strong disorder limit, the avalanche size exponent is close to 2.0, while for the weak disorder limit it is close to 1.5. However, numerical studies (Zapperi et al., 2005a) of larger systems (using the algorithm in Nukala and Šimunović, 2004) showed that the avalanche size distribution has an exponent 2.55 and the cut-off of the size distribution scales with the system size as L^{d_f}, with $d_f = 1.5$, where D represents the fractal dimension of the avalanche (see Figure 10.7).

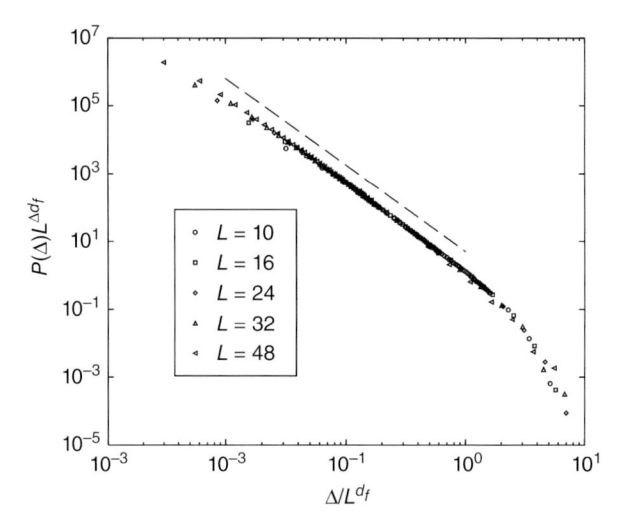

Figure 10.7 The data collapse of the avalanche size distribution in the three-dimensional RFM, where the failure thresholds are distributed uniformly in [0:1]. The data collapse is shown for exponent values $\beta = 2.5$ and $d_f = 1.5$. The straight line shows the best fit of the power-law part with exponent 2.55. From Zapperi et al. (2005a), with kind permission from Elsevier.

A crossover behavior is seen when the thickness of the random fuse model is changed (Barai et al., 2013). Unlike the three-dimensional version mentioned earlier, this is the case where one direction is smaller than the other two ($L \times L \times H$). For smaller thickness, the avalanches are much more widely distributed, but for larger thickness instabilities are apparent and the distribution becomes much sharper. The exponent values of the avalanche size distribution shifts from 1.25 to a much higher value as thickness is increased.

10.2.3
Roughness of Fracture Surfaces in RFM

The final crack line or surface that disconnects a random fuse network has interesting roughness properties. The inherent long-range nature of the load transfer prompted active research for many years in the direction of mapping the surface of a broken fuse network with fracture surfaces. Particularly, in Hansen et al. (1991), roughness properties of crack in two-dimensional fuse network were studied. Generally, a roughness exponent close to 0.7 was obtained, which was conjectured to be in the same universality class as the directed polymer in random media, which gives roughness exponent 2/3 (Halpin-Healy, 1989). This claim was supported in subsequent works as well (Seppälä et al., 2000).

Zapperi et al. (2005b) studied the roughness properties of two-dimensional random fuse model for large systems particularly in the view of claims of anomalous scaling properties. The scaling of local width is a way to determine anomalous scaling properties. For a self-affine surface,

$$w(l) = \left\langle \sum_i \left[h_i - \left(\frac{1}{l} \sum_j h_j \right) \right]^2 \right\rangle^{1/2}, \qquad (10.30)$$

where the sum is within the limit l. One expects $w(l) \sim l^\zeta$ for $l \ll L$ and eventually saturating to a value $W = w(L) \sim L^\zeta$ corresponding to the global width. Also, the power spectrum decays as $S(k) \sim k^{-(2\zeta+1)}$. However, in various theoretical and experimental situations, anomalous scaling is seen, where the local and global roughness exponents differ. Particularly, in anomalous scaling, the local width scales as

$$w(l) \sim l^{\zeta_{loc}} L^{\zeta - \zeta_{loc}}, \qquad (10.31)$$

such that the global width scales as $W \sim L^\zeta$. Similarly, the scaling of the power spectrum changes to

$$S(k) \sim k^{-(2\zeta_{loc}+1)} L^{2(\zeta - \zeta_{loc})}. \qquad (10.32)$$

Figure 10.8 shows the variation of local width for diamond and triangular lattices, respectively. To begin with, in the $l \ll L$ limit, curves for different system sizes do not superpose, indicating anomalous scaling. In addition, the power-law fits indicate $\zeta = 0.80 \pm 0.02$ and $\zeta = 0.83 \pm 0.02$ for diamond and

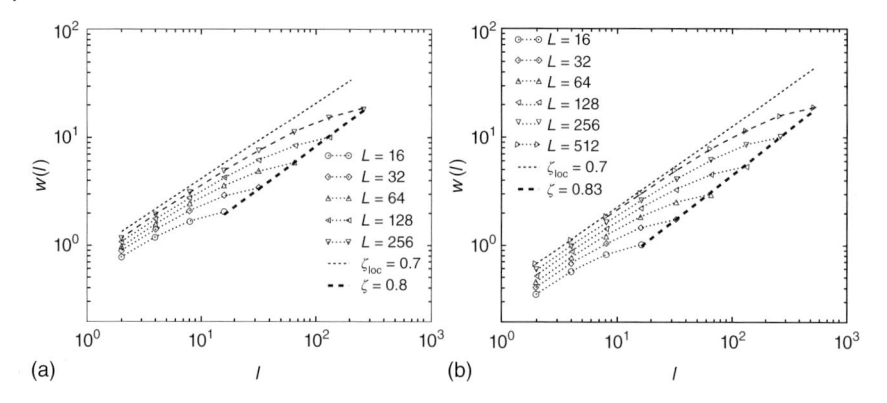

Figure 10.8 Data for local width of diamond (a) and triangular (b) lattices. The local exponents fit with $\zeta_{loc} = 0.7$ and the global exponent ζ is higher than the local one. From Zapperi et al. (2005b), with kind permission from American Physical Society.

triangular lattices, respectively, and $\zeta_{loc} \approx 0.7$ for both the cases. Therefore, a small but definite difference is present between the two roughness exponents. Further analysis was done in terms of the power spectra. The data collapse using $\zeta - \zeta_{loc} = 0.1$ and $\zeta - \zeta_{loc} = 0.13$ for diamond and triangular lattices, respectively fits in a power-law decay with $\zeta_{loc} = 0.7$ and $\zeta_{loc} = 0.74$ for the two lattices, giving $\zeta = 0.8$ and $\zeta = 0.87$, respectively. This is a clear indication of the presence of anomalous scaling in the model. Furthermore, although the local roughness exponent is somewhat close to the exponent for minimum energy surface (2/3), the global roughness exponent is far from it. In addition, in view of the fact that the minimum energy surface is not expected to show anomalous scaling, that the two cases are in the same universality class is ruled out.

The distribution of the crack global width scales as

$$P(W) = P(W/\langle W \rangle)/\langle W \rangle, \qquad (10.33)$$

with $\langle W \rangle \sim L^\zeta$. The width distributions can be collapsed to a single curve for different lattices and its shape fits well with a lognormal distribution (see also Nukala et al., 2008).

Similar analysis was done for three-dimensional fuse model as well. Nukala et al. (2006) found anomalous scaling in three-dimensional fuse model. The global width scaled as $L^{0.5}$. Because of the fact that the broken surface is now two dimensional (see Figure 10.9), the roughness exponents can be measured in both x- and y-directions. Figures 10.10 and 10.11 show the local widths in both directions, measured over $L - 1$ lines and averaged. The systematic variation for $l < L$ in the local widths indicate the presence of anomalous scaling. The global roughness exponents are estimated to be close to 0.52, the local roughness exponent cannot be obtained for the small range of data. However, assuming that the power spectrum scales as $S(k) \sim k^{-(2\zeta_{loc}+1)} L^{2(\zeta-\zeta_{loc})}$, and $\zeta - \zeta_{loc} = 0.1$, one gets good

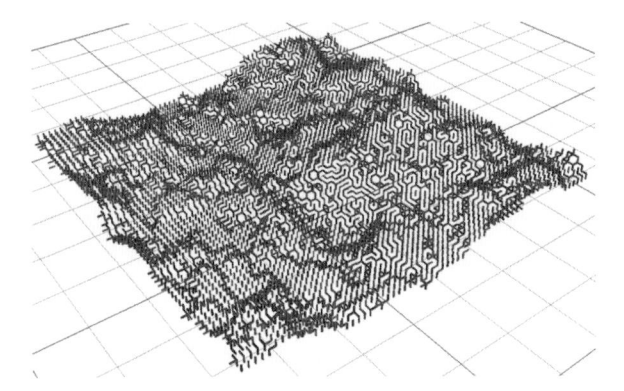

Figure 10.9 The surface of the crack in a three-dimensional random fuse model for $L = 64$. From Nukala et al. (2006), with kind permission from American Physical Society.

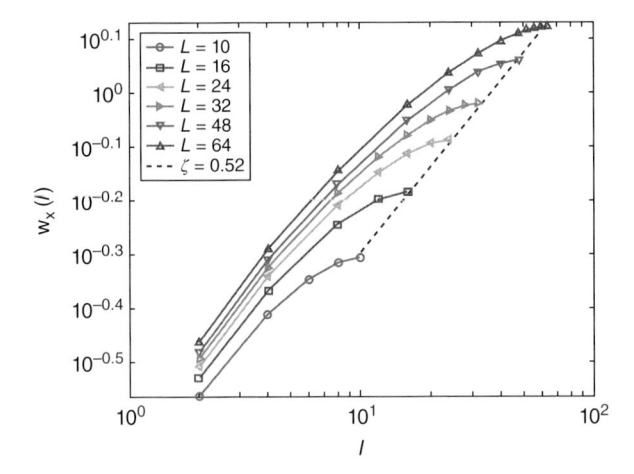

Figure 10.10 The local width in x-direction. The saturation values give estimate for the global roughness exponent to be 0.52. From Nukala et al. (2006), with kind permission from American Physical Society.

data collapse (see Figure 10.12). The power-law fit indicates $\zeta_{loc} = 0.4$, implying $\zeta = 0.5$. Therefore, one finds the same global roughness exponent from both the methods and the value differs significantly from the local exponent, indicating anomalous scaling. The distribution of the global width again fits with lognormal distribution as in two dimensions.

10.2.4
Effect of High Disorder

The degree of disorder often determines the abruptness of failure (Kahng et al., 1988). It is therefore interesting to study the effect of strong disorder in fracture growth and subsequent rupture processes. The effect of disorder and scaling with

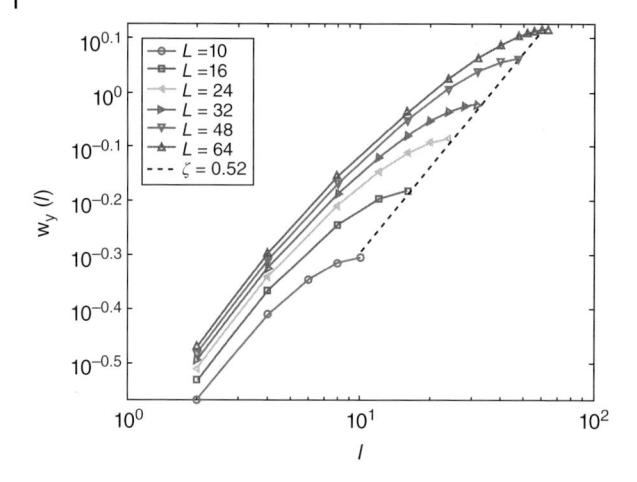

Figure 10.11 The local width in y-direction. The saturation values give estimate for the global roughness exponent to be 0.52. From Nukala et al. (2006), with kind permission from American Physical Society.

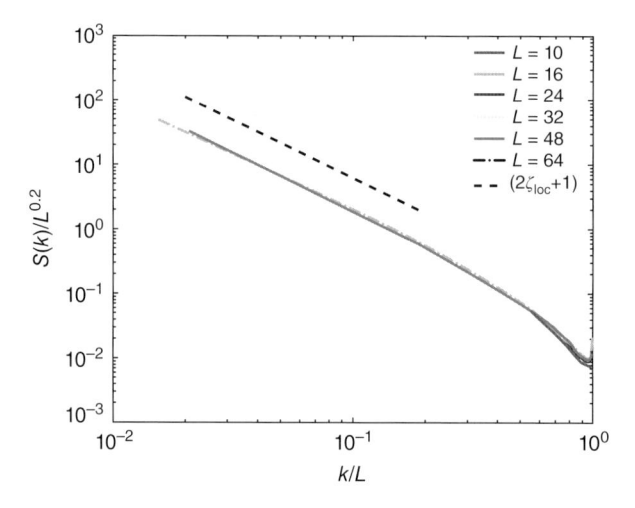

Figure 10.12 The scaling of the power spectrum assuming $\zeta - \zeta_{loc} = 0.1$. The power-law fit indicates $\zeta_{loc} = 0.4$, giving $\zeta = 0.5$. From Nukala et al. (2006), with kind permission from American Physical Society.

system size was studied using a scalar model of crack growth, that is, the random fuse model in Moreira et al. (2012). In the simplest case, all bonds in the fuse model were considered to have identical conductance. However, a bond irreversibly breaks when its current exceeds a preassigned threshold value $I_i = 10^{xR_i}$, where R_i is a random number uniformly distributed in the interval $[0:1]$. Therefore, the distribution of thresholds is $F_I \sim I^{-1}$ in the range 10^x to 10^{-x}.

In the limit of very large disorder ($x \to \infty$), the redistribution and consequently the fluctuations introduced in currents are practically irrelevant. The bonds break in the inverse rank of their thresholds, as long as they are participating in the transport. Now, since the thresholds are randomly distributed spatially, it is equivalent to randomly remove bonds from the system. In this way, it is similar to random percolation. However, when a cavity is formed, which is connected to outside just by a single bond, then all bonds within the cavity are equipotential and hence cannot burn.

Three quantities were measured in this case: the backbone mass M_b, the mass of the largest broken cluster M_f, and the mass of all broken bonds M_t (see Figure 10.13). It is seen that $M_t \sim L^{d_t}$ with $d_t = 2.00 \pm 0.01$, suggesting that the bonds are burned homogeneously throughout the system. The backbone mass scales as $M_b \sim L^{d_b}$ with $d_b = 1.22 \pm 0.01$, which is similar to the strands of invasion percolation fronts (Cieplak et al., 1996) and many other similar cases. In the largest cluster, however, the scaling is $M_f \sim L^{d_f}$ with $d_f = 1.86 \pm 0.01$, which is smaller than the similar statistics for percolating lattice, where it is 1.8958 (Stauffer and Aharony, 1994; Roux et al., 1988). This difference is due to the presence of unbroken bonds within the cavities present in the fuse network. Even though it cannot change the backbone dimension, it can affect the largest cluster size. A plot of M_f/M_p against L (see inset of Figure 10.13), where M_p is the mass of the spanning cluster in percolation, clarifies the difference in scaling prominently. That quantity decays with L with an exponent value $d_f - d_p = -0.03$.

If the disorder distribution is narrow, then the backbone mass and size of largest cluster grow linearly with system size ($d_b = d_f = 1$). As can be seen

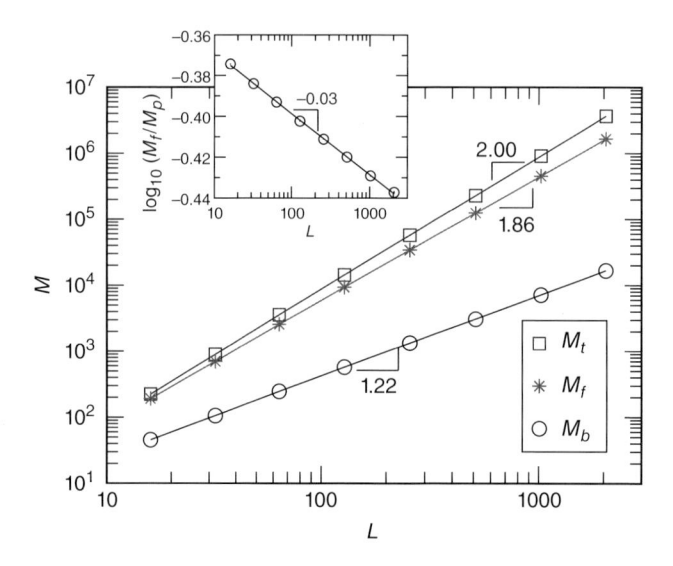

Figure 10.13 The scaling of total number of broken bonds (M_t), mass of backbone (M_b), and mass of the largest broken cluster (M_f) with system size L showing superlinear growth. Inset shows the scaling of M_f/M_p, where M_p is the largest cluster in percolation, showing that the two quantities scale differently with system size. From Moreira et al. (2012), with kind permission from American Physical Society.

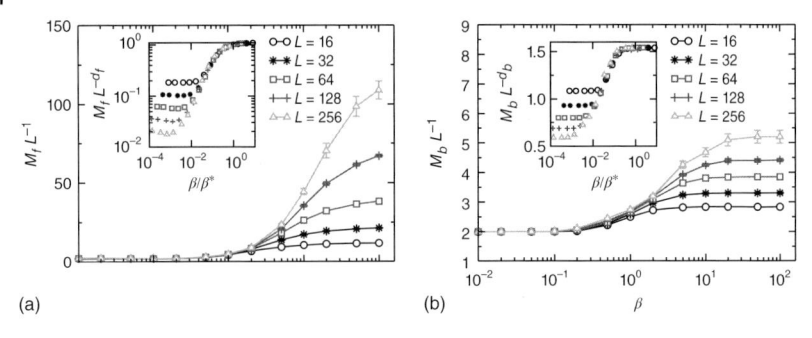

Figure 10.14 The variations of the masses of the largest cluster (a) and those of the backbone (b) with the disorder parameter x. From Moreira et al. (2012), with kind permission from American Physical Society.

from Figure 10.14, M_f and M_b both grow linearly with system size and they are independent of x when $x < 0.1$. As x is increased, in the intermediate range $0.1 < x < x^*(L)$, the quantities still grow linearly with L, however now they are dependent on x. Even up to this point, the weak disorder regime continues. It is only when $x > x^*(L)$, that M_b and M_f grow superlinearly and strong disorder regime begins. One expects that the system scales as in the weak disorder regime beyond a length scale ξ, which must diverge in the strong disorder limit as $\xi \sim x^{1/\eta}$. This implies $x^*(L) \sim L^\eta$, with $\eta = 0.9 \pm 0.01$. Therefore, the role of disorder is to set a length scale below which the system behaves as if there is no disorder and above which effect of disorder can be seen (in superlinear growth of backbone and spanning cluster masses with system size).

10.2.5
Size Effect

The effect of system size in fracture nucleation is interesting. On the one hand, fracture nucleates due to stress concentration and on the other hand, disorder prevents such nucleation. In this competition, it is important to see what happens in the thermodynamic limit. In the context of random fuse model, it was claimed that (Shekhawat et al., 2013) in the limit of large system size, the breakdown process is dominated by first order such as nucleation unless the disorder distribution is very broad. Particularly, for a cumulative disorder distribution of the form $F(x) = x^{x'}$ and $x' > 0$, the limit $x' \to 0$ corresponds to infinite disorder and $x' \to \infty$ corresponds to infinitesimal disorder. It was shown that the upper limit of failure threshold decays with system size as $(x'/2 \log L)^{(x'/2)}$. This suggests that in the thermodynamic limit, the failure process is abrupt and driven by damage nucleation as in the zero disorder limit. However, since the decrease of failure threshold with system size is rather slow, a crossover regime is observed which shows the power-law avalanche statistics. On the other extreme of large disorder, percolation behavior persists. The combined phase diagram is shown in Figure 10.15.

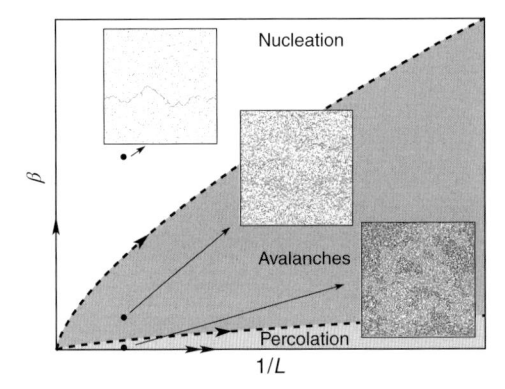

Figure 10.15 The phase diagram for brittle fracture in random fuse model. As the disorder decreases (increasing value of x') or system size increases, the rupture process is driven by crack nucleation. For large disorder, percolation behavior dominates. The crossover region shows scale-free avalanche size distributions. But as is shown, the scale-free behavior persists in the thermodynamic limit only for infinite disorder. From Shekhawat et al. (2013), with kind permission from American Physical Society.

10.3
Dielectric Breakdown Problem

As opposed to the fuse problem, the dielectric breakdown problem has a small fraction of conductors. The dielectric parts can withstand certain voltage e_c across it beyond which it gives in to conduction. A global failure is said to have occurred when the conducting path spans the entire system and it becomes conducting.

In 2d, there exists a duality relation between the fuse and dielectric breakdown problem (Mendelson, 1975; Bowman, 1989). Initially, the system has a conductor fraction $p < p_c$ and consequently the sample is nonconducting. Now, the induction vector **D** and the field **E** follow

$$\nabla \cdot \mathbf{D} = 0$$
$$\nabla \times \mathbf{E} = 0$$
$$\mathbf{D}(r) = \epsilon(r)\mathbf{E}(r), \tag{10.34}$$

where $\epsilon(r)$ is the dielectric constant of the local insulating portion and **E** is irrotational. Writing $\mathbf{E} = -\nabla\phi$, Eq. (10.34) gives

$$\frac{\partial}{\partial x}\left[\epsilon(r)\frac{\partial \phi}{\partial x}\right] + \frac{\partial}{\partial y}\left[\epsilon(r)\frac{\partial \phi}{\partial y}\right] = 0, \tag{10.35}$$

where ϕ is the scalar potential. Now, consider the dual situation where the conductors are replaced by dielectric and vice versa. The sample is now conducting and the relevant quantities are the current density **i** and electric field $\overline{\mathbf{E}}$. They obey

$$\nabla \cdot \mathbf{i} = 0$$
$$\nabla \times \overline{\mathbf{E}} = 0$$
$$\mathbf{i}(r) = \sigma(r)\overline{\mathbf{E}}(r) \tag{10.36}$$

Since **i** has zero divergence, it can be written as a curl of a vector potential **V**. $\mathbf{V}(\mathbf{V} = V_z = \psi(x,y))$ is chosen such that only the z-component of **i** vanishes. Taking the local conductivity of the form $\sigma(r) = 1/\epsilon(r)$, Eq. (10.36) gives

$$\frac{\partial}{\partial x}\left[\epsilon(r)\frac{\partial \psi}{\partial x}\right] + \frac{\partial}{\partial y}\left[\epsilon(r)\frac{\partial \psi}{\partial y}\right] = 0. \tag{10.37}$$

Comparing this with Eq. (10.35), one gets

$$\frac{\partial \psi}{\partial x} = \frac{\partial \phi}{\partial x}, \quad \frac{\partial \psi}{\partial y} = \frac{\partial \phi}{\partial y}. \tag{10.38}$$

The components of the current density in the dual system then become

$$i_x = \partial \psi/\partial y = E_y, \quad i_y = -\partial \psi/\partial x = -E_x. \tag{10.39}$$

Therefore, it is seen that the magnitudes of the current density (in the dual system) and electric field (in the original sample) are same and are perpendicular in direction. Similar relation can be shown between $\overline{\mathbf{E}}$ in the fuse network and **D** of the dielectric problem (Chakrabarti and Benguigui, 1997; Sahimi, 2002). Using this duality, Eqs. (10.16) and (10.19) may be used in substituting I_0 and I_f by V_0 and V_b, respectively and also $(1-p)$ by p.

For the discrete lattice version of the dielectric problem, percolation theory can be used as in the case of the fuse network. In the following, we discuss the dilute and critical limits of the problem using percolation theory.

10.3.1
Dilute Limit ($p \to 0$)

In this case, the enhancement in the field is due to the presence of a long (of size n) defect of conductors in an otherwise dielectric medium, perpendicular to the direction of the field Beale and Duxbury (1988)

$$E_e = E(1 + kn), \tag{10.40}$$

where k is the enhancement factor, as before. The probability of having defect of linear size n scales as

$$P(n) \sim p^n L^d. \tag{10.41}$$

The most probable size follows

$$n_c = -\frac{d}{\ln p}\ln L. \tag{10.42}$$

Therefore, the field enhancement adjacent to the most probable defect is

$$E_e = E\left[1 + K_d\left(-\frac{\ln L}{\ln p}\right)\right]. \tag{10.43}$$

Using this, the breakdown voltage can be estimated as

$$V_b = \frac{E_0 L}{\left[1 + K_d\left(-\frac{\ln L}{\ln p}\right)\right]}, \tag{10.44}$$

where E_0 is the breakdown field for the pure sample.

10.3.2
Close to Critical Point ($p \to p_c$)

As before, the relevant length scale in this limit is the percolation correlation length ξ_p. This means that the typical separation between defects (conductors) is of the order ξ_p, implying the average field to be $V_1/\xi_p = V/L$, where V and V_1 are respectively the externally applied voltage and potential difference between any two defects. If a is the lattice constant, then the maximum possible value of the field can be V_1/a. But, as soon as the field value reaches e_c, the bond breakdown field, local failure occurs. The average field value for such failures is $E_b = (a/\xi_p)e_c$. But, near the percolation point, $\xi_p(p) \propto a(p - p_c)^{-\nu}$, therefore the breakdown field on average behaves as

$$E_b \sim (p_c - p)^{\nu}. \tag{10.45}$$

in all dimensions. An alternative to arrive at this result is the minimum gap $g(p)$. Since E_b is proportional to the gap (see Section 10.1.6), and the gap in turn goes as ξ_p^{-1} (see Eq. 10.20), the above result is obtained. Since the breakdown field goes as the inverse of the linear scale of the most dangerous defect, it will go to zero in the percolating critical point, where the correlation length, a measure of such defect, diverges.

10.3.3
Influence of Sample Size

Let us take two conducting regions, approximately of size l and being placed one after the other in the direction perpendicular to the applied field E in a hypercubic lattice. The enhanced field in the vicinity of the defects will be El. In the low defect concentration regime (far from percolation), the probability of appearance of such clusters is $(1/\xi_p)\exp(-l/\xi_p)$. The largest cluster will therefore scale as $l_{max} \sim \xi_p \ln(L^d)$. In this limit (low disorder), the first breakdown field E_b^1 is also the global breakdown field, and it scales as $1/l_{max}$ (Beale and Duxbury, 1988). The breakdown field has the scaling form

$$E_b \sim \frac{(p_c - p)^{\nu}}{\ln L}. \tag{10.46}$$

as was obtained by Beale and Duxbury (1988). Clearly, for a larger sample, the chance of having a weaker defect is higher. That is why the breakdown field decreases. On the other hand, when the percolation limit is approached, percolation fluctuations win over the extreme statistics Bergman and Stroud (1992). Very near the percolation point, the following scaling form was proposed by Chakrabarti and Benguigui (1997)

$$E_b \sim \frac{(p_c - p)^{\nu}}{\ln(L/\xi_p)}. \tag{10.47}$$

Table 10.2 Theoretical estimates for the dielectric breakdown exponent ϕ_b.

Dimension	Lattice percolation	Continuum percolation
2	$v(=4/3)$	$v+1(=7/3)$
3	$v(\simeq 0.88)$	$v+1(\simeq 1.88)$

A unified expression for the breakdown field in the dielectric problem may be written as

$$E_b = E_0 \frac{[(p_c - p)/p_c]^{\phi_b}}{1 + K \left[\frac{\ln(L/\xi_p)}{-\ln p} \right]}, \quad (10.48)$$

where ϕ (same as v for the percolation on a lattice) depends on the dimension and the type of the system. For theoretical estimates of its value for different cases, see Table 10.2.

10.3.4
Distribution of Breakdown Field

The cumulative distribution function $F_L(E)$ for the breakdown field for a system of size L is usually fitted with the Weibull form (Wiederhorn, 1984; Weibull, 1961)

$$F_L(E) = 1 - \exp(-rL^d E^m) \quad (10.49)$$

where r and m are constants. However, it was also argued (Duxbury et al., 1987, 1986; Duxbury and Leath, 1987; Beale and Duxbury, 1988) that the distribution follows Gumbel form

$$F_L(E) = 1 - \exp\left[-AL^d \exp\left(-\frac{K}{E}\right)\right]. \quad (10.50)$$

Its derivation is possible (Beale and Duxbury, 1988) using scaling arguments of the percolation statistics (Stauffer, 1979; Essam, 1980; Kunz and Souilard, 1978). The Weibull and Gumbel distributions are similar for large values of the Weibull exponent m. Nevertheless, Eq. (10.50) fits better with simulations (Beale and Duxbury, 1988). It was also argued (Sornette, 1988) that Eq. (10.50) is not suited for continuum system, and Weibull distribution is more appropriate there.

10.3.5
Continuum Model

As before, here the dielectric sample contains conducting regions of spherical or circular (for 3d and 2d samples, respectively) shapes that may overlap. The breakdown voltage is proportional to the width δ of the region between the defects. Following the link-blob picture, breakdown must be preceded by about ξ_p^{-1} number of breaking of links. Following the logic described in Section 10.1.5, $E_b \sim \delta_{\min} \xi_p^{-1} \sim L_c^{-1} \xi_p^{-1}$ or

$$E_b \sim (p_c - p)^{\nu+1} \tag{10.51}$$

very close to the percolation point (Chakrabarti et al., 1988; Lobb et al., 1987). Hence, in the continuum limit, the system is weaker than the corresponding discrete case, since here the conductivities of the links are an increasing function of p but for the discrete case, there was no dependence.

10.3.6
Shortest Path

As before, consider a random walker continuing its journey following a self-avoiding walk and converting each insulator it crosses to a conductor. This creates a conducting path across the sample. If the number of conversions is n_0, then the shortest path for a given defect configuration is the one with minimum value of n_0. This is also called the minimum gap $g(p) = \langle n_0 \rangle / L$.

The functional dependence of $g(p)$ over p was studied in details (Stinchcombe et al., 1986; Duxbury and Leath, 1987) for both 2d and 3d. As the critical point is approached, $g(p)$ decreases from 1 to 0 with the exponent ν.

10.3.7
Numerical Simulations in Dielectric Breakdown

10.3.7.1 Stochastic Models

A random growth model was proposed by Sawada et al. (1982) to mimic the breakdown properties of the disordered dielectrics. A pattern can grow in either of the following two ways: the tip grows with a probability p_0 and a new tip is formed (branching) with a probability p_n (with $p_n < p_0$). However, such a simplified picture does not capture the experimental results (Christophorou, 1982).

To model the discharge pattern in gases, a stochastic model was proposed in (Niemeyer et al., 1984), where the growth pattern in turn influences the local electric field. The fractal properties of the dielectric breakdown processes were reproduced in the simulation of this model. Here, the growth starts from the center where one electrode is placed, and the other electrode is placed far away on the boundary of a circle. In every step, one bond in the nearest neighbors of the broken pattern fails and is added to the broken pattern and the voltage is shorted with that in the central electrode. The growth probability p depends on the local field and in turn the local field controls the breakdown pattern via the relation

$$p(i,j \to i',j') = \frac{(V_{i',j'})^\eta}{\sum (V_{i',j'})^\eta}, \tag{10.52}$$

where i,j and i',j' are the lattice coordinates. The electric potential is obtained from the solution of the Laplace equation

$$V_{i,j} = \frac{1}{4}(V_{i+1,j} + V_{i-1,j} + V_{i,j+1} + V_{i,j-1}) \tag{10.53}$$

with the boundary condition that $V = 0$ for all points on the growing pattern and $V = 1$ outside the outer circle. The pattern follows a fractal nature governed by

$$N(r) \sim r^{d_f}, \tag{10.54}$$

where d_f denotes the Hausdorff dimension and $N(r)$ is the total number of discharge points. The value of the Hausdorff dimension is 2, 1.89 ± 0.01, 1.75 ± 0.02, and 1.6 for $\eta = 0, 0.05, 1$, and 2, respectively. The structure becomes more linear for higher η values. For $\eta = 1$, d_f value is in good agreement (~ 1.7) with experiments (Christophorou, 1982). The pattern is very similar to that of the Lichtenberg figure.

Note that similar pattern is also observed for diffusion-limited aggregation model (DLA) Witten and Sander (1981) (For review, see Meakin, 1998). Many other models (see Sahimi, 2002 for details) have been proposed, but a complete understanding is still lacking.

10.3.7.2 Deterministic Models

A deterministic model was introduced by Takayasu (1985) to get the dendritic fractal pattern (seen in lightning) by putting an a priori fluctuation in the bond resistance values ($r_i = \theta r$; $\theta \in [0, 1]$). Also, when a potential difference across a nonconducting element reaches a preassigned value, say v_c, its resistance r_i changes to δr_i, where δ is a small positive number. The value of the resistance never changes again. This is the notion of breakdown in the model. Here, also a breakdown may cause other breakdowns and a pattern emerges, which is anisotropically fractal with a dimension $d_f = 1.58 \pm 0.12$ in 2d. It is lower than the random percolation cluster fractal dimension (1.89), due to its anisotropy.

The stochastic model considered by Niemeyer et al. (1984) was made deterministic in Family et al. (1986) by associating a random breaking coefficient $\theta (\theta \in [0, 1])$ with every insulating bond. Two limits of this model were considered: in one limit, in each time step one interface bond ij that has the highest value for θV_{ij}^{η} fails, and in the other limit an interface bond fails following a probability $\theta V_{ij}^{\eta}/p_{max}$. Here, η is a tunable parameter and p_{max} is the largest value of the quantity θV_{ij}^{η} between all interface bonds at an instance. The resulting pattern is a fractal with dimension that depends on η. In the first case, highly anisotropic pattern with fractal dimension close to 1.2 appears in 2d, but the pattern in the second case is remarkably isotropic. These are similar to the ones obtained in Niemeyer et al., (1984), apart from a different fractal dimension 1.70 ± 0.05 (for $\eta = 1$) in 2d. The values of d_f are similar to those of DLA in a square lattice (Meakin, 1985) and those in case of the model in a homogeneous medium (Family et al., 1986).

The theoretical predictions regarding the defect and system size dependence of the dielectric breakdown have been confirmed by direct simulation of the breakdown process. Here, the voltage across the defects is obtained by numerically solving the Laplace equation (10.53). The dielectric having the largest voltage drop (which is equal to or greater than v_c) is converted to a conductor and the voltages are recalculated. This process is continued until there remains to element the voltage drop across which exceeds v_c. Then the external voltage is slowly increased. This process is continued until there is a system spanning conducting path. The breakdown field is given by V_b/L, where V_b is the applied breakdown voltage. It is seen that the field required for the first failure E_b^1 is also the breakdown field E_b.

The variation of E_b^1 and $g(p)$ for different p below the critical point was studied by Manna and Chakrabarti (1987). It was seen that both the quantities go to zero at p_c with the exponent value close to unity, which was argued to be the correlation length exponent ν of the percolation point (see Figure 10.16). The departure was attributed to the finite size effect. Bowman (1989) reported that E_b^1 becomes zero with an exponent 1.1 ± 0.2 in 2d and 0.7 ± 0.2 in 3d, which are consistent with the corresponding correlation length exponent values 4/3 and 0.88 for 2d and 3d, respectively.

A functional dependence of the first breakdown field E_b^1 was given by Beale and Duxbury (1988) in the form

$$E_b^1 \sim \frac{1}{A(p) + B(p)\ln L}. \tag{10.55}$$

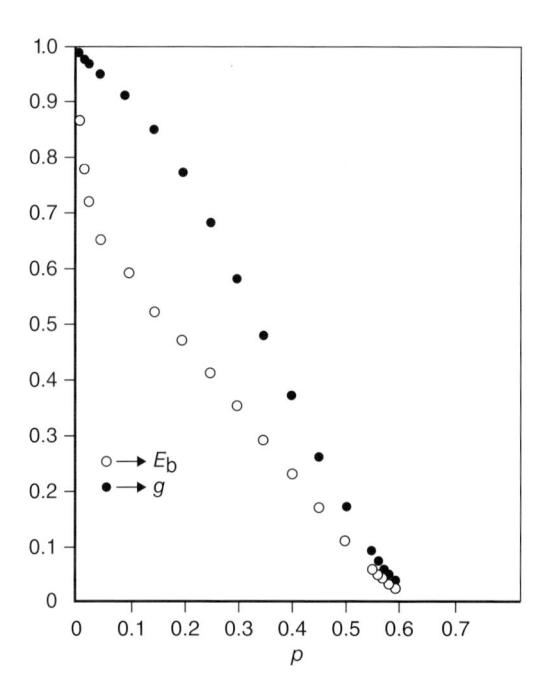

Figure 10.16 The variations of first breakdown field E_b and the minimum gap $g(p)$ with initial impurity probability p are shown. The two quantities behave similarly very near to the percolation threshold p_c, whereas far from p_c they behave differently. From Manna and Chakrabarti (1987), with kind permission from American Physical Society.

As the critical point was approached, $A(p)$ and $B(p)$ varied like $(p_c - p)^{-\nu}$ (in the lattice of size $L = 50, 70,$ and 100) as is expected from Eq. (10.46). By plotting $\ln\{B(p)\ln(p)\}$ against $-\ln(p_c - p)$, it was seen that $\nu = 1.46 \pm 0.22$ in reasonable agreement with the theoretical value 4/3. For the cumulative distribution of failure, their data supported Gunbel form Eq. (10.50). The exponent ϕ_b in Eq. (10.48) was found to be close to 1.0 by Manna and Chakrabarti 1987 and Beale and Duxbury (1988).

Acharyya and Chakrabarti (1996c) studied the rate of breakdown for $p < p_c$ and gradually approaching the percolation point. They found that the rate diverges at the breakdown voltage V_b.

It is indicating that the global breakdown is a highly correlated process close to V_b. A quantity called breakdown susceptibility χ ($= dn/dV$) was introduced, where $n(V)$ denotes the average number of failures for a voltage V for a given p. Obviously, for large V, $n(V)$ reaches L^d. χ shows a maximum (see Figure 10.17) at some voltage V_b^{eff}, different from V_b^{fin}. But V_b^{eff} tends to V_b^{fin} as the system size is increased. Also, it appears to diverge with system size, giving a way to predict an imminent failure from the response of the system. The cluster statistics was studied by Acharyya and Chakrabarti (1996a) up to the breakdown voltage V_b and it was seen that the rate of various quantities, such as the total number of conducting bonds and average cluster size, tends to diverge at the global breakdown point.

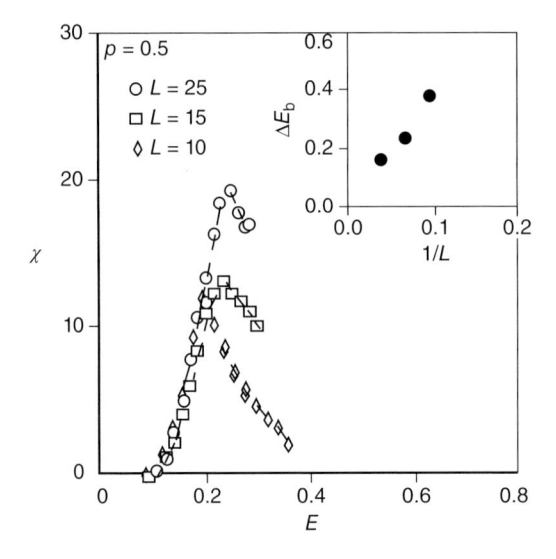

Figure 10.17 The plot of dielectric breakdown susceptibility against applied voltage for several sample sizes is shown. The inset shows that the difference of the maximum susceptibility field and the minimum sample spanning cluster field decreases with increasing sample size. From Acharyya and Chakrabarti (1996b), with kind permission from American Physical Society.

10.4
Summary

We have discussed the electrical failure problems analogous with the mechanical fracture. Particularly, we focused on the dielectric breakdown and random fuse models. The competition between the extreme statistics and the percolation statistics was discussed. While for low disorder concentration extreme statistics (with Gumbel distribution for the cumulative failure probability) dominates, for disorder concentration close to the percolation threshold, percolation statistics wins.

In the random fuse model, the avalanche statistics and the roughness properties of the fractured surfaces were discussed (Section 10.2.3). The roughness properties show anomalous scaling.

11
Earthquake as Failure Dynamics

In this chapter, we intend to discuss statistical physical models of earthquake phenomenon. Earthquakes are the mechanical failure phenomena known to occur in the largest length scale. Because of their catastrophic nature, understanding earthquakes has been a long-standing goal for scientists across various disciplines, ranging from engineers and seismologists to physicists, leading to the development of a rich and diverse literature for years. While it is interesting to study the different approaches toward this problem, detailed account of these studies is beyond the scope of this textbook. Here, we discuss the statistical physical models of earthquakes, focusing mainly on two types of studies. One is the equation of motion-based spring-block type models of earthquake, visualizing the fault system as massive blocks connected by linear springs and studying the frictional instabilities within the system. And the other is the much more simplified cellular automata-type models that only retain a threshold-activated dynamical failure process and rely on the self-evolution of the system toward a "critical" state in reproducing scale-free laws observed empirically.

In this chapter, we first discuss the empirical laws and their generalizations seen in earthquakes. Then, we go over to equation of motion-based spring-block type models with nonlinear frictional properties that capture those laws, emphasizing the roles of heterogeneities in such models. Then, we go over to the self-organized critical models of earthquakes, also looking into the roles of heterogeneity. We also consider the models that deal with the fractal geometry of the fault systems and relate the energy release with the overlap of two such fractal surfaces. Finally, we attempt to link these approaches by studying a minimal version of spring-block type models, which is essentially equivalent to self-organized dynamical models of elastic manifolds driven through a medium with quenched disorders.

11.1
Earthquake Statistics: Empirical Laws

The motivation for the statistical physical modeling of earthquakes comes from the fact that many of its empirical laws have similarity with those seen in systems with many components in a critical state due to cooperative interactions among

Statistical Physics of Fracture, Breakdown, and Earthquake: Effects of Disorder and Heterogeneity, First Edition.
Soumyajyoti Biswas, Purusattam Ray, and Bikas K. Chakrabarti.
© 2015 Wiley-VCH Verlag GmbH & Co. KGaA. Published 2015 by Wiley-VCH Verlag GmbH & Co. KGaA.

EARTHQUAKE MAGNITUDE, INTENSITY, ENERGY, AND ACCELERATION
(Second Paper)
By B. Gutenberg and C. F. Richter

ABSTRACT

This supersedes Paper 1 (Gutenberg and Richter, 1942). Additional data are presented. Revisions involving intensity and acceleration are minor. The equation $\log a = I/3 - \frac{1}{2}$ is retained. The magnitude-energy relation is revised as follows:

$$\log E = 9.4 + 2.14 M - 0.054 M^2 \qquad (20)$$

A numerical equivalent, for M from 1 to 8.6, is

$$\log E = 9.1 + 1.75 M + \log(9 - M) \qquad (21)$$

Equation (20) is based on

$$\log(A_a/T_a) = -0.76 + 0.91 M - 0.027 M^2 \qquad (7)$$

applying at an assumed point epicenter. Eq. (7) is derived empirically from readings of torsion seismometers and USCGS accelerographs. Amplitudes at the USCGS locations have been divided by an average factor of 2½ to compensate for difference in ground; previously this correction was neglected, and $\log E$ was overestimated by 0.8. The terms M^2 are due partly to the response of the torsion seismometers as affected by increase of ground period with M, partly to the use of surface waves to determine M. If M_S results from surface waves, M_B from body waves, approximately

$$M_S - M_B = 0.4 (M_S - 7) \qquad (27)$$

It appears that M_B corresponds more closely to the magnitude scale determined for local earthquakes.

A complete revision of the magnitude scale, with appropriate tables and charts, is in preparation. This will probably be based on A/T rather than amplitudes.

Figure 11.1 Beno Gutenberg (1889–1960) was a seismologist who made very important contributions to the statistics of earthquake occurrences. He and his colleague C. Richter in 1942 proposed the empirical law relating the frequency and magnitude of earthquake occurrences, later came to be known as the Gutenberg–Richter law. Gutenberg was born in Darmstadt, Germany, and received his doctorate from Gottingen University. He moved to Caltech in 1930 to become the founding director of the Seismological Laboratory. There he made the observation of the statistical law mentioned above, which remains one of the inspiration to capture the basic features of earthquake via simple models. He also worked on determining the depth of the core mantle boundary and other interior properties of earth. Abstract from Gutenberg and Richter (1956), with kind permission from Seismological Society of America.

those components. Particularly, the laws governing the frequency distribution of earthquakes and also their temporal occurrence are scale-free in nature, which is the common signature of criticality. These laws are:

Gutenberg–Richter (GR) law: The Gutenberg–Richter law (Gutenberg and Richter, 1944), named after the seismologists B. Guttenberg (see Figure 11.1) and C. Richter (see Figure 11.2), relates the cumulative number of earthquakes with the earthquake magnitude. It states that the probability of the occurrence of an earthquake of magnitude greater than or equal to M is given by

$$\log P(M) = A - b \log M, \qquad (11.1)$$

where A and b are constants. Also, the magnitude of an earthquake is related to the energy released E by a log-linear relation as

$$\log E = C + D \log M \qquad (11.2)$$

which together with the previous relation gives another form of the GR law as

$$P(E) \sim E^{-\beta} \qquad (11.3)$$

with $\beta \approx 1$. This law is seen to be valid widely over many fault systems across the world over long intervals (see Figure 11.3).

Another widely seen statistical law about earthquake is the Omori law (1836) (see Figure 11.4): This law states that the rate of aftershocks (smaller earthquakes

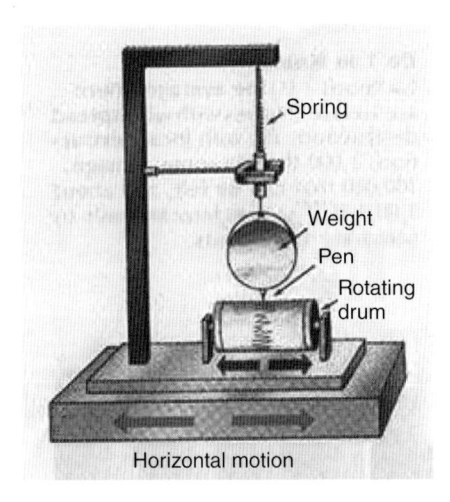

Figure 11.2 Charles Francis Richter (1900–1985) was born in Ohio, USA. He was a seismologist in the California Institute of Technology. He worked with Beno Gutenberg and also pioneered the so-called Richter magnitude scale (1935) in quantifying the intensity of earthquakes. The Richter magnitude is related to the logarithm of amplitude of the waves recorded in seismograph (figure shown). Each whole number increase in the magnitude corresponds to about 32 times increase in the energy released. Richter is also known for the Gutenberg–Richter law of earthquakes, which quantifies the spatial clustering of earthquakes. [http://www.unc.edu/~kthuw/measuringquakes.html; Richter (1935).]

after a major one or the main shock) decays with time following a power-law. Particularly,

$$N(t) = \frac{K}{(C+t)^p}, \qquad (11.4)$$

where K and C are constants. This form of the Omori law is called modified Omori law (Omori, 1895). The value of the exponent varies from 0.7 to 1.5 in different places. This law is seen to be valid for a long time (months) after a main shock.

11.1.1
Universal Scaling Laws

The scale-free nature of the dynamics of earthquake in space and time and also the fractal character of the epicenters led to the speculation of unifying these scaling similar to the ones seen for critical phenomena. This is a step closer to model earthquake as a self-organized critical event (to be discussed later).

Bak *et al.* (2002) and Christensen *et al.* (2002) studied a unifying scaling law for the distribution of waiting times between the earthquakes (see also Corral, 2006a). They considered earthquake catalog during the period 1984–2000 as a region of California ($20°N - 45°N$ latitude and $100°W - 125°W$ longitude), containing 335 076 events. In order to study the spatiotemporal complexity, they divided the region into $L \times L$ size grids. Then, the distribution of the waiting

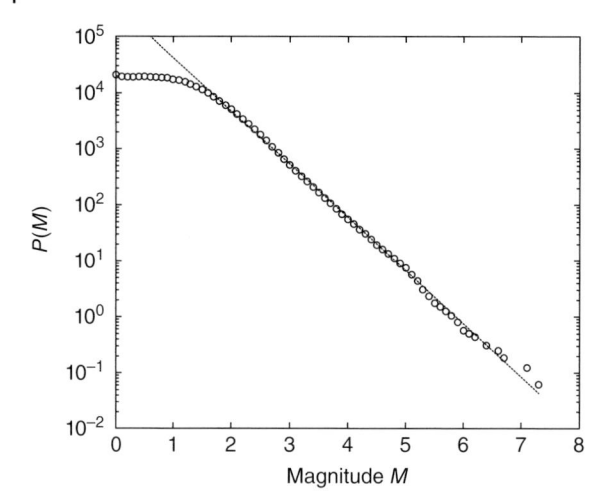

Figure 11.3 Earthquake magnitude distribution from California catalog showing GR exponent 1. From Bak et al. (2002), with kind permission from American Physical Society.

time $Q_{M_0,L}(t_w)$ was measured, where the waiting time t_w is the time between two earthquakes of magnitude greater than $M = \log(M_0)$ occurring within the area $L \times L$. The distribution shows an initial power-law followed by a faster decay (see Figure 11.5), which can be nicely scaled in the form

$$Q_{M_0,L} = t_w^{-P} f(t_w M_0^{-b} L^{d_f}). \tag{11.5}$$

The scaling exponents have the values $p \approx 1.0$, $b \approx 1.0$, and $d_f \approx 1.2$. This shows a power-law decay for a long time (eight decades) with exponent value close to 1 and then a sharper decay (see Figure 11.6).

Bak et al. (2002) and Christensen et al. (2002) also noted that the quantity $M_0^{-b} L^{d_f}$ is the measure of the average number of earthquakes per unit time in the area $L \times L$ having magnitude greater than $M = \log(M_0)$. Therefore, $t_w M_0^{-b} L^{d_f}$ is the average number of such earthquakes. The scaling states that the distribution depends on this number only and when a certain value is exceeded, the events become less correlated (the sharper tail in the distribution). However, it does not correspond to a characteristic scale, since the scale depends on the choice between L and M_0 and therefore is not a unique quantity. The scaling in fact indicates that crust operates in the slowly driven SOC-like state, where the individual avalanches do not overlap. However, by increasing the scale size, if spatially uncorrelated avalanches are sampled, the scaling function shows a sharp decay.

The waiting time distribution studies mentioned above are the blocks of size $L \times L$. Corral (2003) pointed out that such coarse-grained picture counts times from different parts of the blocks together. However, the seismic activities on different parts can vary significantly. Corral (2003) proposed a local measure of waiting time $Q_{xy}(t_w)$, concerning only about the waiting time at coordinate x,y.

On the After-shocks of Earthquakes.

by

F. Ōmori, Rigakushi.

I. General Considerations.

§ 1. A strong earthquake is almost invariably followed by weaker ones and when it is violent and destructive the number of minor shocks following it may amount to hundreds or even thousands. When after-shocks are not reported to have happened it is probably because they were deemed unimportant to record. Or it may be that the seat of origin of the earthquake being very deep or far out under the ocean-bed, the after-shocks did not reach the observer.

Complete records of after-shocks were obtained, I believe, for the first time in the cases of the three recent great earthquakes in Japan; namely, those of Kumamoto in 1889, of Mino and Owari in 1891, and of Kagoshima in 1893. The discussion of these records forms the subject of the present paper.

§ 2. The numbers, daily, monthly, etc., of the after-shocks of these three earthquakes, together with other matters relating thereto are contained in Tables I—XXII at the end of the paper; the shocks being distinguished as "violent," "strong," and "weak" (or "feeble") according to their intensity; the total or aggregate intensity of a number of shocks is obtained by multiplying each shock by one of the coefficients 3, 2, 1, according to its intensity, and taking the sum of the numbers so obtained. The after-shocks of the Mino-Owari earth-

Figure 11.4 Fusakichi Omori (1868–1923) was a Japanese seismologist particularly remembered for his study and quantification of the rate of aftershocks in an earthquake, later came to be known as the Omori law. He was the chairman of seismology at the Imperial University of Tokyo and the president of the Japanese Imperial Earthquake Investigation Committee. He made several investigations worldwide to measure seismic activities following a major earthquake, such as earthquakes in Mino and Owari provinces in 1891 and San Francisco earthquake 1906. In 1894, he proposed the empirical law stating that the rate of aftershocks decays with time almost as inverse of time (Omori, 1895). He also studied volcano eruptions in Japan. The above figure (text) is from Omori (1895); taken from Japanese Institutional Repositories Online.

The global distribution $Q(t_w)$ (also depends on L and M, but is not explicitly mentioned) is a mixture of these local distribution as

$$Q(t_w) \propto \sum_{x,y} \int Q_{xyt}(t_w) r(x, y, t) dt, \qquad (11.6)$$

where $r(x, y, t)$ is the rate of occurrence of earthquakes in the region (x, y) at time t. It was then assumed that the dependence of space and time enters $Q(t_w)$ only through the rate $r(x, y, t)$

$$Q_{xyt}(t_w) = Q(t_w | r(x, y, t)). \qquad (11.7)$$

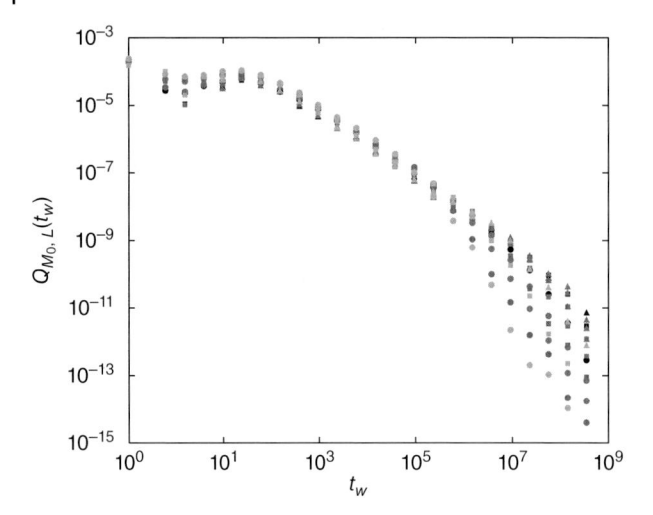

Figure 11.5 The distribution of the waiting time $Q_{M_0,L}(t_w)$ of earthquakes of magnitude greater than $M = \log(M_0)$ occurring within the area $L \times L$. From Bak et al. (2002), with kind permission from American Physical Society.

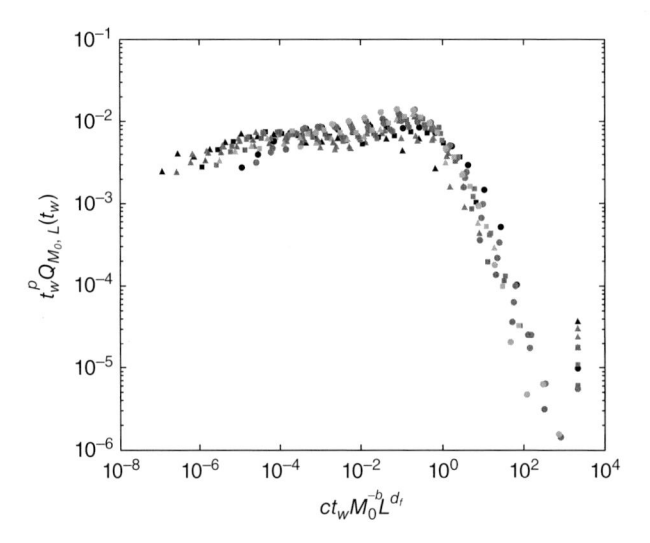

Figure 11.6 The scaled form of the data in Figure 11.5 following Eq. (11.5). The exponent values $b \approx 1.0$, $p \approx 1.0$, and $d_f \approx 1.2$ were used. From Bak et al. (2002), with kind permission from American Physical Society.

Therefore,

$$Q(t_w) = \int_0^\infty Q(t_w|r) \frac{r\rho(r)}{\mu} dr, \tag{11.8}$$

where $\rho(r)$ is the probability density of r and $\mu = \langle r \rangle$ is the normalization factor.

11.1 Earthquake Statistics: Empirical Laws

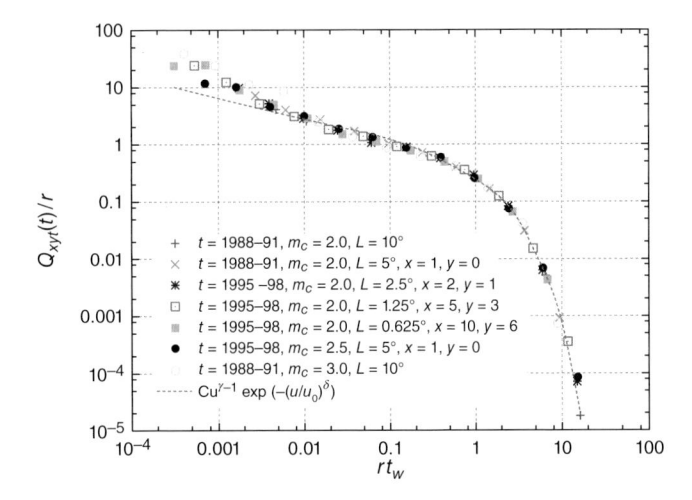

Figure 11.7 The scaling of local distribution of waiting time following Eq. (11.9). From Corral (2003), with kind permission from American Physical Society.

As can be seen from Figure 11.7, the distribution Q_{xyt} has the scaling form

$$Q(t_w|r) \approx r f(r t_w), \qquad (11.9)$$

with the scaling function of the form

$$f(x) = C \frac{1}{x^{1-\gamma}} e^{-(x/x_0)^\delta}, \qquad (11.10)$$

with $\gamma = 0.63 \pm 0.05$, $\delta = 0.92 \pm 0.05$, $x_0 = 1.50 \pm 0.15$, and $C = 0.5 \pm 0.1$, that is, a gamma distribution.

Now, looking for the distribution $\rho(r)$ (Corral, 2003), one can fit it in the form

$$\rho(r) = D\theta \frac{(\theta r)^{\alpha'}}{[1 + (\theta r)^c]^{(\alpha' + \beta')/c}}, \qquad (11.11)$$

which gives $\rho \sim 1/r^{1-\alpha'}$ for $r \ll \theta^{-1}$ and $\rho \sim 1/r^{1+\beta'}$ for $r \gg \theta^{-1}$, with $\alpha' \approx 0$ and $\beta' \approx 1.2$. One can also calculate $\mu = \langle r \rangle \sim L^{d_f}/M_0^b$. Now, knowing the form of the rate distribution function, one can calculate $Q(t_w)$. It can be shown to have the limiting forms

$$Q(t_w) \sim \frac{\theta^{1-\beta'}}{t_w^{2-\beta'}} \quad \text{for} \quad t_w \ll \theta \qquad (11.12)$$

and

$$Q(t_w) \sim \frac{\theta^{1+\alpha'}}{t_w^{2+\alpha'}} \quad \text{for} \quad t_w \gg \theta, \qquad (11.13)$$

which can be seen in Figure 11.8. The fitting gives $\beta' \approx 1.1$ and $\alpha' \approx 0.2$.

Finally, the scaling relation Eq. (11.9) not only includes the GR law (which can be found by calculating the mean then inverting it to get the rate of events), but also says that the GR is obeyed at any time, if the times are properly selected

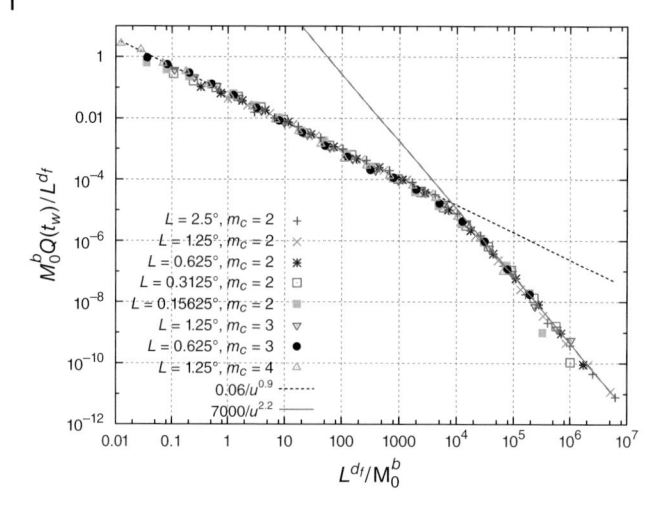

Figure 11.8 The scaled form of the global waiting time distribution. The two exponents are 0.9 and 2.2. From Corral (2003), with kind permission from American Physical Society.

(Corral, 2006a). Events having waiting time t'_w for $M \geq M'_c$ and waiting time t_w for $M \geq M_c$ occur with a GR ratio $10^{-b(M'_c - M_c)}$ if the ratio of the waiting times is given by $10^{b(M'_c - M_c)}$.

11.2
Spring-block Models of Earthquakes

The spring-block representation of a fault system was first introduced by Burridge and Knopoff (1967) (see Figure 11.9). In their table-top experiment (see Figure 11.10 for a schematic diagram), they took eight massive wooden blocks (about 140 g each) and connected them with identical springs, and the system is placed over a rough surface. The last block is connected to a motor via a string, which pulls it slowly. Initially, even though a force is applied on the first block by the chain, it does not slip due to static friction between the block and the rough surface. However, after some time, the block will slip leading to readjustment of position and elastic energies of the springs. After some initial dynamics, the whole system of blocks will be in the intermittent stick-slip motion, giving rise to interesting complex dynamical responses.

Denoting the energy of the system at time t by $E(t)$, one can write

$$E(t) = \frac{k_c}{2} \sum_{i=1}^{N} [U_{i-1}(t) - U_i(t)]^2, \quad (11.14)$$

where $U_i(t)$ is the displacement of the ith block at time t and k_c is the spring constant of the connecting springs. As there is a slow but constant drive (strain rate), the system charges during the time when the blocks stick to the rough surface. Eventually, the stored energy overcomes the barrier due to static friction and

Bulletin of the Seismological Society of America
www.bssaonline.org

Bulletin of the Seismological Society of America June 1967 vol. 57 no. 3 341-371

Copyright © 1967, by the Seismological Society of America

Model and theoretical seismicity

R. BURRIDGE and L. KNOPOFF

▸ Author Affiliations

    DEPARTMENT OF APPLIED MATHEMATICS AND THEORETICAL PHYSICS UNIVERSITY OF CAMBRIDGE England
    INSTITUTE OF GEOPHYSICS UNIVERSITY OF CALIFORNIA LOS ANGELES, CALIFORNIA

ABSTRACT

A laboratory and a numerical model have been constructed to explore the role of friction along a fault as a factor in the earthquake laboratory model demonstrates that small shocks are necessary to the loading of potential energy into the focal structure; a large pa the stored potential energy is later released in a major shock, at the end of a period of loading energy into the system. By the viscosity into the numerical model, aftershocks take place following a major shock. Both models have features which describe shocks in the main sequence, the statistics of aftershocks and the energy-magnitude scale, among others.

Figure 11.9 Leon Knopoff (1925–2011): was a geophysicist working in University of California, Los Angeles, who worked in the area of physics and statistics of earthquakes, prediction of earthquakes, along many other areas. He, along with R. Burridge proposed the spring-block model for earthquake in 1967, which later went on to be named after them. In the model (both laboratory experiment and numerical study), they considered a system of massive wooden blocks connected by springs and pulled over a rough surface. They showed that the system gives intermittent dynamics with large main shocks followed by aftershocks having power-law size distributions, analogous to what are seen in real earthquake statistics. Their model is still being actively studied as a model for earthquake. [Abstract from Burridge and Knopoff, 1967]

Rough surface

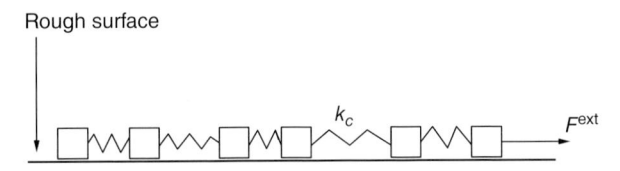

Figure 11.10 A schematic diagram of the spring-block model of earthquake. Massive blocks are connected by linear springs and driven over a rough surface.

collective motion of the blocks takes place, leading to system-wide readjustment of the position and hence the sharp decay in the stored energy. Then the system charges again for another cycle. Note that small readjustments take place during the charging, leading to smaller bursts of energies. In fact, throughout these quasi-random cycles of the charging and discharging, the number of events in which the energy release is E or higher, indeed follows a power-law $P(E) \sim E^{-\beta}$, with $\beta \approx 1$. This is analogous to the Gutenberg–Richter law seen in earthquakes.

11.2.1
Computer Simulation of the Burridge–Knopoff Model

The upsurge in the numerical studies of the Burridge–Knopoff model started with the studies of Carlson and Langer (Carlson and Langer, 1989a,b) (for a review, see Carlson et al., 1994). In one dimension, the model consists of N identical blocks. Each block is connected with its two nearest neighbors by identical linear springs of spring constant k_c. In addition, each block is connected with a moving plate via a spring with spring constant k_p. In addition to the forces due to the linear springs, a block is acted upon by the friction between its bottom surface and the rough surface over which it is pulled. The equation of motion for the ith block reads

$$m\ddot{U}_i = k_p(vt - U_i) + k_c(U_{i+1} + U_{i-1} - 2U_i) - \Phi_i, \tag{11.15}$$

where U_i is the displacement of the ith block, v is the velocity with which the upper plate is moved, and Φ_i is the nonlinear friction term. This equation can be cast into a dimensionless form as follows: Let time be rescaled by the characteristic time period of the springs $\tilde{T} = \sqrt{m/k_p}$ and displacement be measured in units of a characteristic length $L^* = \Phi_0/k_p$, where Φ_0 is a reference value in the friction force. The dimensionless stiffness parameter is $l = \sqrt{k_c/k_p}$. Now, the dimensionless equation becomes

$$\ddot{u}_i = Vt - u_i + l^2(u_{i+1} + u_{i-1} - 2u_i) - \phi_i, \tag{11.16}$$

where $u_i = U_i/L^*$, $V = v\tilde{T}/L^*$, and $\phi_i = \Phi_i/\Phi_0$. The friction law is the only source of nonlinearity in this model. In general, a velocity weakening friction law is chosen. This form of the friction force assumes that the friction is a unique function of the instantaneous velocity $\phi_i = \phi_i(\dot{u}_i)$. The functional form, of course, is to be velocity weakening. The form that is widely considered is as follows

$$\phi(\dot{u}_i) = (-\infty, 1] \quad \text{for} \quad \dot{u}_i \leq 0$$
$$= \frac{1-\delta}{1+2\alpha\dot{u}_i/(1-\delta)} \quad \text{for} \quad \dot{u}_i > 0 \tag{11.17}$$

This Burridge–Knopoff model is widely studied numerically in both one and two dimensions and with short- and long-range interactions. In general, it is seen that although the smaller events follow a GR law behavior, the larger events often significantly differ from a power-law (see Kawamura et al., 2012 for a detailed discussion).

In simulations, the measured quantity is the magnitude of an event, defined as

$$M = \ln\left(\sum_i \Delta u_i\right), \tag{11.18}$$

where Δu_i is the displacement of the blocks moving in an event (and the sum over it gives the seismic moment). The magnitude distribution of an event essentially depends upon the velocity weakening parameter α of the friction law. When $\alpha = 1$, the magnitude distribution is a power-law over a long range and the behavior of the model can be called "critical." However, as the value of α is increased over 1, qualitatively different features emerge. Now, the larger events occur and those events show a characteristic peak, which may be called a "supercritical" behavior (see Figure 11.11). On the other hand, the distribution property changes when $\alpha < 1$ values are considered. In this case, the large events are suppressed and the situation is called a "subcritical" behavior (see Figure 11.12). Therefore, the "critical" region in the one-dimensional BK model comes for a narrow range of the velocity weakening parameter of the friction law.

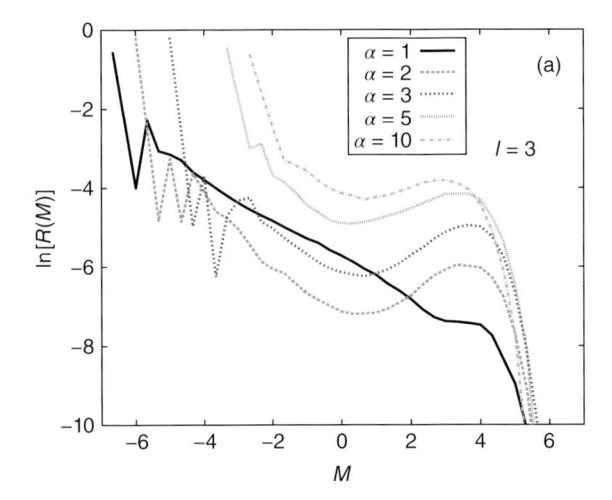

Figure 11.11 The magnitude distributions for one-dimensional BK model for different values of $\alpha \geq 1$. For large values of α, supercritical behavior is seen. From Mori and Kawamura (2006), with kind permission from John Wiley and Sons.

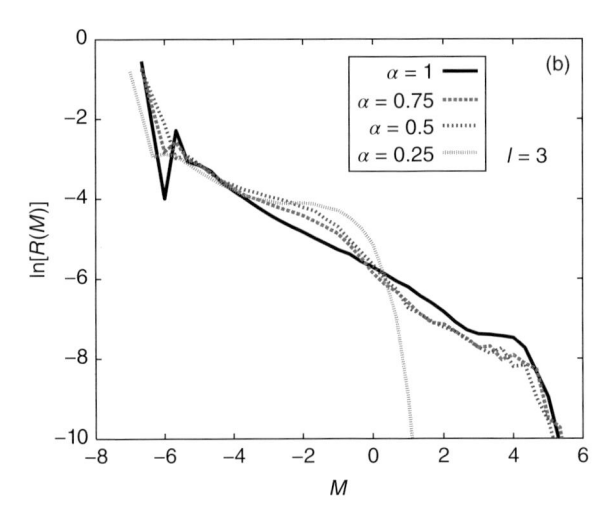

Figure 11.12 The magnitude distributions for one-dimensional BK model for different values of $\alpha \leq 1$. For small values of α, subcritical behavior is seen. From Mori and Kawamura (2006), with kind permission from John Wiley and Sons.

The real faults are two dimensional. Hence it is important to see how the features of the BK law change when studied in two dimensions. It is seen that, generally, the statistical properties of the two-dimensional BK model are similar to those of the one-dimensional version. However, in terms of phase diagram, it is richer than the one-dimensional version. As before, when the velocity weakening parameter α is small, the large events are again suppressed, showing what is called a "subcritical" behavior. However, as the value of α is increased above $\alpha_{c1} \approx 0.5$, suddenly large events occur, although the intermediate-sized events (say, between 2 and 6) are much smaller in number. This sudden occurrence of large events along with small events is a signature of first-order transition. Further increase in the α value gives rise to pronounced peak structures for larger magnitudes and near power-law for the lower magnitudes. This is the supercritical behavior. A near-critical behavior is seen for $\alpha \approx 13$ and further increase of α suppresses large size avalanches and shows subcritical behavior. Therefore, in the phase diagram, the "critical" region is rather narrow and is to be fine-tuned. Even then the behavior is not truly critical, since there is a cut-off in the power-law, which is not the finite size effect.

A more realistic version of the BK model is the one with long-range interactions. This is because, the faults are three dimensional and they extends into the orthogonal direction to the direction of propagation. Hence, the blocks should have a long-range connectivity as well, which in reality may be obtained by the interactions through bulk. There are several such attempts (Rundle et al., 1995). Xia et al. (2005, 2008) consider R connected neighbors for each block, with a spring constant scaled by $1/R$. This reduces to the mean-field limit when $R \to \infty$. In addition, models with distance-dependent interactions were studied. In particular, Mori and Kawamura (see Mori and Kawamura, 2008b) studied BK model with interaction strength decaying as $1/r^3$. The qualitative behaviors remain the same with those with short-range interactions. In Figure 11.13, the phase diagram separating the subcritical, near-critical, and supercritical states

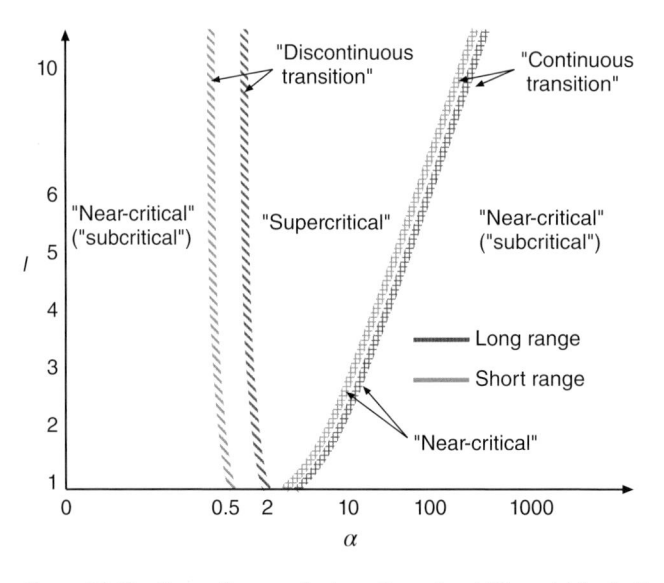

Figure 11.13 Phase diagrams for two-dimensional BK model for both short- and long-range interactions are shown for friction parameter α and elastic parameter l. From Mori and Kawamura (2008a), with kind permission from American Physical Society.

11.2.2
Train Model of Earthquake

The train model of earthquake (de Sousa Vieira, 1992) deals with the system studied experimentally by Burridge and Knopoff. It considers a system of N blocks (each of mass m) arranged in an array and each block is connected to its two neighbors by a linear spring of force constant k. The system is placed on a rough surface, and nonlinear velocity weakening friction law is considered between the surface of the block and the rough surface. This chain of blocks is now pulled at one end at a constant velocity V. The equation of motion reads

$$m\ddot{U}_j = k(U_{j+1} + U_{j-1} - 2U_j) - F(\dot{U}_j) \tag{11.19}$$

and $U_0 = Vt$. The last term is the nonlinear velocity weakening friction force as before. The equation can again be cast into a dimensionless form as follows: $\tau = \omega_p t$, $\omega_p^2 = k/m$, and $u_j = kU_j/F_0$, then the equation takes the form

$$\ddot{u}_j = u_{j+1} + u_{j-1} - 2u_j - \Phi(\dot{u}_j/v_c) \tag{11.20}$$

with $v = V/V_0$, $v_c = V_c/V_0 = (2\alpha)^{-1}$ and $V_0 = F_0/\sqrt{km}$. The only two parameters of the model are α and v. The model is completely deterministic and no randomness is needed to get the SOC-like features. The difference between this model and the earlier one is that for the ith block to slip, all blocks up to that point have to slip. On driving, the first block will slip first. That will induce further slips on the next block and gradually a system-wide "charging" will take place and then a global failure will take place.

For earthquake models, an important quantity is the moment M_0 of an event. It is defined as

$$M_0 = \sum_j \Delta u_j \tag{11.21}$$

where the sum is over the blocks that slipped in the event. With this, one can study the scaling of average moment with the number of blocks n slipping together. Numerically, a scaling relation of the form

$$\overline{M(n)} \propto n^{\beta'} \tag{11.22}$$

holds, with $\beta' \approx 2.0$.

The statistics of the event sizes are collected once the model reaches the stationary state, that is, an occurrence of at least one system-wide event. The probability of an event of moment M_0 greater than M_0' gives a power-law distribution analogous to the Guttenberg–Richter law

$$P(M_0 > M_0') = AM_0'^{-b}. \tag{11.23}$$

For a wide range of the parameter v, the exponent value remains robust $b \approx 0.6$.

11 Earthquake as Failure Dynamics

This model obeys the sum rule

$$\frac{1}{N}\int_{M_{min}}^{M_{max}} M_0 P(M_0) dM_0 = v, \qquad (11.24)$$

where M_{min} and M_{max} are the minimum and maximum sizes of the events, respectively. This is essentially a conservation that ensures the average velocity of the blocks is equal to the pulling velocity. With the form of $P(M_0)$ given in Eq. (11.23), for the integration to have convergence in the limit $M_{min} \to 0$, one must have $b \le 1$, which is seen in the numerical results.

Another quantity studied in this model is the probability distribution of events involving n blocks. This also follows a power-law

$$D(n) \sim n^{-\theta} \qquad (11.25)$$

with $\theta \approx 2.2$ (see Figure 11.14). There is no appreciable variation of the scaling exponents β', b, and θ for large values of the friction parameter α. However, for smaller values, all exponents show monotonic decay.

Some analytical results of this model can be obtained for the displacement of the center of mass. As before, if we consider n blocks slips as a whole, then its center of mass coordinate can be written as

$$R_n = \frac{1}{n}\sum_{j=1}^{n} u_j \qquad (11.26)$$

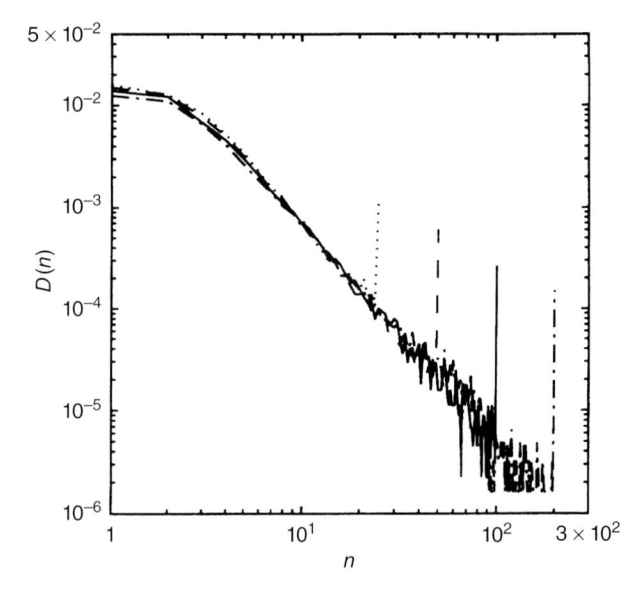

Figure 11.14 The probability distribution involving slip of n blocks is shown for different system sizes $N = 25$, 50, 100, and 200. From de Sousa Vieira (1992), with kind permission from American Physical Society.

and using the equation of motion, one gets

$$\ddot{R}_n = -\Omega_n^2 R_n + 1 - \Phi(\dot{R}_n/v_c) + v\tau, \tag{11.27}$$

where $\Omega^2 = 2/n$. Taking the linear approximation $\Phi(\dot{R}_n/v_c) \approx 1 - 2\alpha \dot{R}_n$ for the friction term, the solution becomes (with initial conditions $R(\tau = 0) = \dot{R}(\tau = 0) = 0$)

$$R_n(\tau) = \frac{v \exp(\alpha\tau)}{2i\Omega_n^4 \Lambda_n}[\Omega_-^2 \exp(i\Lambda_n\tau) - \Omega_+^2 \exp(-i\Lambda_n\tau)]$$
$$\frac{2\alpha v}{\Omega_n^4} + \frac{v\tau}{\Omega_n^2}, \tag{11.28}$$

where $\Omega_\pm = \alpha \pm i\Lambda_n$ and $\Lambda_n = (\Omega_n^2 - \alpha^2)^{1/2}$. For $\alpha \approx 0$, the solution is approximately

$$R_n(\tau) = \frac{v}{\Omega_n^2}\left[\tau - \frac{\sin(\Omega_n \tau)}{\Omega_n}\right]. \tag{11.29}$$

The duration of an event and displacement of center of mass are

$$\delta\tau = 2\pi/\Omega_n \quad \text{and} \quad \delta R_n = 2\pi v/\Omega_n^3. \tag{11.30}$$

Clearly, the maximum velocity attained by the blocks can be greater than the pulling velocity and the average velocity is $\delta R_n/\delta\tau = nv/2$, which is greater than v for $n > 2$.

11.2.3
Mapping of Train Model to Sandpile

It has been conjectured before that train model of earthquake, as discussed earlier, is in the same universality class with the Oslo rice pile SOC model (Paczuski and Boettcher, 1996). To this end, here we consider the effect of random pinning in place of velocity-dependent nonlinear friction term and its effect on avalanche statistics.

The effect of velocity-independent friction law was studied in Chianca et al. (2009), where the discrete train model was studied using stochastic friction. Each block is acted upon by three forces: the elastic forces from the two nearest neighbors and a frictional force between the block and the rough surface below. Whenever the sum of these forces is greater than zero for a block, it moves up to the point where the force again balances, and the total force is never negative in between. The friction force in this model is considered as random pinning. Particularly, the friction between two surfaces increases in the presence of an asperity. In this model, the rough surfaces are modeled by a string of bits having values 0 and 1. Each bit represents an arbitrarily small area. When the friction force at one point is strong (larger asperity) on average, then its bit is taken as 1 and otherwise 0. Therefore, when two surfaces come into contact, the contribution to friction comes from the overlapping 1 bits (see Figure 11.15).

The dynamics of the model is the same as the usual train model, except for the fact that the motion is now discrete and friction force is random. Therefore,

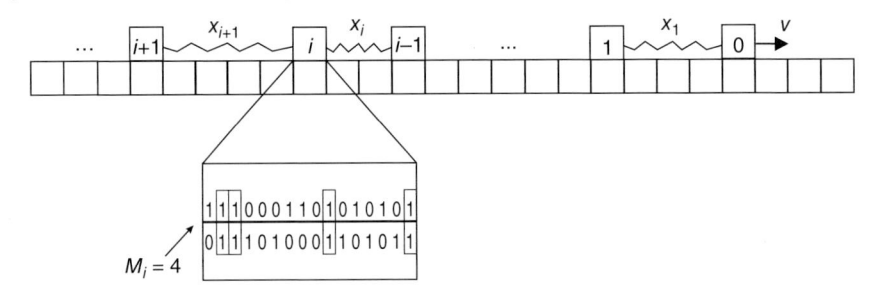

Figure 11.15 Schematic diagram for the discrete train model with stochastic friction with 32 bits for surfaces. The overlap magnitude is $F_i = 4$. From Chianca et al. (2009), with kind permission of European Physical Journal (EPJ).

one beginning with a global equilibrium when total force on each block is zero. Then, one end of the string is moved until instability sets in (at least one more block slips). When a block (say, ith) becomes unstable, it moves a distance $\Delta u = u_i - u_{i+1} \leq F_i$, where i1th block is u_i units to the right and $i+1$ th block is u_{i+1} units to the left of the ith block and F_i is the maximum friction between the block and the surface below it. All the blocks are scanned and moved from right to left until the instability stops (no further sliding). This completes one avalanche. The size of an avalanche is taken as the number of distinct blocks moving in an avalanche. An equivalent definition can be (without changing essential results) the total number of slides, where one block may be counted more than once if it moves more than once.

The random bits chosen for the surfaces are completely uncorrelated. Then the avalanche statistics can be studied after waiting for the transients to go. The cumulative avalanche size distribution $P(s)$ of avalanches of size greater than s follows a power-law with exponent value 0.55 ± 0.04. This suggests that the complex nonlinear velocity weakening friction force is not the necessary ingredient for the complex behavior of the train model. This can be obtained by a simple stochastic friction force term.

11.2.3.1 Mapping to Sandpile Model

The discrete train model with stochastic friction described here can be mapped to Oslo rice-pile model. The model in one dimension consists of L sites. Each site has a height variable h_i attached to it, which denotes the number of grains in that site. Grains are slowly added at the 0th site. The local slope of the sandpile model $z_i = h_i - h_{i+1}$ is mapped to the net elastic force on the ith block in the train model, that is, $u_i - u_{i+1}$ (see Figure 11.16). A site here becomes unstable when the local slope crosses a threshold z_i^c, which depends on the site. In the case of an instability, a grain is moved to the lower slope direction, thereby decreasing the local slope by two units. The latter site can now become unstable, and the avalanche may continue for a while. Whenever a grain topples, a new site-dependent threshold z_i^c is assigned to that site. This threshold plays the role of random friction of the

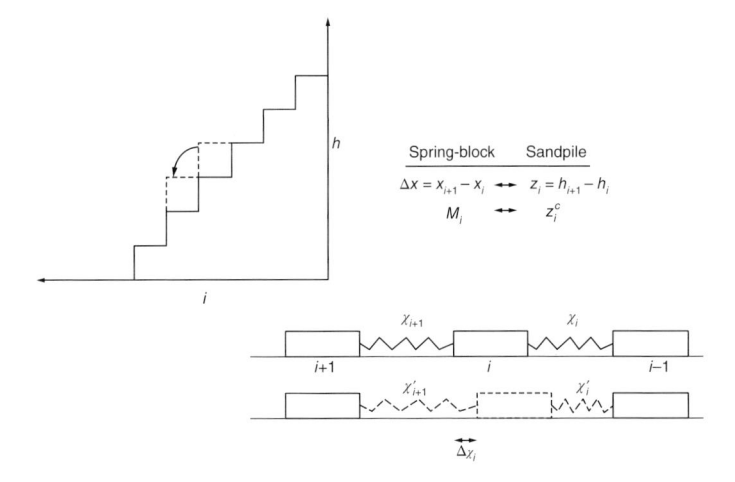

Figure 11.16 Schematic correspondence between the discrete train model and sandpile model. From Chianca et al. (2009), with kind permission of European Physical Journal (EPJ)

train model. Particularly, when an instability occurs

$$h_i \to h_i - 1, h_{i+1} \to h_{i+1} + 1$$
$$z_i \to z_i - 2, z_{i\pm 1} \to z_{i\pm 1} + 1. \tag{11.31}$$

This dynamics is continued until $z_i < z_i^c$ for all i. The whole event when toppling occurs is called an avalanche (see Figure 11.17) and no new grain is added during an ongoing avalanche. With the height of the grain columns to the displacements of the neighboring blocks in the train model, the avalanche dynamics in the two models become equivalent. The results of the train model with 32 bits for each block compare very well with those of the sandpile when z_i^c is uniformly distributed in [1,16] (with integer values). When z_i^c is taken as the integer values in [1,2], it is known as the Oslo rice pile model. The avalanche size distribution (noncumulative) is shown for the train model in Figure 11.17.

11.2.4
Two-fractal Overlap Models

So far, we have discussed models related to the self-organized stick-slip motion of spring-block systems. The intermittency and thereby the avalanche statics appeared due to the nonlinear friction term. However, here we discuss a class of models that directly relates the surface properties of the faults to the observed statistical laws of earthquakes. The fact that the fault surfaces are fractal in nature is known (see e.g., Brown and Scholz, 1985; Power et al., 1987). Therefore, some attempts were made to model the earthquake statistics through the overlap of two fractal surfaces. The highly nonuniform overlaps of the two surfaces and thereby the release of the stress developed can be related to the GR law. Particularly, random fractional Brownian profiles (the so-called Self-affine Asperity Model)

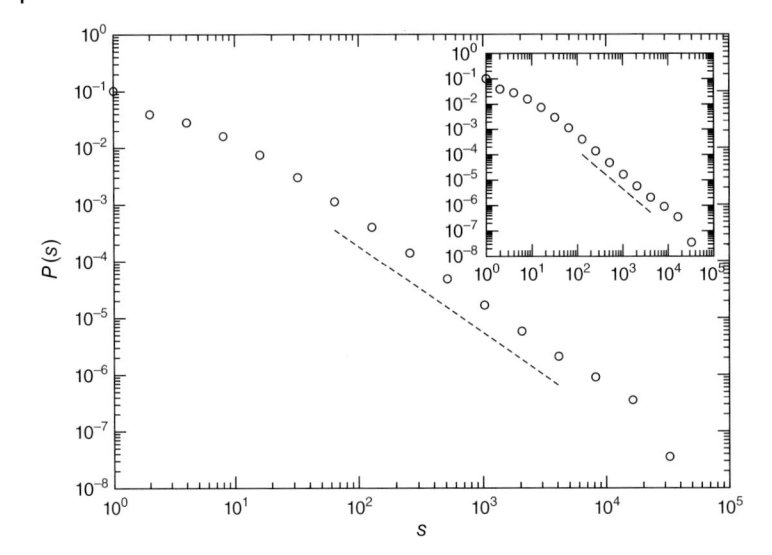

Figure 11.17 The noncumulative avalanche size distribution for the train model with 32 bits and sandpile model (inset) with z_j^c in [1,16]. Exponent value is 1.52. From Chianca et al. (2009), with kind permission of European Physical Journal (EPJ).

(De Rubeis et al., 1996) were used in modeling the rough fault surfaces. The GR exponent was related to the geometry of the fault. A subsequent, more general version also obtains Omori law (Hallgass et al., 1997), although with a much different value of the exponent than is generally found.

The simplest model of this type is the one where a Cantor set slides over its replica (Chakrabarti and Stinchcombe, 1999). This deterministic model surely is oversimplified in modeling the entire complexity of the situation, however this captures the essential part that the two surfaces are fractal and their overlap statistics are going to be manifested in the macroscopic observables. On the other hand, the simplicity of the model makes it fully analytically tractable, and the results are in reasonable agreement with real earthquake data.

11.2.4.1 Model Description

Even though earthquakes are caused due to slip of adjacent fault planes, as it is a purely surface phenomenon and the motion is generally in a given direction, the process of release of stored energy can be analyzed effectively in one dimension. Therefore, a fractal embedded in one dimension is the suitable framework for such analysis. This is also consistent with the fact that projection of any fractal surface in one dimension will be a Cantor set, although a random one. Due to analytical tractability, Chakrabarti and Stinchcombe (1999) considered a deterministic version of the Cantor set. Cantor set is a prototype example of fractal (see Appendix E). In order to construct a triadic Cantor set, in the first step the middle third of a base interval [0,1] is removed. In the successive steps, the middle thirds of the remaining intervals ([0,1/3] and [2/3,1] and so on) are removed. After

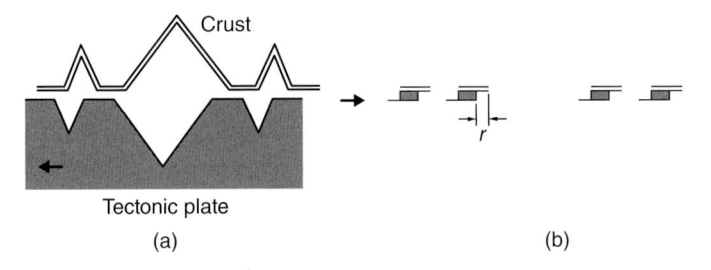

Figure 11.18 (a) Schematic representation of the rough earth surface and the tectonic plate. (b) The one-dimensional projection of the surfaces form overlapping Cantor sets.

n such steps, the remaining set is called a Cantor set of generation n. When this process is continued ad infinitum, that is, in the limit $n \to \infty$, it becomes a true fractal. Note that this model becomes similar to BK model if the spring constants are finite rather than absolutely rigid, as considered here.

In this model, the solid–solid contact surfaces of both the earth's crust and the tectonic plate are considered as average self-affine surfaces (see Figure 11.18). The strain energy grown between the two surfaces due to a stick period is taken to be proportional to the overlap between them. During a slip event, this energy is released. Considering that such slips occur at intervals proportional to the length corresponding to that area, a power-law for the frequency distribution of the energy release is obtained.

This model was initially analyzed in the continuum limit using renormalization group approach (Chakrabarti and Stinchcombe, 1999). Subsequently, it was also solved in the discrete limit (Bhattacharyya, 2005). Here, we discuss the main results and their comparison with real earthquake data, while the detailed calculations are included in Appendix F.

11.2.4.2 GR and Omori Laws

The distribution of the overlap magnitude $F(M)$ is found to be of the form

$$F(M) = \frac{3}{2\sqrt{n\pi}} \exp\left[-\frac{9}{4}\frac{(M-n/3)^2}{n}\right], \qquad (11.32)$$

where $M = \log_2 s_n$. The cumulative distribution function can be shown to be of the form

$$F_{cum}(M) = \frac{1}{3}\sqrt{\frac{n}{\pi}} \exp\left[-\frac{9}{4}\frac{(M-n/3)^2}{n}\right](M-n/3)^{-1}. \qquad (11.33)$$

On simplification for large M, one gets (see Appendix F)

$$\log F_{cum}(M) = A - \frac{3}{4}M, \qquad (11.34)$$

where A is a constant depending upon n. This is the Gutenberg–Richter law with exponent 3/4. Note that the value arises from the choice of the form of the Cantor set. For a Cantor set of dimension $\log(q-1)/\log q$, the exponent value will be $q/(q+1)$. This implies that the value of the exponent will depend on the

11 Earthquake as Failure Dynamics

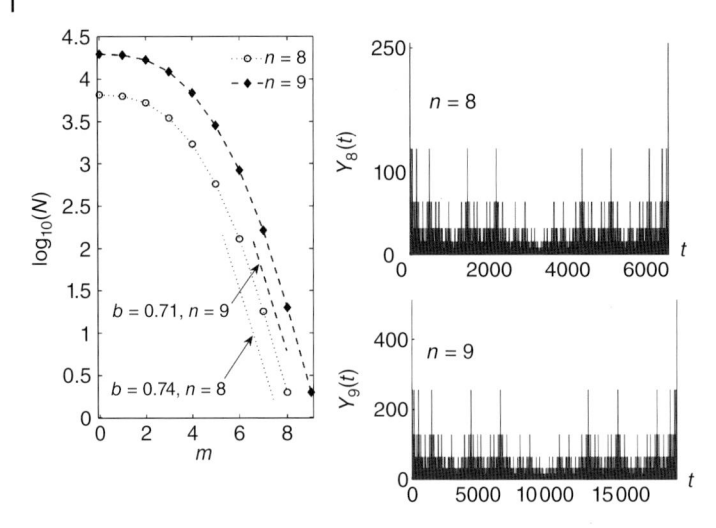

Figure 11.19 The frequency magnitude plot (right) and its cumulative distribution for the two-fractal overlap model for generations 8 and 9. From Bhattacharya et al. (2011), with kind permission from IOP (UK).

local fractal characteristics of the fault. This is indeed consistent with the fact that the exponent value is not completely universal, but varies from place to place. In Figure 11.19, the GR laws for generations 8 and 9 are shown.

The structure of Omori law arises naturally in this model. The magnitude time series is a nested structure of geometric progression. However, if one considers the minimum threshold to be 1, then the situation is trivial, since there is an event at every time step. Interesting situation arises when the magnitude cut-offs are placed at higher threshold values, for example, the second highest threshold value $n-1$. It can be shown that the cumulative number of events goes as $N(t) = \log_3(t)$, and for this model we get $p = 1$, which is indeed close to what one observes in real data. Also note that while measuring aftershocks, there is also a natural lower cut-off arising from the low recording capacity of the measuring device. Therefore, even in that sense, imposing an artificial lower cut-off in the model was justified. Figure 11.20 shows the cumulative number of aftershocks and its fit with theoretical prediction.

The variation in the macroscopic laws with local fault properties is also manifested in another measurement, which is the time integral of the aftershock magnitudes:

$$Q(t) = \int_0^t M(t')dt', \tag{11.35}$$

where t denotes the time since the main shock. The analysis reported in Bhattacharya et al. (2011) indicates a linear relation

$$Q(t) = St, \tag{11.36}$$

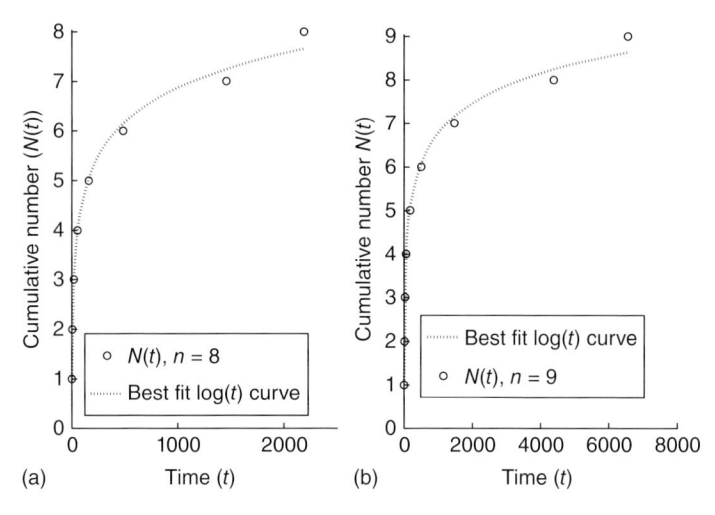

Figure 11.20 Omori law from the two-fractal overlap model for generations 8 and 9 and its best fit with logarithmic function. From Bhattacharya et al. (2011), with kind permission from IOP (UK).

where S denotes the slope. It was observed that the values of the slope are different in different fault zones. Furthermore, the linearity was not affected by shifting the time origin, and the slope also remained unchanged. The linear relation obtained from real data set is shown in Figure 11.21. On interpreting the slope, it was first assumed that the interoccurrence times and their magnitudes are independent in long time. Then, for a large sequence one has

$$Q(t) = \int_0^t M(t')dt' \approx \int_0^t M_{avg}(t')dt' \approx M_{avg} t \qquad (11.37)$$

where M_{avg} is the average magnitude calculated from the GR distribution, which is

$$S = M_{min} + \frac{0.43}{b}, \qquad (11.38)$$

where M_{min} is the lower cut-off. This expression connects the slope with GR exponent, which in turn is related to the fractal nature of the fault. In this way, the slope provides information about the different fractal nature of different faults.

This model was applied in studying the seismic activity rates in two zones in Turkey (Sezer, 2012). The rates differ strongly with the predicted change in the fractal dimension.

11.3 Cellular Automata Models of Earthquakes

As mentioned earlier, the scale-free distribution of different quantities in the case of earthquakes leads to the consideration of the fault system always being in a

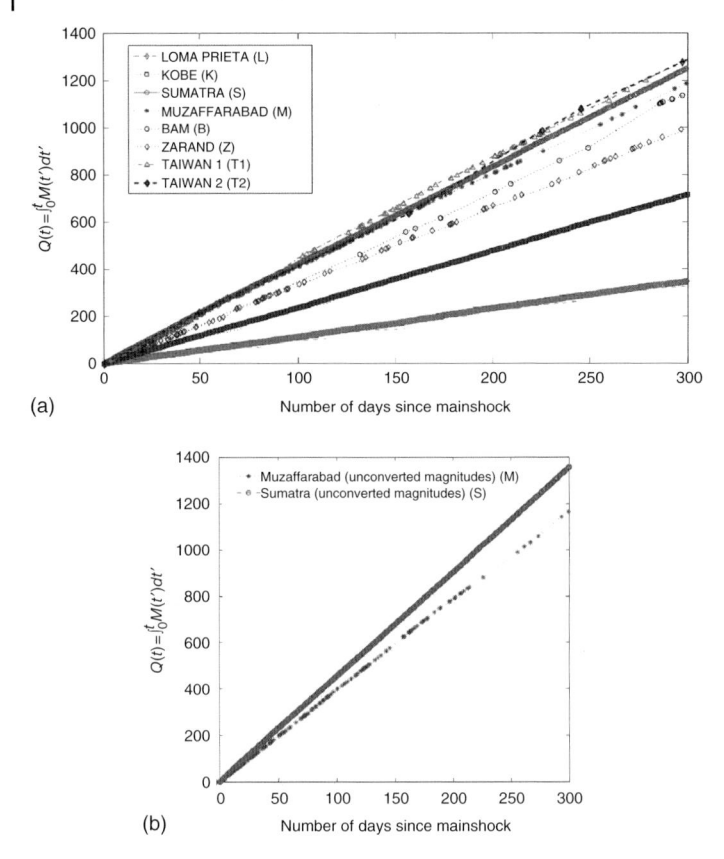

Figure 11.21 The time cumulative of aftershocks and its linear fit for data from different places as indicated. From Bhattacharya et al. (2011), with kind permission from IOP (UK).

"critical" state. Even though this point of view is sometimes contradicted due the fact that earthquake size distribution is not always scale-free, but sometimes a characteristic size and characteristic time are also seen. However, as long as one considers the 'critical' picture for earthquake, the stick-slip motion of the spring-block system captures some of the essential features of earthquake.

However, the qualitative features seen in the BK model can also be obtained from a much simpler model, which is the cellular automata model of self-organized criticality (SOC), introduced by Bak et al. (1987) (BTW model; see Figure 11.22).

11.3.1
Bak Tang Wiesenfeld (BTW) Model

Here a two-dimensional lattice is considered (say, square lattice). With each site, a variable $F_{i,j}$ is associated that mimics the stress at that point. It is assumed to increase at a constant rate with time at random locations. Once at some point

JOURNAL OF GEOPHYSICAL RESEARCH, VOL. 94, NO. B11, PAGES 15,635–15,637, NOVEMBER 10, 1989

Earthquakes as a Self-Organized Critical Phenomenon

PER BAK AND CHAO TANG

Brookhaven National Laboratory, Upton, New York

The Gutenberg-Richter power law distribution for energy released at earthquakes can be understood as a consequence of the earth crust being in a self-organized critical state. A simple cellular automaton stick-slip type model yields $D(E) \sim E^{-\tau}$ with $\tau \sim 1.0$ and $\tau \sim 1.35$ in two and three dimensions, respectively. The size of earthquakes is unpredictable since the evolution of an earthquake depends crucially on minor details of the crust.

Figure 11.22 Per Bak (1948–2002): received his PhD from the Technical University of Denmark in 1974. While working at the Brookhaven National Laboratory with Chao Tang and Kurt Wiesenfeld, he proposed the idea of self-organized criticality (SOC) (Bak *et al.*, 1987) in 1987. Essentially, the concept is that due to slow external drive and in the presence of dissipation, the critical point of the system can become an attractive fixed point. Hence, the system remains at the critical point without fine-tuning any parameter (as is the case for general criticality). Hence, the system shows scale-free response while perturbed. They proposed a simple sandpile model (named BTW model after them) where the sand grains are added to the system at a slow rate, but once the system reaches the SOC state, the amount of sand displaced can be of any size. This idea was later applied to many areas including the scale-free avalanche dynamics of earthquakes. [Abstract from Bak and Tang, 1989]

$F_{i,j} \geq F_0$ that site topples, and the stress on its four nearest neighboring sites is increased:

$$F_{i\pm1,j} = F_{i\pm1,j} + 1, F_{i,j\pm1} = F_{i,j\pm1} + 1, F_{i,j} = 0. \tag{11.39}$$

The boundaries are open, and the energy is dissipated through the boundaries. If this dynamics is continued, then the average stress at each point increases. Finally, the average stress reaches a saturation value ($F_c \approx 2.12$ for square lattice; see Figure 11.23) and the avalanches, that is, the readjustments of the stress values once a site becomes unstable, are of all sizes (limited by the system size only). The system is then said to have reached a self-organized critical state, since there is no external tuning parameter by which the system is brought into the critical state. In natural occurrences of scale-free avalanches (e.g., earthquakes, landslides, and forest fires), there is no external tuning parameter. Therefore, application of the idea of SOC in such cases is very appealing.

Since its introduction, BTW model was extensively studied both numerically and analytically. It was noted that the model is "Abelian" (i.e., the final configuration does not depend on the order in which the sites are toppled). Using this property, some static characteristics of the model (height correlations, height probabilities, etc.) could be calculated exactly (Dhar and Ramaswamy, 1989; Dhar, 1990; Majumdar and Dhar, 1991) (see also Dhar, 1999). However, no exact calculation exists for avalanche statistics of the model. There are, however, approximate schemes (Pietronero *et al.*, 1994; Ivashkevich, 1996) for avalanche size predictions, which do not match with numerical estimates.

11 Earthquake as Failure Dynamics

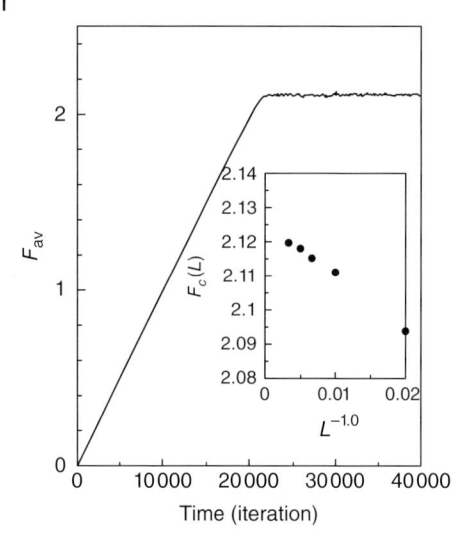

Figure 11.23 The growth and subsequent saturation of the average stress value ($L = 100$) of the BTW model. The inset shows the finite size scaling of the saturation value F_c. From Pradhan and Chakrabarti (2001), with kind permission from American Physical Society.

Several quantities show scale-free avalanches in this model once it reaches the SOC state (Lübeck and Usadel, 1997). They are number of topplings (Δ) (the avalanche size); number of distinct sites toppled (D), since one site may be toppled more than once; duration of an avalanche (τ) (the number of scans required to complete an avalanche); and the radius of gyration of the avalanche cluster (r). These quantities show scale-free statistics

$$P_x(x) \sim x^{-\beta_x} \tag{11.40}$$

with $x = \Delta, D, \tau,$ or r.

The exponent values (Lübeck and Usadel, 1997) are $\beta_\Delta = 1.293 \pm 0.009$, $\beta_\tau = 1.480 \pm 0.011$, and $\beta_r = 1.665 \pm 0.013$. The exponent β_D, however, shows (see Figure 11.24) significant system size dependence.

It can be noted (Lübeck and Usadel, 1997) that the exponent value for the avalanche duration is close to 3/2. As we see, this is a feature for other sandpile models (Manna, Zhang) as well. A possible explanation of this can be given through the analogy between avalanche propagation and random walk as described below. The number of unstable sites at each time step $n(t)$ can be thought of as a random walker. Obviously, $n(t = 0) = 1$; while the transition rates can be written as $p(n, n')$. The avalanche stops when the walker returns to origin ($n = 0$). If one assumes that the transition probabilities are homogeneous $p(n, n') = p(n - n')$, symmetric $p(\nabla n) = p(-\nabla n)$ and the sequence of ∇n is uncorrelated, the avalanche duration is simply the first return probability of a random walker to the origin. This can be calculated easily, considering that the

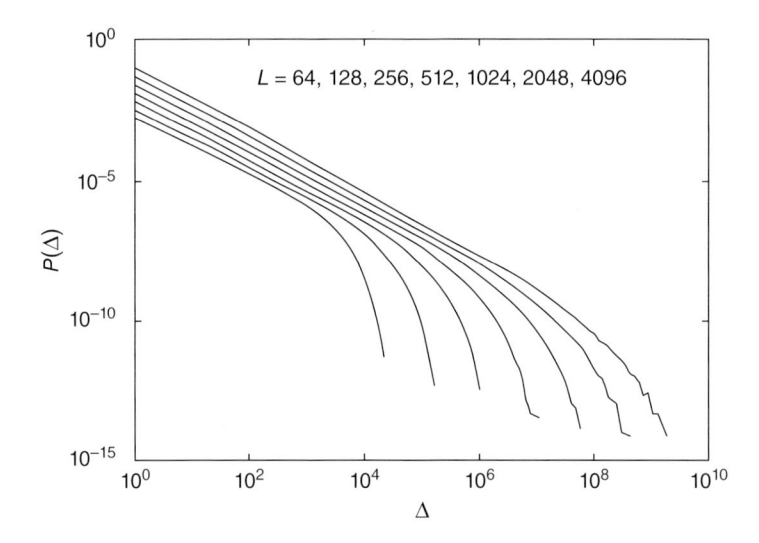

Figure 11.24 The probability distributions of avalanches for different system sizes. From Lübeck and Usadel (1997), with kind permission from American Physical Society.

probability distribution of position of a random walker is given by

$$P(x,t) = \frac{1}{\sqrt{4\pi Dt}} \exp -x^2/4Dt, \tag{11.41}$$

where D is the diffusion constant. The first passage probability to a point x_1 is given by the diffusion flux at that point

$$F(t) = D\frac{\partial P(x,t)}{\partial dx}\bigg|_{x=x_1} = \frac{x_1}{\sqrt{4\pi Dt^3}} \exp(-x_1^2/4Dt), \tag{11.42}$$

which in the long time limit $Dt \gg x_1^2$ is of the form

$$F(t) \sim t^{-3/2}. \tag{11.43}$$

Figure 11.25 shows the probability distribution $p(\nabla n)$ and the correlation function

$$C(\nabla t) = \frac{\langle \nabla n(t) \nabla n(t+\nabla t) \rangle}{\langle \nabla n^2 \rangle}. \tag{11.44}$$

The distribution function is symmetric for large system size limit, which is a requirement for the random walker to return to origin and not drift off. However, the correlation function does not become zero after $\nabla t = 0$, but shows an oscillation (see Figure 11.25). Therefore, the walk is not fully uncorrelated, but only approximately so. Therefore, β_τ is almost 3/2.

The scaling relations between the exponents can be arrived at if one assumes that area, size, duration, and radius scale in power-laws with each other. For dynamical scaling $\tau \sim r^z$, by using $P_\tau(\tau)d\tau = P_r(r)dr$, one gets

$$z = \frac{\beta_r - 1}{\beta_\tau - 1}. \tag{11.45}$$

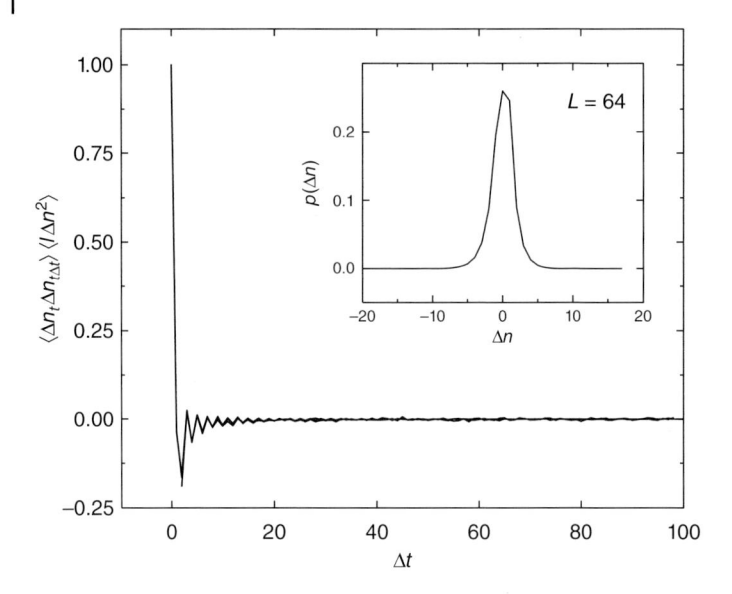

Figure 11.25 The probability distribution function $p(\nabla n)$ (inset) and the correlation function $C(\nabla t)$ for different system sizes. From Lübeck and Usadel (1997), with kind permission from American Physical Society.

However, one can show that $z = (d+2)/3$ (Diaz-Guilera, 1994). In addition, the compactness of the clusters implies $D \sim r^2$. Therefore, one can estimate $\beta_\tau = 3/2$, $\beta_r = 5/3$, and $\beta_D = 4/3$.

11.3.2
Zhang Model

The BTW model, as discussed earlier, is a discrete model in terms of stress values at each site. A continuum version was subsequently proposed in Zhang (1989). The stress values $F_{i,j}$ are now continuous variables. As before, there is a threshold stress F_0 and if at any site the stress value exceeds that limit, the stress value at that site is set to zero and the stress of that site is equally redistributed among its nearest neighbors. Note that this model is the zero dissipation limit of the OFC model discussed in Section 11.3.5. The simulation starts from an empty lattice, and a randomly chosen site is given an extra stress δF. The average stress of the system increases at first and eventually reaches a constant value (see Figure 11.26) after of the order of $L^2/\delta F$ perturbations (Lübeck, 1997).

If one measures the distribution of the stress values within the full range $[0 : F_0]$ ($F_0 = 1$, say), then it shows distinct peaks at specific locations (see Figure 11.27). The number of peaks is actually the number of nearest neighbors. The maximum value of each peak scales with system size as follows

$$p_{\max}(F) \sim L^y \qquad (11.46)$$

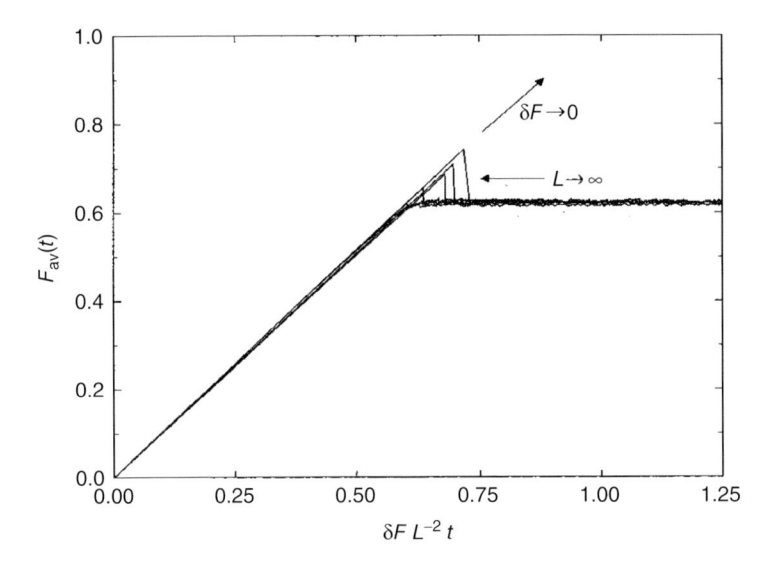

Figure 11.26 The average stress as a function of the rescaled time $\delta F L^{-2} t$ for $L \leq 512$ and different values of δF. From Lübeck (1997), with kind permission from American Physical Society.

with $y \approx 0.6$. Since the distribution of energy has to be normalized, one would also expect the peaks to get sharper with increase in system sizes. If one plots $L^{-y} p(F)$ against $L^y (F - F_{\max}(L))$, a scaling collapse is obtained (see inset of Figure 11.27).

The discrete stress values can be calculated by the following way (Lübeck, 1997): Assume that the stress values are already discrete with allowed values

$$F \in \{0, F_1, 2F_1, 3F_1, \ldots, nF_1, \ldots\}. \tag{11.47}$$

Then the maximum value of n must satisfy

$$n_{\max} F_1 \leq F_0 \leq (n_{\max} + 1) F_1. \tag{11.48}$$

Only the sites with stress $F = (n_{\max} + 1) F_1$ topple, and the stress is redistributed among the nearest neighbors. If the number of nearest neighbors is $\bar{z}$, then one must have

$$\frac{(n_{\max} + 1) F_1}{\bar{z}} = F_1. \tag{11.49}$$

This implies that the number of peaks should be equal to the number of nearest neighbors ($n_{\max} + 1 = \bar{z}$), a relation also proposed by Diaz-Guilera (1992).

Now, the separations in the peak can also be calculated (Lübeck, 1997). Starting from a stable configuration, stress is added until a site becomes critical (in the limit $\delta F \to 0$). Therefore, each lattice site gains a stress value $\Delta F = F_0 - n_{\max} F_1$. The stress of any site is now $F = nF_1 + \Delta F$. The critical site topples, adding stress $F_0 / \bar{z}$ to its neighbors. If the new energy of the neighbors is the next allowed energy, then

$$nF_1 + F_0 - n_{\max} F_1 + \frac{F_0}{\bar{z}} = (n+1) F_1. \tag{11.50}$$

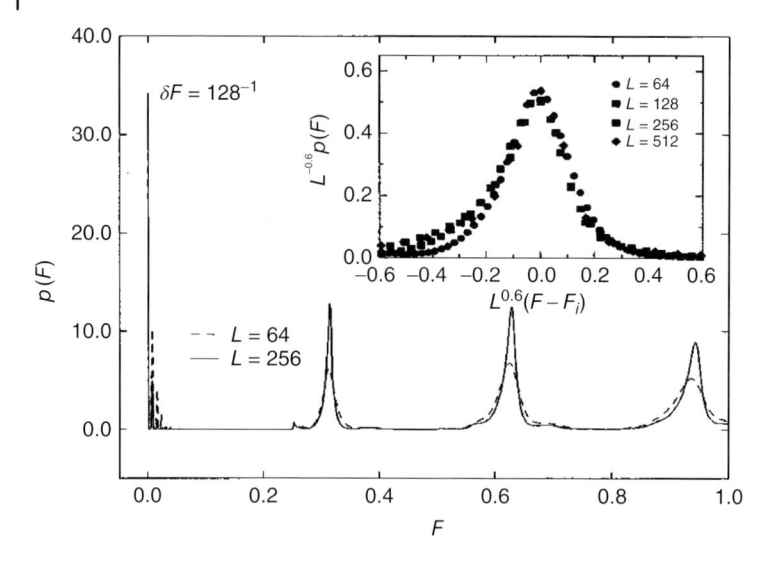

Figure 11.27 The probability distribution of the stress values in the steady state showing the peak values. Inset shows the scaling of the third peak. From Lübeck (1997), with kind permission from American Physical Society.

Simplification gives

$$F_1 = \frac{F_0}{n_{max}+1} \frac{\bar{z}+1}{\bar{z}} = F_0 \frac{\bar{z}+1}{\bar{z}^2} \tag{11.51}$$

Therefore, one can conclude that discretization depends only on the number of nearest neighbors and not on the dimension of the lattice. The statistical weights of the four peaks can be measured. These weights differ from those of BTW, which are exactly known (Priezzhev, 1994).

In addition to the static properties mentioned above, this model shows scale-free avalanche statistics. The quantities measured are the same as those for the BTW model mentioned earlier. Interestingly, the exponent values depend on the size of the input stress δF (see Figure 11.28). However, in the limit $\delta F \ll F_0$, the exponent values are independent of δF in the large system size limit.

11.3.3
Manna Model

The above-mentioned sandpile models follow deterministic dynamics. Manna (1991) introduced a stochastic version of the sandpile model. In two dimensions, the dynamics of the model can be defined as follows: As before starting from an empty lattice stress is added by unit amount in a randomly chosen site. However, as soon as the stress value reaches $F_0 = 2$, the site topples and the two units of the stress (sand grain) are redistributed between two randomly chosen nearest neighbors of the site. In this way, stochasticity is introduced in the dynamics of the model.

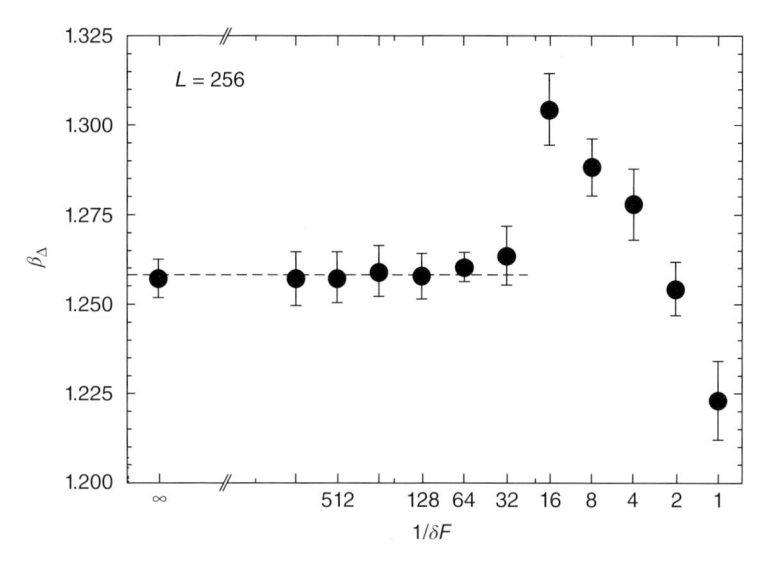

Figure 11.28 The variations in β_Δ for a fixed system size $L = 256$ with different input stress δF. Note that in the limit $\delta F \ll F_0$, the exponent value is independent of δF. From Lübeck (1997), with kind permission from American Physical Society.

With a slow rate of adding the stress, the average value of the stress first increases, until it reaches a critical value F_c (see Figure 11.29). This critical height shows finite size scaling of the form

$$F_c(L) = F_c(\infty) + CL^{-1/\nu} \tag{11.52}$$

with $\nu \approx 1.0$ and $h_c(\infty) \approx 0.716$ (see inset of Figure 11.29). Once the critical state is reached, a further small perturbation to the model gives response in all scales, that is, the avalanche behavior. As before, the size (Δ), duration (τ), and area (D) of the avalanche can be measured. These distributions follow the finite size scaling of the form (Chessa et al., 1999)

$$P(x) \sim x^{-\beta_x} \mathcal{G}(x/x_c), \tag{11.53}$$

where $x = \Delta, D$, and τ and x_c is the cut-off to the power-law and diverges as $x_c \sim L^{y_x}$. The set of exponent values $\{\beta_x, y_x\}$ specifies the universality class. One can directly measure these exponents using the finite size scaling mentioned above, as was done in Manna (1991) and some other subsequent works. But the accuracy is hard to improve in those methods.

A moment analysis similar to the one in Janssen (1998) was done for Manna model in Chessa et al. (1999). The qth moment of x on a lattice of size L was defined as

$$\langle x^q \rangle_L = \int x^q P(x) dx. \tag{11.54}$$

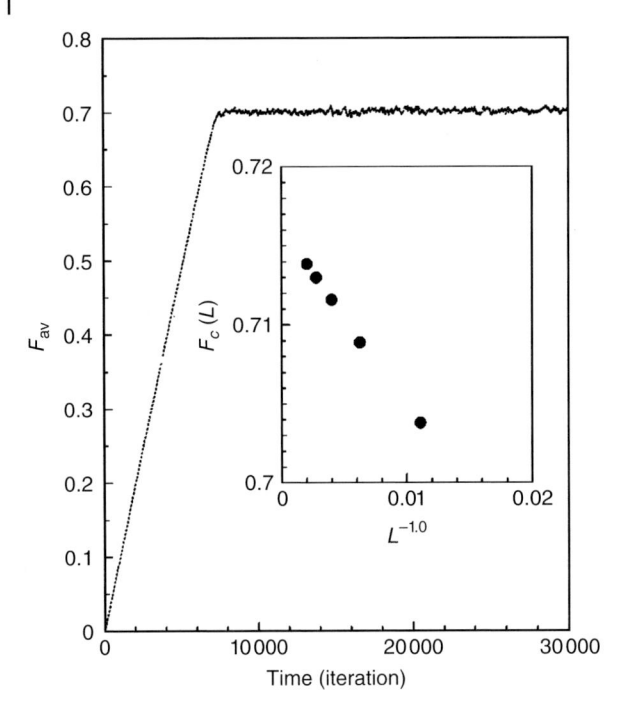

Figure 11.29 The growth and subsequent saturation of the average stress value in the Manna model. Inset shows the finite size scaling of the saturation value F_c. From Pradhan and Chakrabarti (2001), with kind permission from American Physical Society.

Now, using Eq. (11.53) and substituting the transformation $z = x/L^{y_x}$ in the asymptotic limit $x \to \infty$, one gets

$$\langle x^q \rangle_L = L^{y_x(q+1-\beta_x)} \int z^{q+\beta_x}(G)(z) dz \sim L^{y_x(q+1-\beta_x)}. \tag{11.55}$$

Therefore, the L dependence comes out as $\langle x^q \rangle_L \sim L^{\alpha_x(q)}$, where

$$\alpha_x(q) = y_x(q + 1 - \zeta_x), \tag{11.56}$$

implying $\alpha_x(q + 1) - \alpha_x(q) = y_x$. Therefore, the slope of the $\alpha_x(q)$ versus q curve gives the cut-off exponent y_x. The plots of the moments for $P_\Delta(\Delta)$, $P_\tau(\tau)$, and $P_D(D)$ are shown in Figures 11.30(a), 11.31(a), and 11.32(a), respectively. Above $q \approx 0.7$, these curves are linear, giving the slope as $y_\Delta = 2.73 \pm 0.02$, $y_\tau = 1.50 \pm 0.02$, and $y_D = 2.02 \pm 0.02$. These estimates compare well with other works (Manna, 1991). Substituting $q = 1$ in Eq. (11.56), one gets the scaling relation $(2 - \beta_x)y_x = \alpha_x(1)$. From the values of y_x obtained from the slopes and the values of $\alpha_x(1)$ computed, one can estimate the β_x values. These values are $\beta_\Delta = 1.27 \pm 0.01$, $\beta_\tau = 1.50 \pm 0.01$, and $\beta_D = 1.35 \pm 0.01$. Finally, these sets of values $\{y_x, \beta_x\}$ can be used to obtain a data collapse for different system sizes, as shown in Figures 11.30(b), 11.31(b), and 11.32(b).

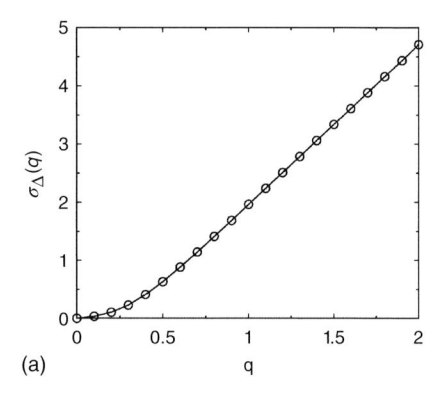

Figure 11.30 (a) The moment values are plotted for the size distribution of toppling numbers $\sigma_\Delta(q)$. The linear part gives the cut-off exponent $y_\Delta = 2.73$. (b) Using the values $\beta_\Delta = 1.27$ and $y_\Delta = 2.7$, a data collapse for different system sizes is seen. From Chessa et al. (1999), with kind permission from Elsevier.

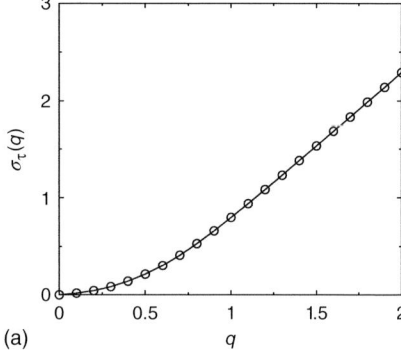

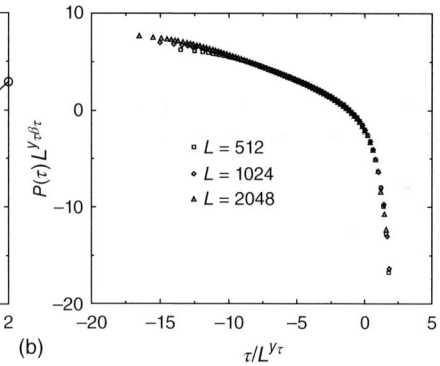

Figure 11.31 (a) The moment values are plotted for the size distribution of duration of avalanche $\sigma_\tau(q)$. The linear part gives the cut-off exponent $y_\tau = 1.50$. (b) Using the values $\beta_\tau = 1.5$ and $y_\tau = 1.5$, a data collapse for different system sizes is seen. From Chessa et al. (1999), with kind permission from Elsevier.

11.3.4
Common Failure Precursor for BTW and Manna Models and FBM

As discussed earlier, the sandpile models, when driven slowly, gradually approach the critical state starting from an empty lattice. In the time of going toward the critical state, the average stress F_{av} per site increases. The avalanches get spatially correlated and involve more number of sites and take longer time to die. These signify a gradual building of both spatial and temporal correlations in the model, which ultimately becomes system-wide when the SOC state is reached.

From the point of view of earthquakes, a large avalanche signifies a main shock. Therefore, it is important to identify the signatures of precursors of such failures. A common response in terms of precursors in the BTW and Manna models was

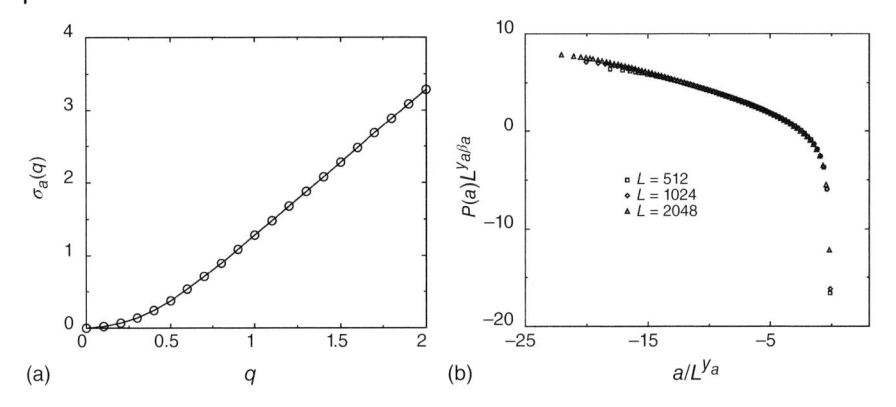

Figure 11.32 (a) The moment values are plotted for the size distribution of area of avalanche $\sigma_D(q)$. The linear part gives the cut-off exponent $y_D = 2.02$. (b) Using the values $\beta_D = 1.35$ and $y_D = 2.0$, a data collapse for different system sizes is seen. From Chessa et al. (1999), with kind permission from Elsevier.

studied by Pradhan and Chakrabarti (2001). In addition to the sandpile models, similar precursor was also noted for the random fiber bundle models. While the global load sharing fiber bundle model is not SOC (no dynamics beyond $\sigma_0 > \sigma_f$), its gradual approach to catastrophic failure and growing temporal correlation are similar to the sandpile models.

Let us first discuss the two sandpile models. Starting from the empty lattices, stress was increased at randomly chosen sites. Before reaching saturation, the average stress increases with time. Measurements were performed at some stress value F_{av} below the critical saturation value F_c. For a given value of F_{av}, a site was randomly chosen and a perturbation was applied such that the site topples. Mainly three quantities were then measured (Pradhan and Chakrabarti, 2001): the total number of topplings (Δ), total duration of the avalanche (τ), and the correlation length (ξ), which is measured as the distance to the farthest site toppled from the point of perturbation.

Subsequently we also discuss the precursors to failure in fiber bundle models, where breakdown susceptibility and relaxation time were measured. As the model is mean field-like, the notion of any length scale divergence is not present for this version of the model.

11.3.4.1 Precursor in BTW Model

The initial condition for the system is chosen such that the average stress has reached a value $F_{av} < F_c$ and all sites are stable, that is, $F_{i,j} < 4$ for all i, j. Now, a centrally located site is randomly chosen and its stress value is increased by 4, that is, $F_{i_0,j_0} \to F_{i_0,j_0} + 4$. This is bound to induce a toppling at that site and possibly other subsequent ones. As mentioned earlier, three quantities (Δ, τ, and ξ) were measured for different values of F_{av}.

As F_{av} approaches the critical value F_c, the average (over 10^5 initial conditions) size of toppling diverges as $\Delta \sim (F_c - F_{av})^{-\delta'}$, with $\delta' \approx 2.0$ (see Figure 11.33(b)).

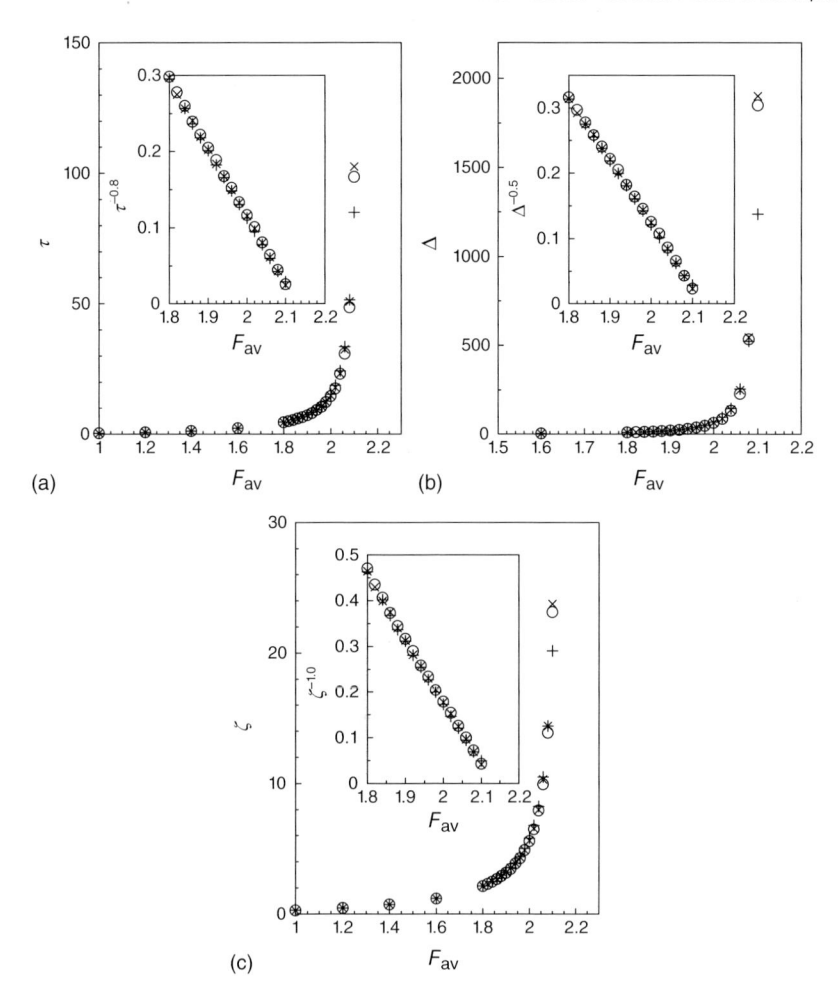

Figure 11.33 The behavior of precursors in BTW model for different system sizes: $L = 100$ (plus), 200 (cross), and 300 (open circle). No prominent system size dependence was found. (a) The divergence of the relaxation time $\tau \sim (F_c - F_{av})^{-1.2}$, inset shows $\tau^{-0.8}$ versus F_{av} which becomes a straight line and can be extrapolated to estimate $F_c \approx 2.13$. (b) Similar divergence and scaling plot (inset) for the number of toppling sites. The exponent value is -2.0. (c) Similar plot for the correlation length, the exponent value is -1.0. From Pradhan and Chakrabarti (2001), with kind permission from American Physical Society.

Similarly, the time (number of scans over the lattice) up to which this toppling continues also diverges as $\tau \sim (F_c - F_{av})^{-z}$, with $z \approx 1.2$ (see Figure 11.33(a)). Finally, the average distance between the initially perturbed site (i_0, j_0) and the farthest site toppled (i_f, j_f) in each realization gives the measure of growing correlation length ξ. This length also diverges as the critical stress is approached: $\xi \sim (F_c - F_{av})^{-\nu}$, where $\nu \approx 1.0$ (see Figure 11.33(c)).

11.3.4.2 Precursor in Manna Model

Like the BTW model, similar precursors can be measured for the Manna model. After the system reaches an average value F_{av}, and all sites are stable, that is, $F_{i,j} < 2$ for all i, j, a centrally located site is chosen and its stress value is increased by 2. That bound to induce a toppling. Then, as before, three quantities (Δ, τ, and ξ) related to toppling were measured for different values of F_{av}.

As the average stress approaches the critical value F_c, the above-mentioned quantities show divergence like that seen in BTW model. Particularly, the number of affected sites scales as $\Delta \sim (F_c - F_{av})^{-\delta'}$ with $\delta' \approx 2.0$, the duration of the toppling $\tau \sim (F_c - F_{av})^{-z}$ with $z \approx 1.2$, and finally the correlation length $\xi \sim (F_c - F_{av})^{-\nu}$ with $\nu \approx 1.0$ (see Figure 11.34). It is interesting to note that in terms of precursor responses, both BTW and Manna models show divergence in some quantities with the same exponent values.

11.3.4.3 Precursor in Fiber Bundle Model

Let us recall that for a global load sharing fiber bundle model with threshold distribution uniform in the range $[0:1]$, the fraction of unbroken fibers at time t, for an applied load per fiber value σ, satisfies (see Appendix C)

$$U_t(\sigma) = 1 - \frac{\sigma}{U_{t-1}(\sigma)}, \tag{11.57}$$

which can be translated to the differential equation

$$\frac{dU(t)}{dt} = -\frac{U(t)^2 - U(t) + \sigma}{U(t)}. \tag{11.58}$$

This has the relevant fixed point solution

$$U(\sigma) = 1/2 + (\sigma_c - \sigma)^{1/2}, \tag{11.59}$$

where $\sigma_c = 1/4$. A departure from the fixed point $U(t) = U + \epsilon$ can be shown to decay as

$$U(t) = U + \text{const} \times \exp(-t/\tau), \tag{11.60}$$

with

$$\tau = \frac{(\sigma_c - \sigma)^{-1/2}}{4}. \tag{11.61}$$

Clearly, this is the relaxation time divergence, giving $z = 1/2$ for this model. Also, from Eq. (11.59), the breakdown susceptibility $\chi = dm/d\sigma$ can be estimated to diverge as $\chi \sim (\sigma_c - \sigma)^{-1/2}$, where $m = N[1 - U(\sigma)]$ is the total number of fibers broken due to stress σ.

11.3.5
Olami–Feder–Christensen (OFC) Model

In this model, the effect of dissipation is explicitly considered to mimic the realistic situation of friction in BK model (Olami et al., 1992). Like the BTW model,

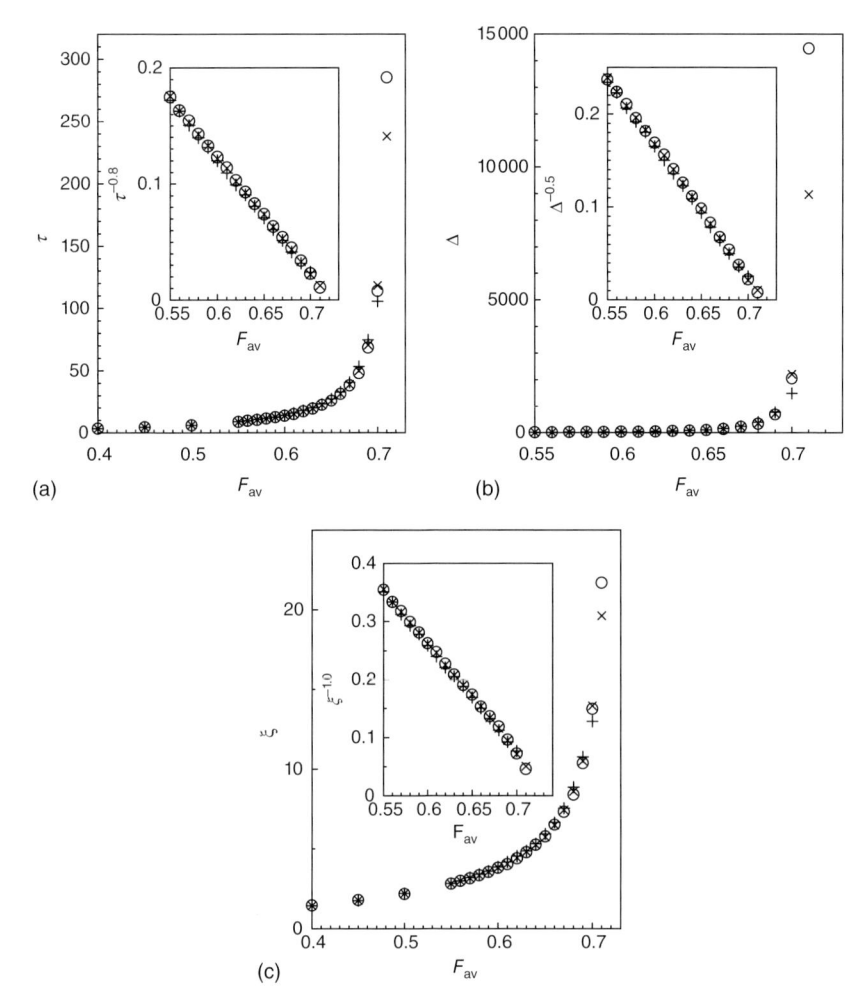

Figure 11.34 The behavior of precursors in Manna model for different system sizes: L = 100 (plus), 200 (cross), and 300 (open circle). No prominent system size dependence was found. (a) The divergence of the relaxation time $\tau \sim (F_c - F_{av})^{-1.2}$, inset shows $\tau^{-0.8}$ versus F_{av} which becomes a straight line and can be extrapolated to estimate $F_c \approx 2.13$. (b) Similar divergence and scaling plot (inset) for the number of toppling sites. The exponent value is −2.0. (c) Similar plot for the correlation length, the exponent value is −1.0. From Pradhan and Chakrabarti (2001), with kind permission from American Physical Society.

here also a stress variable F_i is assigned to every site in the lattice. The values of the stresses are chosen randomly between [0 : 1]. Then the stress values are increased slowly and uniformly until at any site it reaches the critical value $F_0 = 1$. Then, the δF_i stress is added to each of its four neighbors and F_i is set to zero. Therefore, essentially $1 - 4\delta$ fraction of the stress is dissipated. Clearly, the model is conservative in the limit $\delta = 0.25$. If the addition of this stress causes further instability, then the avalanche is continued following the same rule. No stress is added during an ongoing avalanche. The event up to the next stoppage of the dynamics constitutes one avalanche. Then, the system is driven again by adding stress.

This dynamics can be justified from the BK model as follows: The total elastic force in the BK model can be written as

$$F_{i,j} = \kappa[(2u_{i,j} - u_{i+1,j} - u_{i-1,j}) + (2u_{i,j} - u_{i,j+1} - u_{i,j-1})] + u_{i,j}. \tag{11.62}$$

When this force exceeds the local static friction, this block slips. Let us assume that in this case $F_{i,j} = 0$ after a slip by a distance $\Delta u_{i,j}$. Therefore, we should have

$$F_{i,j} \to F_{i,j} + (4\kappa + 1)\Delta u_{i,j} = 0. \tag{11.63}$$

The forces on the neighboring blocks become

$$F_{i\pm1,j} \to F_{i\pm1,j} - \kappa \Delta u_{i,j}$$
$$F_{i,j\pm1} \to F_{i,j\pm1} - \kappa \Delta u_{i,j}. \tag{11.64}$$

Solving for $\Delta u_{i,j}$, one gets

$$F_{i\pm1,j} \to F_{i\pm1,j} + \Delta F_{i,j}$$
$$F_{i,j\pm1} \to F_{i,j\pm1} + \Delta F_{i,j}. \tag{11.65}$$

with

$$\Delta F_{i\pm1,j} = \Delta F_{i,j\pm1} = \delta F_{i,j}, \quad \text{where} \quad \delta = \frac{\kappa}{4\kappa + 1} \tag{11.66}$$

Clearly, $\delta < 1/4$ and the model is dissipative. The OFC model attempts to capture this dynamics by a simple cellular automata model.

Even though it is still not clear whether the model is strictly critical in the nonconservative regime ($\delta < 0.25$), it is widely accepted that the scale-free distribution of avalanches is seen for a wide range of parameter values. The exponent value, however, depends on the dissipation (see Figure 11.35).

In addition to the scale-free behavior in the avalanche size distribution, which is analogous to the GR law, the OFC model also shows another widely seen statistical law for earthquakes, that is, the Omori law. It was shown that (Hergarten and Neugebauer, 2002) the OFC model shows Omori law (aftershocks) (Figure 11.36) and inverse-Omori law (foreshocks) (Figure 11.37). The exponent values however, depend again on the dissipation rate and are somewhat smaller than the observed values in nature.

11.3.5.1 Moving Boundary

The OFC model and for that matter in all the previous SOC models, the stress is dissipated through the open boundaries. However, presence of these boundaries brings strong inhomogeneity in the system, that is, the avalanches near the edges are in different footing than those deep within the bulk. Therefore, measurable quantities strongly depend on the distance from the boundary. This effect is manifested in both the conservative (Zhang model) and non-conservative versions of the OFC model, however in the latter case it is stronger (Grassberger, 1994).

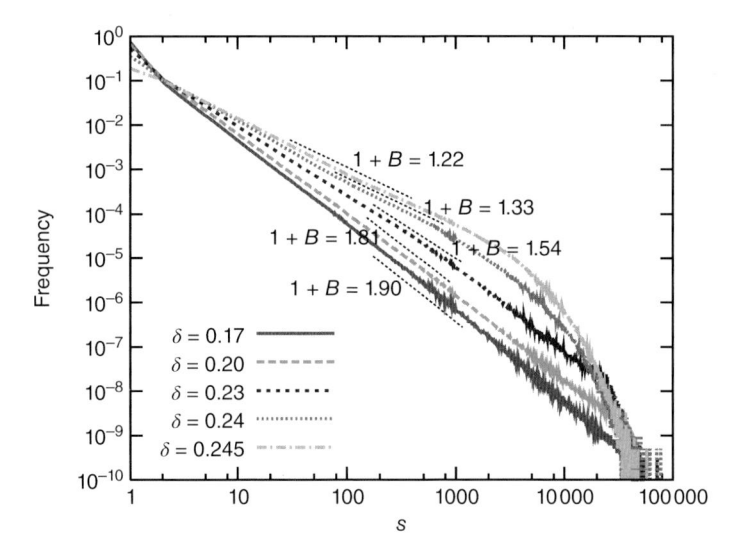

Figure 11.35 The avalanche size distributions in OFC model for different values of dissipation parameter. From Kawamura et al. (2010), with kind permission from American Physical Society.

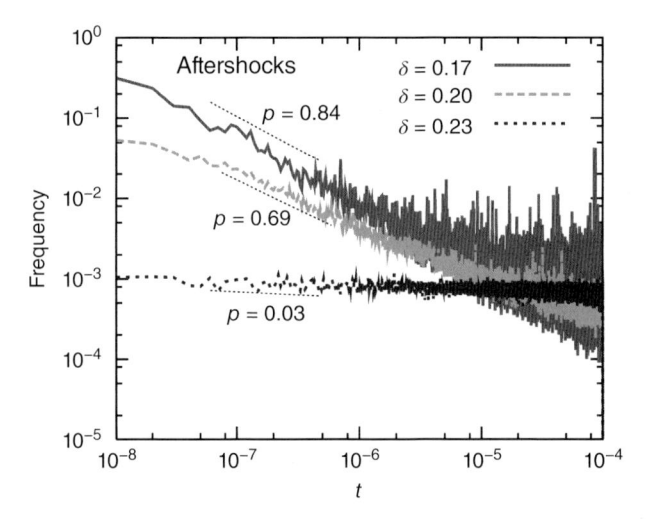

Figure 11.36 The decay of aftershocks with time for different dissipation parameters for the OFC model. Power-law behavior suggests an Omori law-like feature. From Kawamura et al. (2010), with kind permission from American Physical Society.

However, in the case of earthquakes, there are no such fixed dissipative boundaries. The seismic waves propagate in all directions and decay with distance. It is therefore argued (Manna and Bhattacharya, 2006; Bhattacharya and Manna, 2007) that in the models too, all the avalanches come in the same footing with respect to the distance from the bulk.

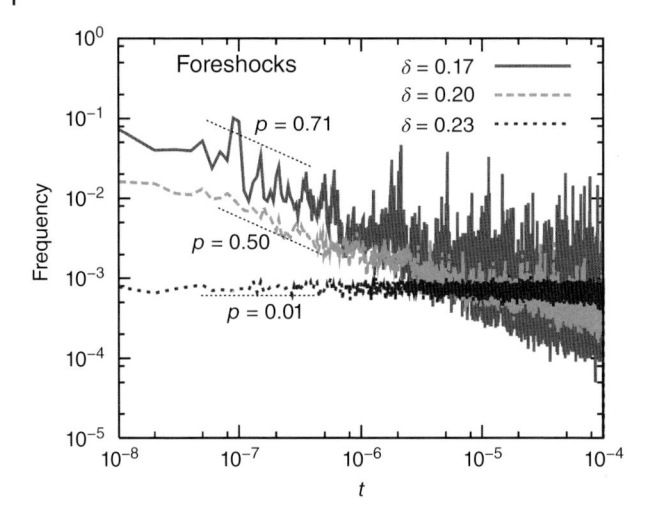

Figure 11.37 The avalanche activities with time before a main shock. The power-law behavior suggests an inverse-Omori law. From Kawamura et al. (2010), with kind permission from American Physical Society.

In Manna and Bhattacharya (2006), a method was proposed to remove this nonuniformity due to fixed boundaries. It is a modification of the OFC model. In the limit of quasi-static drive ($\delta F \to 0$), if the site (i_0, j_0) topples, then the boundary sites are chosen such that the point (i_0, j_0) is at the center of the system. Particularly, the boundary sites are defined in the torus as

$$i = i_0 + L/2 \bmod L \quad \text{and}$$
$$j = j_0 + L/2 \bmod L. \tag{11.67}$$

The boundaries are, as usual, dissipative but no longer fixed and are defined with respect to the starting point of an avalanche. During an ongoing avalanche, the boundaries are fixed. In the rare event of two sites toppling together at the beginning, one of them is randomly selected as the central site.

Since the lattice is now periodic in all directions, all sites are now equivalent. If one compares the profile of average stress per site, average number of avalanche origins per site, and average size of avalanche per site in fixed boundary OFC and moving boundary OFC, the nonuniformity is clear in the former case, while for the latter case all these are uniform with very small fluctuations (see Figure 11.38).

First for the case of zero dissipation (Zhang model), one can write the finite size scaling form of the avalanche distribution as

$$P_\Delta(\Delta, L) \sim L^{-\mu} \mathcal{H}(\Delta/L^\nu) \tag{11.68}$$

where the scaling function $\mathcal{H}(x) \sim x^{-1-\alpha'}$ for $x \to 0$ and for $x \gg 1$, $\mathcal{H}(x)$ decreases faster than a power-law so that $\alpha' = \mu/\nu - 1$. Due to the uniformity of all sites, the avalanche size distribution in this model is remarkably well behaved. The avalanche size distribution and its finite size scaling are shown in Figure 11.39. The exponent values are $\nu = 3.02$ and $\mu = 3.78$ making $\alpha' \approx 0.26$.

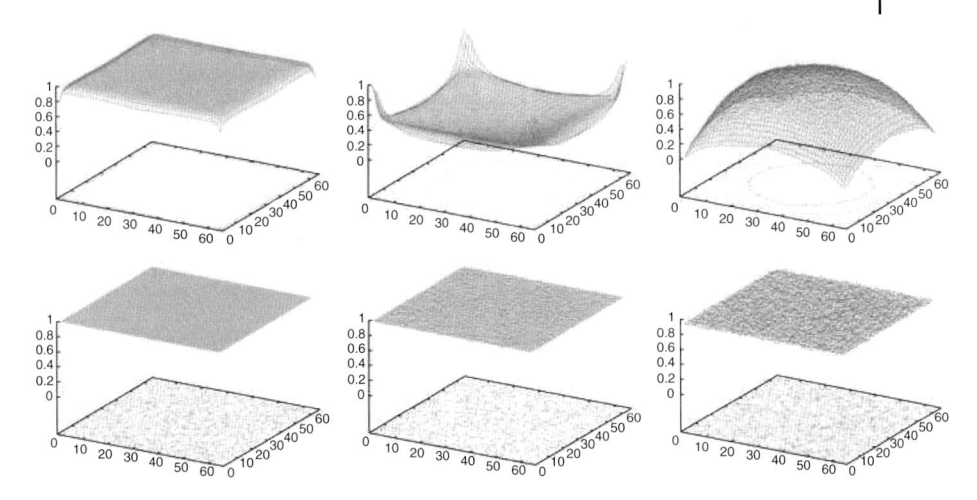

Figure 11.38 Comparison of profiles for conservative OFC model with fixed (top) and moving (bottom) boundaries. (i) Average stress per site, (ii) average number of avalanche origins per site, and (iii) average avalanche size per site. From Bhattacharya and Manna (2007), with kind permission from Elsevier.

As mentioned earlier, another important characteristic of earthquake dynamics is the recurrence time distribution. As this is a quasi-static drive, the waiting time between successive avalanches is exactly equal to δF. It also depends on the lower cut-off s_{down}. The scaling form proposed by Corral (2004, 2003) is

$$Q(t_w, R) \sim R \mathcal{G}(Rt_w) \tag{11.69}$$

where t_w is the waiting time, R is the rate of occurrence of earthquake, and and $\mathcal{G}$ is a universal scaling function having a gamma distribution form. Rescaling the average waiting time to the inverse of rate of occurrence, the scaling collapse for different system sizes and cut-offs is shown in Figure 11.40(a). If the scaled data are fitted with a form

$$\mathcal{G}(x) \sim x^{-a_1} \exp(-a_2 x^{a_3}), \tag{11.70}$$

then the exponent values come out as $a_1 = 0.003$, $a_2 = 1.02$, and $a_3 = 0.99$, which are to be compared with $a_1^c = 0.33$, $a_2^c = 0.63$, and $a_3^c = 0.98$ of earthquake data, respectively (Corral, 2004). Hence, it is seen that the initial power-law decay is absent in this model.

Also to check the scaling form of Bak et al. (2002), one can plot $Q(t_w, L, s_{\text{down}})$ ($s_{\text{down}}^{\alpha'}/L^{d_f}$) versus $t_w L^{d_f}/s_{\text{down}}^{\alpha'}$ and obtain the scaling form

$$Q(t_w, L, s_{\text{down}}) \frac{s_{\text{down}}^b}{L^{d_f}} \sim \mathcal{F}\left(\frac{t_w L^{d_f}}{s_{\text{down}}^{\alpha'}}\right). \tag{11.71}$$

The data collapse is shown in Figure 11.40(b). The exponent values are $d_f = 1.67$ and $\alpha' = 0.29$. The best fit to the form Eq. (11.70) gives $a_1 = 0.001$, $a_2 = 3.21$, and $a_3 = 0.99$. The initial power-law decay is still absent.

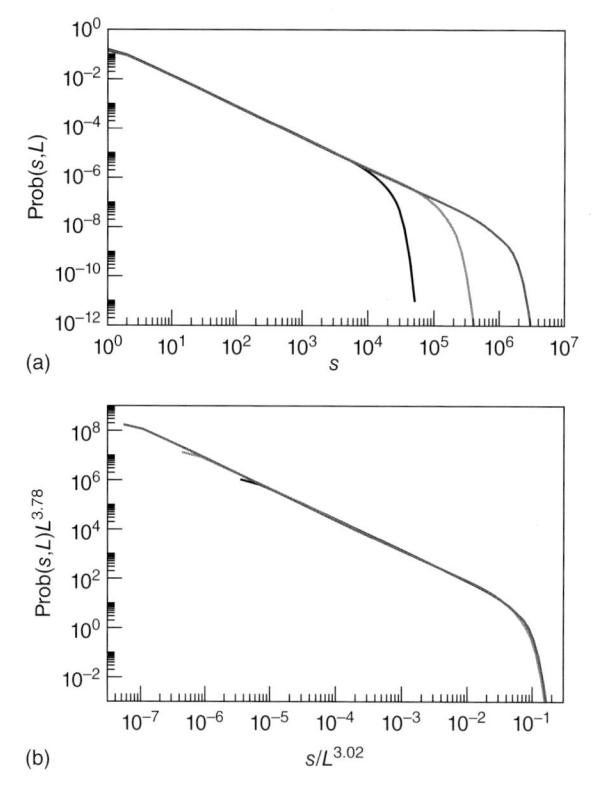

Figure 11.39 (a) The avalanche size distribution for different system sizes. (b) The finite size scaling of the form Eq. (11.68). The exponent values are $v = 3.02$ and $\mu = 3.78$. From Manna and Bhattacharya (2006), with kind permission of European Physical Journal (EPJ).

This model, however, becomes periodic after some relaxation time for the dissipative case.

11.4
Equivalence of Interface and Train Models

As discussed earlier, one version of the spring-block model is the train model (de Sousa Vieira, 1992), where the blocks were connected by springs and the entire system is pulled over a rough surface as one block (the engine) at one extreme end is pulled. This version is much simpler (to analyze and simulate) than the Burridge–Knopoff spring-block model (Burridge and Knopoff, 1967), where the blocks are connected by springs to a moving plane, which drags the blocks over a rough surface. In these models, the surfaces of the blocks and the rough surface over which the system is pulled have frictional contact and the blocks are moved to the positions of zero force once they started moving after they overcame the static frictional force (threshold), and subsequently continued their motion against the

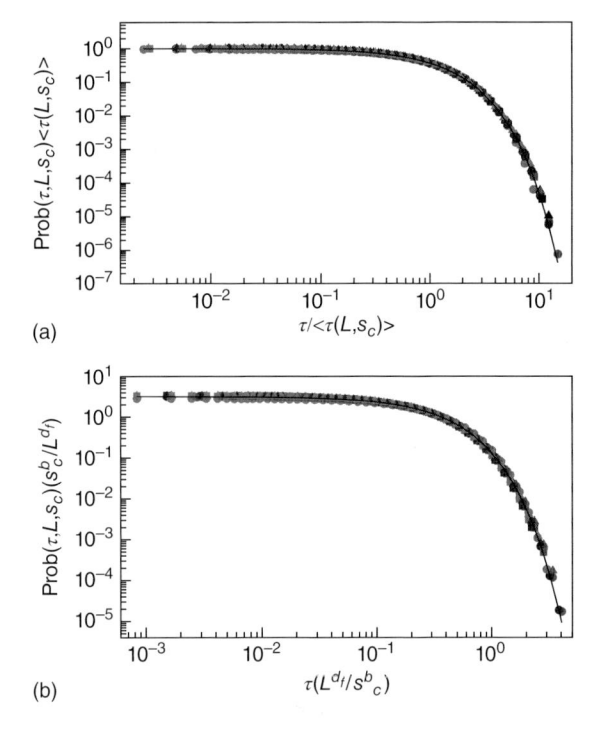

Figure 11.40 Scaling of the waiting time distribution for different system sizes and lower cut-off using (a) Corral method and (b) BCDS method. The exponent values are $d_f = 1.67$ and $\alpha' = 0.29$. From Manna and Bhattacharya (2006), with kind permission of European Physical Journal (EPJ).

strongly nonlinear velocity-dependent dynamic friction force. Movement of one block induces forces on the neighboring connected blocks, and this triggers the movements of other blocks and thus the process of stick-slip motion of blocks continues. Though the model reproduces some features of earthquakes, the presence of velocity-dependent frictional force makes analysis and simulation difficult. Further simplification was achieved in the train model by assuming the friction force to be a stochastic function of the position of the blocks instead of being velocity dependent (Chianca et al., 2009). However, introduction of friction in this way by specifying the asperities of the surfaces and blocks at different positions renders the model complicated and difficult to analyze.

The spring-block model, in essence, captured the threshold-activated processes giving rise to the intermittent dynamics. A different approach was considered (Biswas et al., 2013), where the frictional forces between the blocks and the rough surface were replaced by random pinning forces (drawn from some distribution randomly every time). Apart from substituting the friction force as random pinning, another change in this model (compared to the earlier train model) is that the block positions are now discretized, that is, the blocks are now placed at the sites of a lattice and are connected by springs. In one (two) dimension(s), two such chains (planes) of blocks go past each other as the upper chain (plane) was pulled at the boundary. When two blocks from upper and lower planes come on top of each other, a random pinning force F^{pin} comes into play. A block (ith) slides only when the total force on the block F_i^{tot} exceeds the value of F_i^{pin} and goes to the next lattice site. In this model, F_i^{pin} was a random dynamical variable chosen between 0 and 1 every time a block slides and comes on top of another block.

A different definition for avalanches was used to capture both the Gutenberg–Richter and Omori laws, which are the two prototypical laws concerning the statistical features of earthquake dynamics. Note that in the original Burridge–Knopoff model (numerical simulation version (Carlson and Langer, 1989a, 1989b; Carlson et al., 1994)) and in the train models (de Sousa Vieira, 1992), the Omori law could not be reproduced. Each of the slip event in this model is considered to have occurred at different time steps. This could be justified from the fact that in earthquakes, the aftershocks are essentially a consequence of the primary adjustments (slips) of the existing contact points after the major slip and not due to further movement (and thereby stress building) of the tectonic plates, which moves in a timescale orders of magnitude smaller than at least the initial aftershock event's timescale. Using this definition of avalanche, both the well-known statistical laws of earthquakes, namely the Gutenberg–Richter law and the Omori law can be obtained (Biswas et al., 2013).

The model was simulated and finite size scaling analysis was used to determine the exponent values associated with the avalanche statistics. It may be mentioned that the essential ingredients of the dynamics were pinning and randomness. The universality was shown between the models with random threshold and uniform lattice constant and uniform pinning and disordered lattices. It was also seen that the exponent values did not change if the lattice disorder was in the form of a deterministic Cantor set. The model was also extended to two dimensions. Finally, it was shown that this version of the train model is precisely equivalent to the interface propagation model of Edwards–Wilkinson (EW) (Edwards and Wilkinson, 1982) (see Bonamy and Bouchaud, 2011 for a detailed discussion on dynamical transitions in interface models) pulled at the boundaries.

11.4.1
Model

Here we describe the model introduced by Biswas et al. (2013). In one dimension, blocks are arranged on a discrete Lattice in the form of an array. The blocks are connected by Hookean springs (with identical spring constants). The array is being pulled from one side quasi-statically, that is, the block on the extreme right side is pulled until instability sets in, after which the pulling is paused as long as there are movements of the blocks (see Figure 11.41 for a schematic diagram). Once all the movements stop, the rightmost block is pulled again and so on. This puts the system in an intermittent motion after some transients. As in the original train model, three forces act on each block (except for the two blocks at the two extreme ends): two forces from two nearest neighbor springs and another due to friction

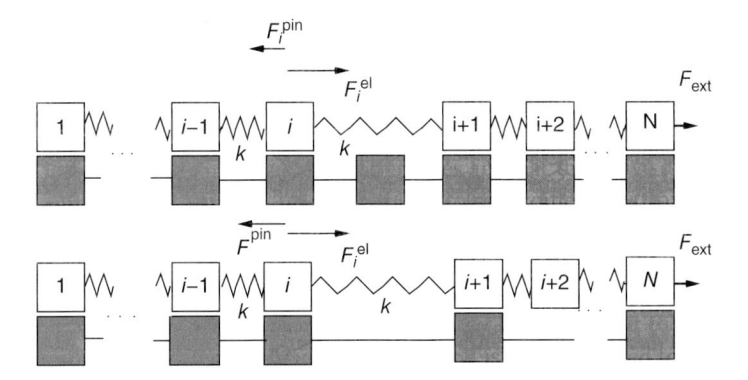

Figure 11.41 Schematic diagram for the discrete spring-block model with pinning. The upper figure depicts the case where there is no disorder in the lattice spacings of the chains (IC) and the lower figure has disorder in the lattice spacing of the chains (DC). The directions of elastic force and pinning force (working as friction) are shown for a typical block. The chains are being pulled from one end quasi-statically (see text). The pinning forces in the IC case are considered to be distributed in a range (typically [0 : 1]) but for the DC case, pinning force has a single value (say, 1); it only comes into play when two blocks come on top of each other and the randomness comes from the randomness in the positions of the blocks in both the chains. From Biswas et al. (2013), with kind permission of European Physical Journal (EPJ).

on that site. In this modified train model, a major simplification is made by replacing friction force by a random pinning force (this is in contrast to how solid–solid frictions are usually modeled, see Appendix G). The value of the friction force may depend on the properties of the two surfaces in contact, and therefore on every new contact, this value gets changed. In this model, the pinning is a random variable chosen between 0 and 1 every time a block slides and come on top of another block of the static lattice below. In a previous attempt (Chianca et al., 2009) to consider friction as random forces, the roughness properties of the surfaces were explicitly considered in obtaining the friction force values. The roughness properties of the surfaces were not explicitly considered in this model and reproduced by random pinning forces only.

The dynamics of the system was carried out as follows: If $r_i(t)$ denotes the position of the ith block of the upper chain at time t (clearly, r_is are now integers, unlike the continuous variable in Eq. (11.20)), then the net force on that block at that time is given by

$$F_i^{tot}(t) = F_i^{el}(t) + F_i^{pin}, \tag{11.72}$$

where F_i^{el} is the net elastic force due to the two neighboring blocks and F_i^{pin} is the pinning force, which acts against the direction of motion (remembering that for the disordered chain (DC), this pinning can be zero if there is no block in the bottom chain at a given point). The expression for the net force is given by

$$F_i^{tot}(t) = k(r_{i+1}(t) + r_{i-1} - 2r_i(t)) + F_i^{pin}. \tag{11.73}$$

Compare this with Eq. (11.20), although this is not exactly an equation of motion and also the position variables are always integers. Now, if $F_i^{tot} > 0$, then the position of the ith block is moved by unit distance provided there is no block at that position. The extreme left block does not have a neighbor on one side, and therefore the equation is to be adjusted accordingly and that on the extreme right is the "engine." Mainly two types of chains were considered: One was like the usual train model, where the blocks were placed an equal distance apart; let us call this the intact chain (IC). A nonzero friction (or pinning) force can only appear when two blocks come on top of each other, otherwise the friction is zero. The pinning force is a random number drawn from the uniform distribution [0:1]. Remembering that the fault surfaces are rough, the case when some of the blocks are removed from both the upper and lower chains was also considered; let us call this a disordered chain (DC). For this case, the pinning force is always unity for all the blocks in the train in contact with the block in the lower chain. As we see, even without a distribution in the friction force, the system shows intermittent dynamics, which suggests that the effect of roughness of the surface on dynamics can be incorporated into the disorders of the chains. Remembering that the rough surfaces are often self-affine, we also consider the case when the blocks in the two chains are arranged in the form of a Cantor set of the same generation; let us call this a self-affine chain. This differs from DC in the sense that the disorders here are correlated.

11.4.2
Avalanche Statistics in Modified Train Model

When the chain is pulled from one side, after some transients, the whole chain starts moving. All statistics are to be taken once the whole chain starts moving. When an instability (a slip event) sets in, pulling is stopped. Successive slips may continue due to the first slip. The number of slip events on every scan of the lattice (with parallel updating) was counted. This number gives the size s of the avalanche and the time was increased by unity after each scan of the blocks of the upper chain. Figure 11.42 shows the avalanche size distributions for one-dimensional IC model. The scaling form assumed here (for this and the subsequent versions of this model) was the following

$$P(s) \sim (sL^a)^{-\beta} f\left(\frac{sL^a}{L^{z'}}\right), \qquad (11.74)$$

where $P(s)$ is the probability of an avalanche of size s. The scaling relation tells us that at constant time t, the average avalanche size $\langle s \rangle$ decreases with L as L^{-a}. The values of the exponent obtained here are: $\beta = 0.70 \pm 0.01$, $a = 0.40 \pm 0.05$, and $z' = 1.20 \pm 0.06$. These values do not depend on the distribution of the pinning force. Note that the exponent values are in contrast to those obtained earlier (Chianca et al., 2009). However, with the earlier definition of avalanches, we get back $\beta \sim 1.5$, with no finite size scaling.

In Figure 11.43, the distribution of avalanche sizes is shown when a fraction (0.5) of blocks is randomly removed from the upper and lower chains. A similar

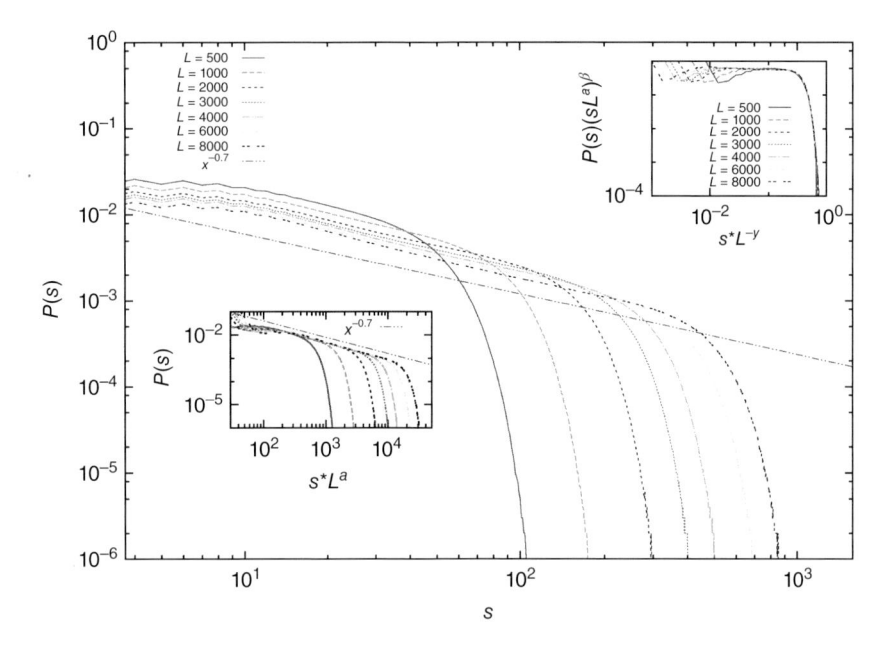

Figure 11.42 The avalanche size distributions for one-dimensional train model with regular spacings of the blocks (IC) are plotted for different system sizes. The inset shows the data collapse with the finite size scaling form assumed (Eq. 11.74). The exponent values are $\beta = 0.70 \pm 0.01$, which is the equivalent of the Gutenberg–Richter law here, and the other scaling exponents are $a = 0.40 \pm 0.05$, $y = z' - a = 0.80 \pm 0.01$. From Biswas et al. (2013), with kind permission of European Physical Journal (EPJ).

finite size scaling study was done as for IC. The exponent values come out to be very close to those obtained earlier. It was checked that the values do not depend on the fraction of the blocks removed, or even for different distribution functions for the interblock spacings (for random removal, it will be exponential; we have checked for uniformity, power-law distributions, etc.).

Next, the particular case when the disorders in the two chains are self-similar and arranged in the form of Cantor sets of some finite generation was considered. Rough surfaces were modeled with Cantor sets earlier in the context of earthquakes (Chakrabarti and Stinchcombe, 1999). That the distribution of friction values becomes nontrivial even though a single pinning value is taken in every contact is also known (Eriksen et al., 2010) (see also Lima et al., 2000 for stochastic friction coefficient). The earlier studies are in fact the limiting cases of the this one when the springs are absolutely rigid. The avalanche statistics is plotted in Figure 11.44. Note that while constructing the Cantor set, in each generation, the middle third of the chain was removed. Therefore, while performing the finite size scaling, the system size used was 2^G, where G is the generation number. A single value for the pinning force as in the DC case was considered. The exponent values obtained remain similar to the values obtained for IC and DC cases. In this sense, the avalanche statistics are universal.

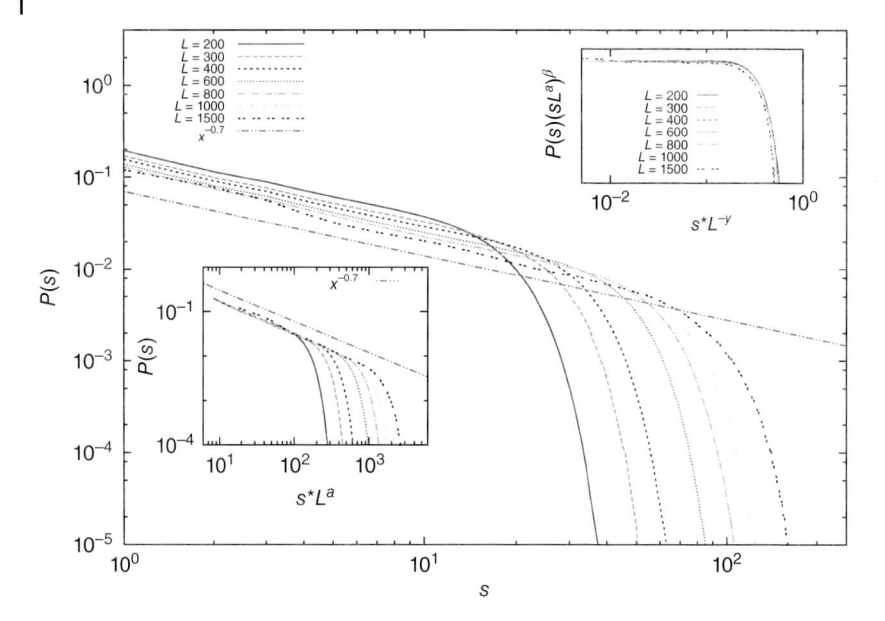

Figure 11.43 The avalanche size distributions for one-dimensional train model with a fraction (0.5) of blocks removed from both upper and lower chains (DC) are plotted for different system sizes. The inset shows the data collapse with the finite size scaling form assumed (Eq. 11.74). The exponent values are $\beta = 0.70 \pm 0.01$, which is the equivalent of the Gutenberg–Richter law here, and the other scaling exponents are $a = 0.40 \pm 0.05$, $y = z' - a = 0.80 \pm 0.01$. From Biswas et al. (2013), with kind permission of European Physical Journal (EPJ).

Next, the analogue of Omori law was studied in this context. First, an upper cut-off (s_u) of the avalanche size was defined. The avalanches above this size were called a main shock. A lower cut-off (s_l) was also defined, below which an avalanche was not measured. The probability that an avalanche of size s_l or above has occurred after time t after a main shock was measured. This was found to decay as a power-law with time

$$n(t) \sim t^{-p} \tag{11.75}$$

when t is large and the exponent value $p = 0.85 \pm 0.05$ is not sensitive to the cut-off values (see Figure 11.45 for IC, Figure 11.46 for DC models and Figure 11.47 for SAC model).

Next, the model was considered in two dimensions. The blocks were initially arranged in a regular square lattice ($L \times L$) and the system was pulled from one side (all L blocks) quasi-statically. The definition of avalanches remained the same. The case of randomly distributed pinning force (uniformly) and the regular lattice (IC) was considered. The exponent values for the avalanche statistics are very different from those for the one-dimensional case. As can be seen from Figure 11.48, the exponent values for two dimensions become $a = 0.07 \pm 0.01$, $\beta = 1.15 \pm 0.02$, and

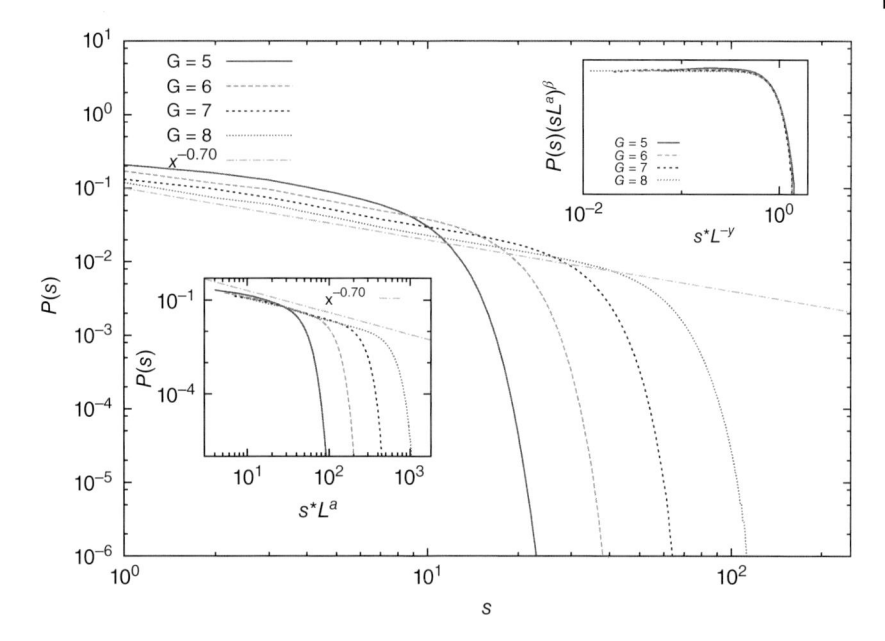

Figure 11.44 The avalanche size distributions for one-dimensional train model with the blocks arranged in the form of Cantor sets are plotted for different generation numbers. The inset shows the data collapse with the finite size scaling form assumed (Eq. 11.74). The exponent values are $\beta = 0.70 \pm 0.01$, which is the equivalent of the Gutenberg–Richter law here, and the other scaling exponents are $a = 0.40 \pm 0.05$, $y = z' - a = 0.80 \pm 0.01$. From Biswas et al. (2013), with kind permission of European Physical Journal (EPJ).

$y = 1.20 \pm 0.01$. However, the analogue of Omori law (see Figure 11.49) remains almost similar, although the power-law fit is not very good in this case.

11.4.3
Equivalence with Interface Depinning

The avalanche statistics of the discrete train model discussed here is similar to the statistics of the interface depinning problem. Particularly, this linear elastic model with threshold-activated dynamics is formally similar to the Edwards–Wilkinson (EW) equation with quenched randomness (Edwards and Wilkinson, 1982). The EW equation for interface depinning reads

$$\frac{\partial h(x,t)}{\partial t} = \nabla^2(h(x,t)) + \eta(h(x,t)) + f_{\text{ext}} \tag{11.76}$$

where $h(x,t)$ is the height of the interface (from some arbitrary reference) on position x at time t, η is the quenched noise, and f_{ext} is the applied force, which is uniform at all points. This equation shows a depinning transition in the sense that if one measures the steady-state velocity of the interface for different values of

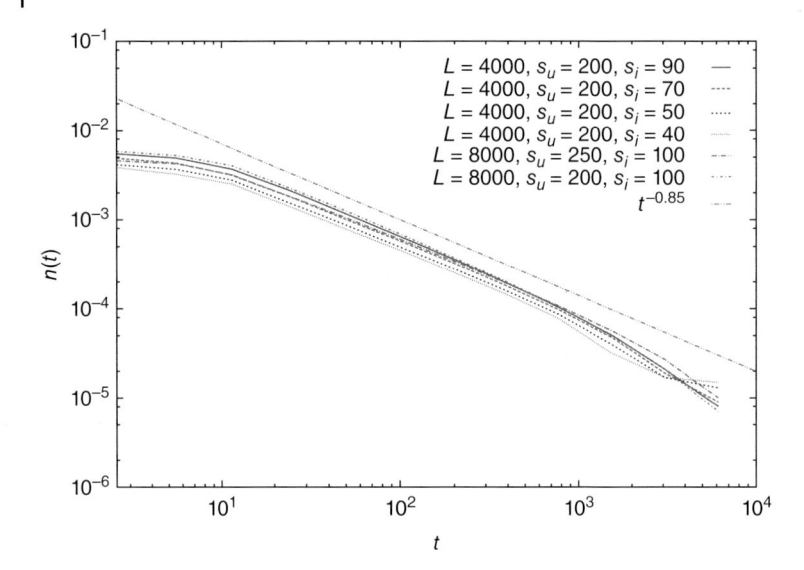

Figure 11.45 The probability that an avalanche of size s_l or above takes place after a time t of a main shock (an event of size equal to or higher than s_u) is plotted for IC. This shows a power-law dependence, and the exponent value is 0.85 ± 0.05. As can be seen, this value is insensitive to the cut-off values imposed. The power-law dependence is analogous to the Omori law. From Biswas et al. (2013), with kind permission of European Physical Journal (EPJ).

external force, one gets it in the form

$$v_{sat} \sim (f_{ext} - f_c)^\theta \quad \text{when } f_{ext} > f_c$$
$$v_{sat} = 0 \text{ otherwise.} \tag{11.77}$$

where θ is the velocity depinning exponent. Depinning transition in bead spring model was studied before (Cule and Hwa, 1996). However, that model was not discretized and the driving was applied on each bead, where Omori law-like features cannot be observed.

In addition to this steady-state exponent θ, one can also study the avalanche dynamics in this model when it is driven quasi-statically at the boundaries. The fact that the boundary-driven interface dynamics and train model can be in the same universality class was conjectured before (Paczuski and Boettcher, 1996) (see also Malthe-Sørenssen, 1999). Here, it was shown that the discrete version of the train model with random pinning forces is exclusively the boundary-driven interface model.

The avalanche statistics, with avalanches defined as before, in the interface problem is the same as in the case of train model (in one dimension). In Figure 11.50, the avalanche size distribution is shown and its finite size scaling gives the same exponents. The rate of events after a large shock with similar statistics (see Figure 11.51) was also measured. As mentioned earlier, the steady-state order parameter exponent θ in both train and EW models were measured

11.4 Equivalence of Interface and Train Models

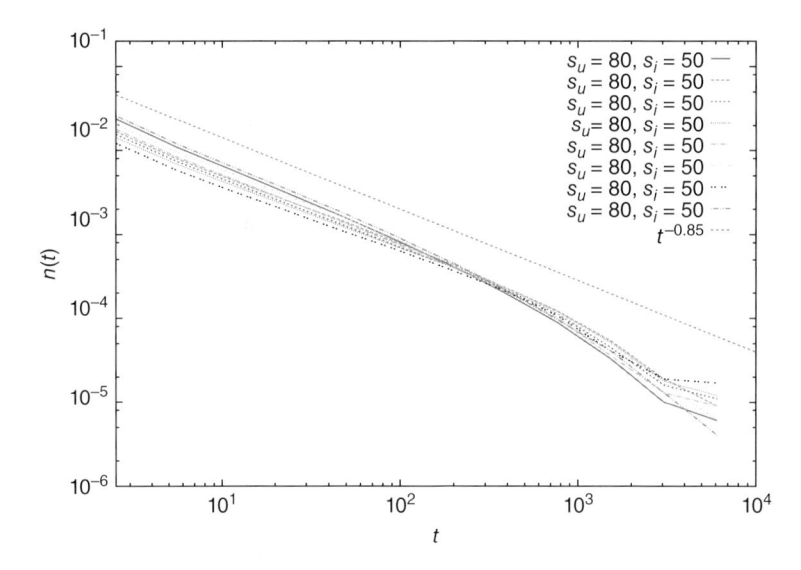

Figure 11.46 The probability that an avalanche of size s_l or above takes place after a time t of a main shock (an event of size equal to or higher than s_u) is plotted for DC. This shows a power-law dependence, and the exponent value is 0.85 ± 0.05. As can be seen, this value is slightly sensitive with the cut-off values imposed externally. This power-law dependence is analogous to the Omori law. From Biswas et al. (2013), with kind permission of European Physical Journal (EPJ).

too. The results are shown in Figure 11.52. The proximity in the values of the exponents suggests that these are manifestations of the same model.

Furthermore, two-dimensional EW model was studied to compare it with the two-dimensional train model. The two-dimensional surface was again driven from the boundaries. The avalanche statistics and rate of activities are plotted in Figures 11.53 and 11.54. Again, these are similar to the values obtained in train model. Therefore, we see that the steady-state and dynamical behaviors of the discrete train model with random pinning are the same as those of the EW model. In both the cases, the systems are boundary driven and are of linear elastic nature with random pinning.

This connection between the two modeling approaches is consistent with the fact that the empirical laws of statistical seismology have been found to exist in fracture experiments in laboratory scales as well Baró et al. (2013).

11.4.4
Interface Propagation and Fluctuation in Bulk

As mentioned earlier, the drive in two-dimensional EW model (and also in two-dimensional train model) is applied along one side. Therefore, one may ask how the disturbances propagate through the bulk. Considering the case of two-dimensional EW model (of course everything can be translated to the train model in two dimensions as well), mainly two aspects of this disturbance propagation

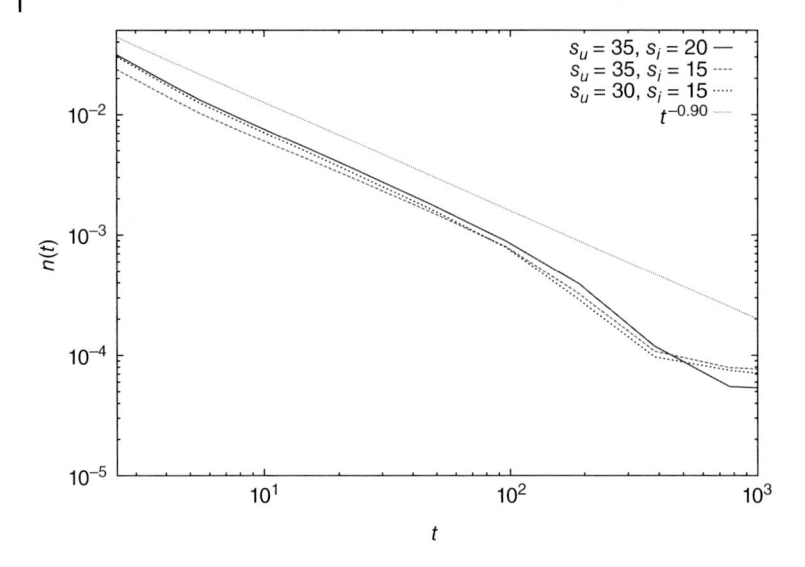

Figure 11.47 The probability that an avalanche of size s_l or above takes place after a time t of a main shock (an event of size equal to or higher than s_u) is plotted for SAC. This shows a power-law dependence and the exponent value is 0.85 ± 0.05. As can be seen, this value is slightly sensitive with the cut-off values imposed externally. This power-law dependence is analogous to the Omori law. From Biswas et al. (2013), with kind permission of European Physical Journal (EPJ).

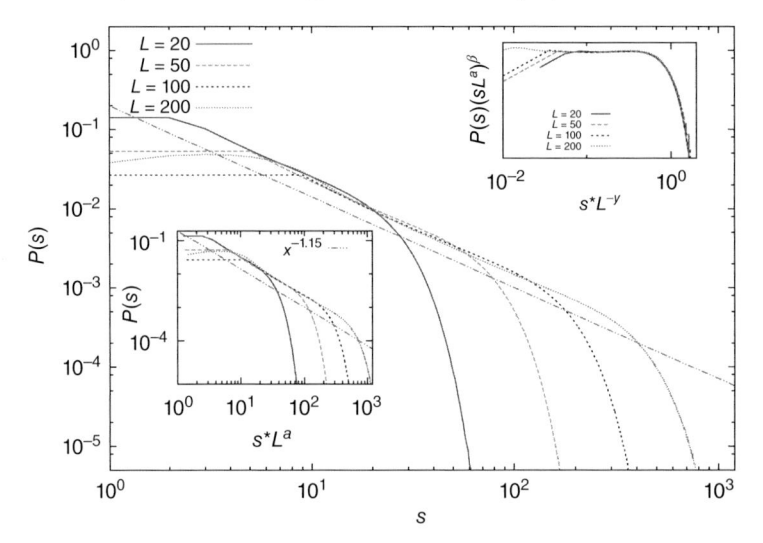

Figure 11.48 The avalanche size distributions for train model with the blocks arranged in the form of two-dimensional lattice are plotted for different system sizes. The inset shows the data collapse with the finite size scaling form assumed (Eq. 11.74). The exponent values are $\beta = 1.15 \pm 0.01$, which is the equivalent of the Gutenberg–Richter law here, and the other scaling exponents are $a = 0.07 \pm 0.01$, $y = z' - a = 1.20 \pm 0.01$. From Biswas et al. (2013), with kind permission of European Physical Journal (EPJ).

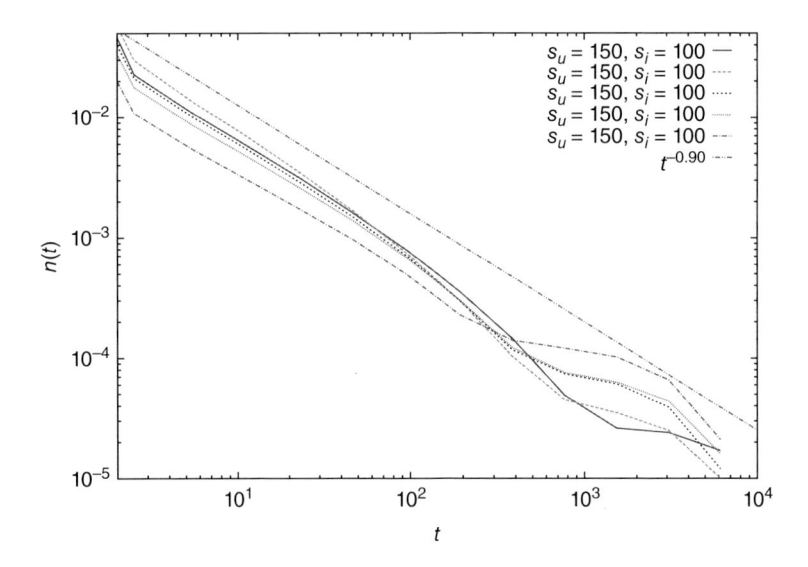

Figure 11.49 The probability that an avalanche of size s_l or above takes place after a time t of a main shock (an event of size equal to or higher than s_u) is plotted for the two-dimensional train model. This shows a power-law dependence and the exponent value is 0.90 ± 0.05. From Biswas et al. (2013), with kind permission of European Physical Journal (EPJ).

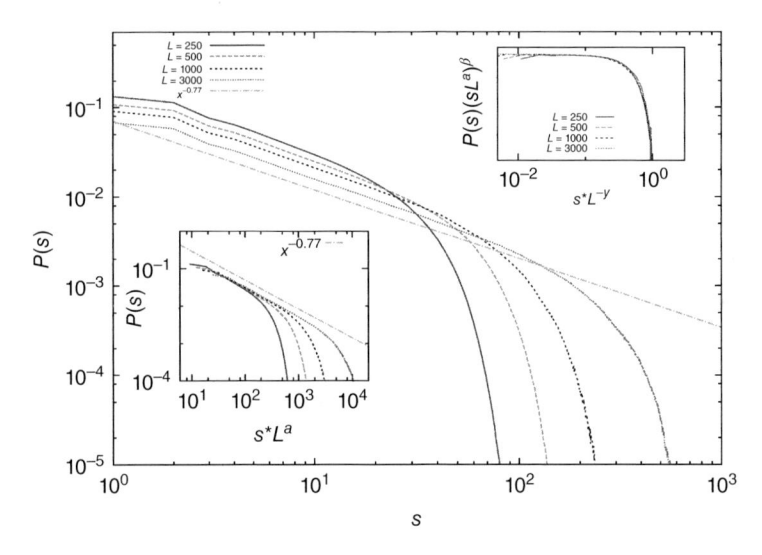

Figure 11.50 The avalanche size distributions for one-dimensional EW model are plotted for different system sizes. The inset shows the data collapse with the finite size scaling form assumed (Eq. 11.74). The exponent values are $\beta = 0.77 \pm 0.02$, which is the equivalent to the Gutenberg–Richter law here, and the other scaling exponents are $a = 0.40 \pm 0.01$, $y = z' - a = 0.82 \pm 0.01$. From Biswas et al. (2013), with kind permission of European Physical Journal (EPJ).

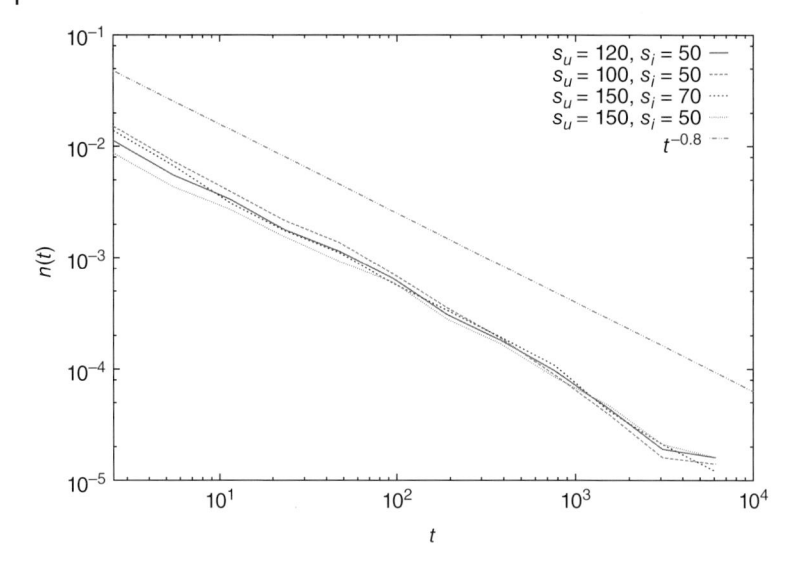

Figure 11.51 The probability that an avalanche of size s_l or above takes place after a time t of a main shock (an event of size equal to or greater than s_u) is plotted for one-dimensional EW model. This shows a power-law dependence and the exponent value is 0.80 ± 0.05. From Biswas et al. (2013), with kind permission of European Physical Journal (EPJ).

were studied: (i) Since the drive is applied only along a side, in the early time dynamics, there will be a "front" similar to the propagation of interface through quenched disordered medium. This is a rough line with an average velocity of propagation that decays in a power-law. In Figure 11.55, the front velocity and the r.m.s. fluctuation of the front are plotted against time. While the velocity decays as $v(t) \sim t^{-0.5}$, the front width does not show appreciable dependence with time, making it similar to a logarithmic growth. (ii) The fluctuation of the system along the direction perpendicular to the drive was also measured. As one goes away from the line which is driven, the widths of each line first increase, reach a maximum, and then decrease to zero in the regions which are yet to experience the drive. Since the front velocity scales as $t^{-0.5}$, the displacement (i.e., the maximum length up to which disturbance has propagated and one gets finite width) scales as $t^{0.5}$. One can therefore scale the displacement axis by $t^{0.5}$ and make the end points coincide (see Figure 11.56). One then finds a symmetric curve, that is, the fluctuation in the bulk is maximum at the halfway of the range up to which it has propagated.

There are two things to be mentioned here: First the "front" may have overhangs. To deal with it, the surface which can be formed by overestimating the overhangs was considered (SOS approximation). One can estimate the effect of overhangs in other ways, and it may differ substantially only near a dynamical transition (depinning), which we are not studying here. Second, while one can study the time dependence by quasi-statically increasing the displacement at one

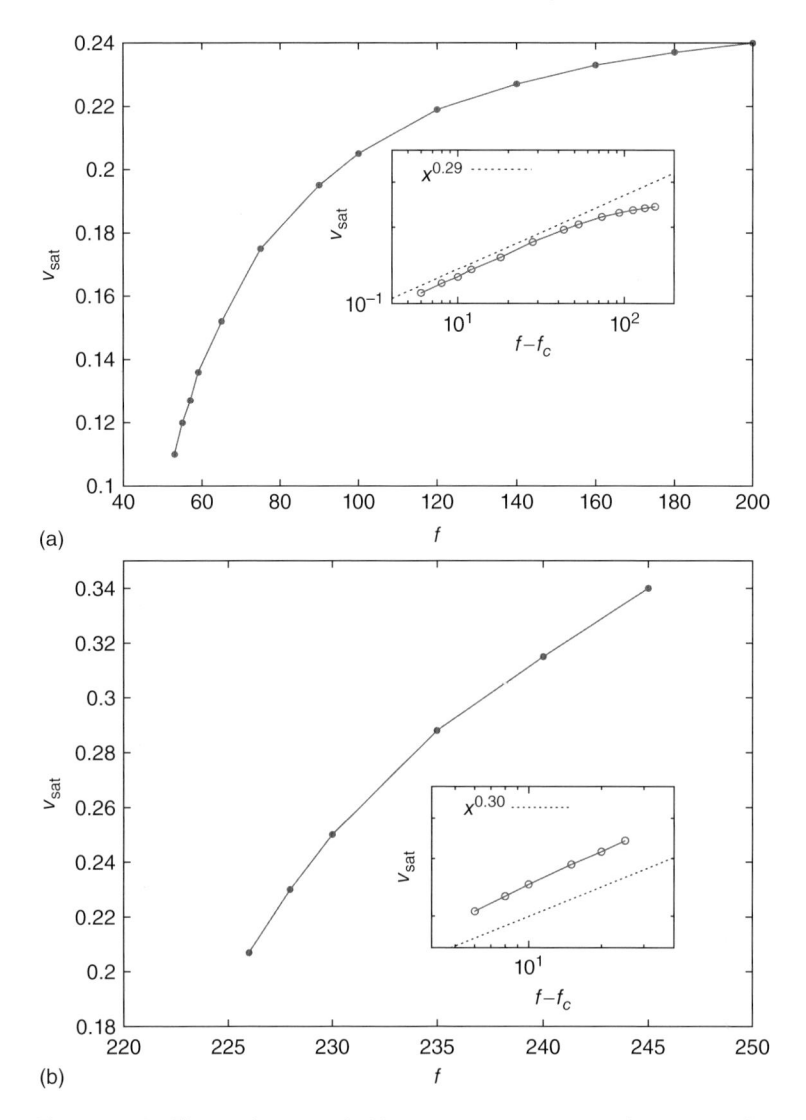

Figure 11.52 The steady-state velocities as a function of external force for train model (a) and EW model (b) in one dimension. In both cases, the velocity increases in a power-law beyond a critical point. The exponents are 0.29 ± 0.01 and 0.30 ± 0.01 for train model and boundary-driven EW model, respectively. From Biswas et al. (2013), with kind permission of European Physical Journal (EPJ).

end and waiting for the system to come to a halt (that makes one time step), one can also make the entire displacement at one go and one time step consists of one scan of the lattice. These two approaches are not the same in terms of avalanche statistics, but in fluctuation in the bulk and velocity of the front, these two turn out to be the same. Hence, the latter was followed, which is computationally cheaper.

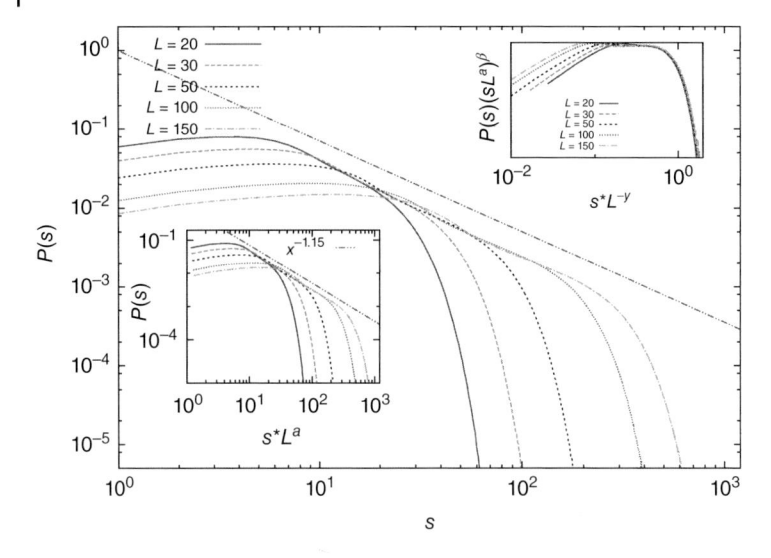

Figure 11.53 The avalanche size distributions for the two-dimensional EW model (driven at the boundaries) are plotted for different system sizes. The inset shows the data collapse with the finite size scaling form assumed (Eq. 11.74). The exponent values are $\beta = 1.15 \pm 0.01$, which is the equivalent to the Gutenberg–Richter law here, and the other scaling exponents are $a = 0.05 \pm 0.01$, $y = z' - a = 1.20 \pm 0.01$. From Biswas et al. (2013), with kind permission of European Physical Journal (EPJ).

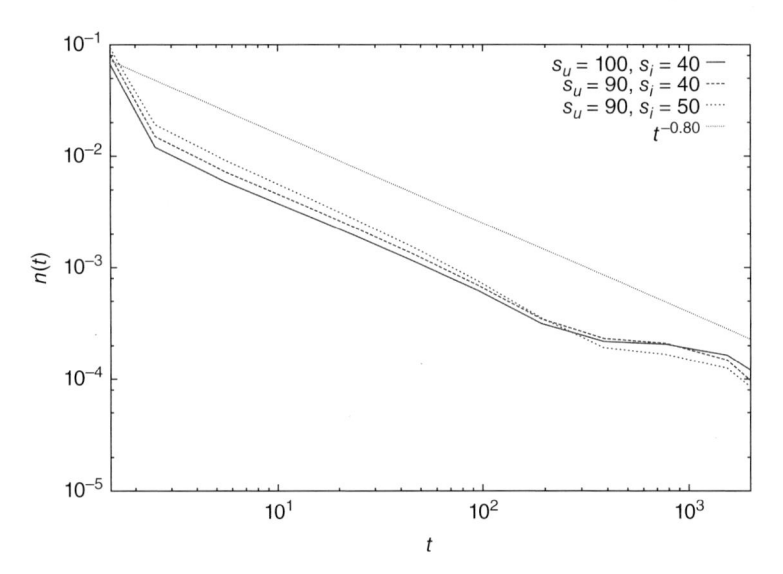

Figure 11.54 The probability that an avalanche of size s_l or above takes place after a time t of a main shock (an event of size equal to or higher than s_u) is plotted for the two-dimensional EW model. This shows a power-law dependence and the exponent value is 0.80 ± 0.02. From Biswas et al. (2013), with kind permission of European Physical Journal (EPJ).

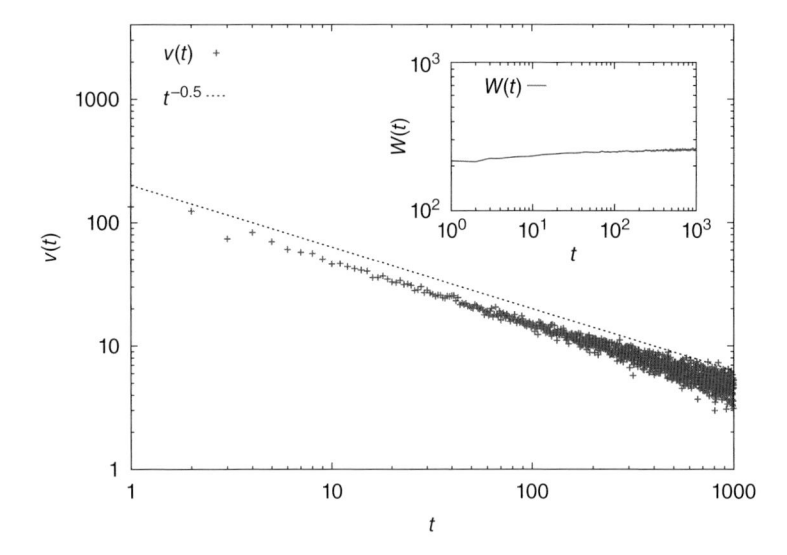

Figure 11.55 The velocity of the front $v(t)$ (up to which disturbance has propagated) is plotted against time for the two-dimensional EW model, showing a $t^{-0.5}$ dependence. The inset shows the width of this front, which is practically independent of time (i.e., logarithmic dependence). From Biswas *et al.* (2013), with kind permission of European Physical Journal (EPJ).

In short, in this section we have discussed a discretized version of the Burridge–Knopoff train model with (nonlinear friction force replaced by) random pinning studied in one and two dimensions. With suitable definitions of avalanches, a scale-free distribution of avalanche and the Omori law-type behavior for aftershocks are obtained. With this simplification, the avalanche dynamics of the model becomes precisely similar (identical exponent values) to the Edwards–Wilkinson model of interface propagation. It also allows the complimentary observation of depinning velocity growth (with exponent value identical with that for Edwards–Wilkinson model) in this train model and Omori law behavior of aftershock (depinning) avalanches in the Edwards–Wilkinson model. The observations of universalities of avalanche dynamics and interface depinning dynamics should shed new light in the existing conjectures regarding the statistical similarity in the dynamics of interface and fracture (Ramanathan and Fisher, 1998), interface and earthquake (Paczuski and Boettcher, 1996), and the recent experimental observations regarding the dynamics of fracture and earthquakes (Baró *et al.*, 2013).

11.5 Summary

In this chapter, we have summarized the statistical models of earthquakes. The empirical laws of earthquakes, the Gutenberg–Richter law, and Omori law are

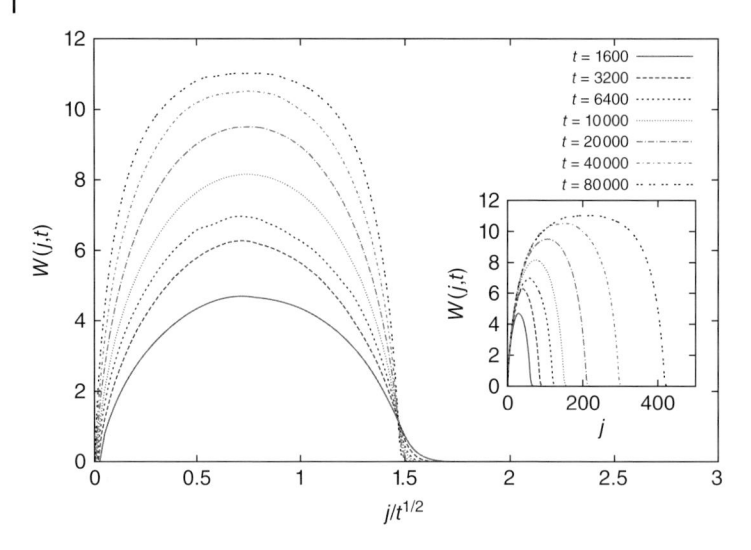

Figure 11.56 The fluctuation within the bulk of the system is measured along the direction perpendicular to the drive in the two-dimensional EW model. After scaling the displacements, one finds a symmetric profile of fluctuation within the bulk. From Biswas et al. (2013), with kind permission of European Physical Journal (EPJ).

known for a long time. Since these are statistical laws and signify the cooperative nature of the system giving scale-free responses in space and time, it has motivated the self-organized critical view of the system of faults.

We have summarized the more realistic spring-block model (Burridge–Knopoff) of earthquakes, where massive blocks of a system are pulled over a rough surface. The adjustments of the blocks give avalanches that obey Gutenberg–Richter law. The model also works in two dimensions. However, in any of the cases, the Omori law is not found.

We have also discussed the more abstract sandpile models that show SOC and are often related to earthquakes. Particularly, the OFC model depicts the effect of dissipation on the dynamics. In a given range of parameter, the model is critical, showing desired statistical features.

We have also discussed a discretized version of the train model (de Sousa Vieira, 1992) of earthquake dynamics that gives both Gutenberg–Richter and Omori laws. The nonlinear velocity weakening friction term is replaced by (random) pinning force threshold, and the movements of the blocks are discretized, in the sense that one block can move one lattice constant at a given time. This allows us to model the system as two chains; one dragged over the other by pulling the upper chain from one end. The blocks in the upper chain, which is being dragged, are connected by linear springs with identical spring constants, while the blocks in the lower chain are fixed in position (see Figure 11.41). The (friction) pinning force is nonzero only in places where two blocks come on top of one another. When both the chains are intact (IC or intact-chain model; blocks are placed at equal distances initially), whenever one block comes on top of another, a random pinning force is drawn from a uniform distribution in [0:1]. In the case of DC (disordered chain) model, both the chains are disordered, that is, some blocks are randomly removed from both the chains, and the pinning (friction) force is taken to be simply a constant (unity) whenever one block comes over another. We make the connection that this version of train model is precisely the same as the Edwards–Wilkinson (EW) model for interface propagation through quenched disordered medium. This is apparent from Eq. (11.73), when the EW front is driven by its two ends in the transverse direction (perpendicular to the initial position of the chain). Both these models have also been studied in two dimensions, where in train model the system is driven along the plane of the surface, for EW model it is driven perpendicular to the surface. We obtained equivalent results. The fact that these two models may be in the same universality class was conjectured before (see, e.g., Paczuski and Boettcher, 1996), but here we show that the dynamics is these two models can be translated to be precisely the same with all its qualitative features intact and identical complimentary features: Power-law of growth of depinning velocity in the train model and Omori law in EW model.

The dynamics of the train model follow Eqs. (11.72) and (11.73). The pulling of the upper chain is stopped whenever a slip event occurs. An avalanche is defined as the number of displacements in one scan of the (entire) upper chain (parallel updates are made). Time is increased by unity after each scan. Further pulling is resumed only when all the blocks have relaxed. This makes an avalanche distinguishable from all the previously occurred avalanches and allows for the study of aftershock statistics. The avalanche statistics changes significantly (see Section III). In both the cases (IC and DC), the avalanche size distribution exponent becomes $\beta = 0.70 \pm 0.01$ and $a = 0.40 \pm 0.01$, $z' = 1.20 \pm 0.01$ (Eq. 11.74). The universality of the avalanche statistics is clear: these exponent values remain unchained irrespective of the details of the models (intact chain, chain with random or Cantor set like disorder or values of spring constants, etc.). Next, we study the aftershock behavior (Omori law) of the avalanches. We first search for upper size (s_u) of the avalanche and choose a lower cut-off s_l. An avalanche of size greater than s_u is then taken as the main shock. We calculate the probability of having an aftershock of size s_l or above after time t of the main shock. This follows a scale-free behavior (see Eq. (11.75)) with exponent value 0.85 ± 0.05, analogous to Omori law. This Omori law was not detected earlier in any spring-block type model. Like the EW model for depinning, we measure the depinned velocity for the train model (see Figure 11.52). The depinning exponent value ($\theta = 0.29 \pm 0.01$) is the same as the EW depinning exponent value. The exponent values mentioned above change significantly in two dimensions (see Figure 11.48), though remain universal (independent of spring constant, pinning force distribution, etc.). The value of the avalanche exponent becomes $\beta = 1.15 \pm 0.01$; the other finite size scaling exponents are $a = 0.07 \pm 0.01$, $z' = 1.27 \pm 0.02$. The Omori law exponent value (0.90 ± 0.05), however, remains close to one-dimensional value. The fluctuation propagation through the bulk in the two-dimensional EW model was also studied. It is found that when the disturbance (of pulling the system by one end line) has not reached the other end, the fluctuation is maximum just halfway the total distance (in units of lattice constants) the disturbance front has propagated (see Figure 11.55). The disturbance-front has a power-law decay in velocity, with exponent value 0.50 ± 0.01 (see Figure 11.56), while the r.m.s. fluctuation grows only logarithmically.

12
Overview and Outlook

As discussed extensively in this book, the fracture and breakdown of inhomogeneous solids are, in general, extreme phenomena. This is in the sense that unlike the linear responses (e.g., elasticity or classical conductivity having self-averaging statistics) which are meaningful in the sense of average over identical samples or systems, breaking phenomena of solids have, in general, non-self-averaging statistics. These statistics are governed by the extreme fluctuations (weakest defects and their tail-end statistics) and consequently some unusual scaling behaviors set in. Of course, with more and more heterogeneity in the solid, such fluctuations and hence the breaking behavior become more and more self-averaging, and standard scaling scenario comes into play. Many of these universal (roughness exponent, avalanche statistics etc.) properties across different (microscopic to geological) length scales have revealed intriguing features in breaking dynamics of solids and have led to major developments recently.

Role of disorders is central in the failure properties of solids. Disorders are generally of two types: lattice defects (point defects) and amorphous defects. The characterizations of lattice disorders and particularly the description of substitutional or point defects within the formalism of percolation theory (in the limit of high disorder) are discussed in Chapter 3. It is these defects or flaws around which stress concentrates. Failure of any of these defects contributes directly to the failure probability of the solid. In other words, survival probability of the entire sample is the joint probability of survival of each of the individual defects (Eq. (4.15)). Hence, it is the weakest (usually the largest) defect, and not the average defect size that contributes to the failure properties, making it follow the extreme statistics rather than the self-averaging one (Section 4.2.1).

With the increase of defect concentration, however, a competition arises between the correlation length of the defects and Lifshitz length. Essentially, the competition is between the nucleation of stress near largest defect (usually making that part most vulnerable), and the fact that since the disorder is very high an even weaker point may be present further away. This fact is also reflected in the electrical breakdown analogy discussed later in Sections 10.2.4 and 10.2.5. The percolation limit of the disorder makes it possible to connect the mechanical responses of the solids to the failure strength through scaling analysis when typical defect size is given by percolation correlation length (Eq. (4.19)), giving

Statistical Physics of Fracture, Breakdown, and Earthquake: Effects of Disorder and Heterogeneity, First Edition.
Soumyajyoti Biswas, Purusattam Ray, and Bikas K. Chakrabarti.
© 2015 Wiley-VCH Verlag GmbH & Co. KGaA. Published 2015 by Wiley-VCH Verlag GmbH & Co. KGaA.

bounds for the exponent values. These bounds are also supported by experimental observations (Benguigui et al., 1987).

The deformation of stress field and thereby the propagation of fracture leave their imprint in the fractured surface left behind. This is seen as the roughness properties of the fractured surfaces (discussed in Chapter 5). The intriguing feature is its universality. It was first noticed by Mandelbrot et al. (1984). This suggests a universal mechanism of fracture nucleation and propagation near fracture front. This is a widely investigated and debated topic. Observations suggest a crossover behavior in the roughness properties with increasing length. While in the short length scale the propagation is coalescence of defects ahead of crack front, in the large length scale (with respect to fracture process zone) the behavior is more like an elastic string driven through a disordered medium (discussed in Section 8.1).

In addition to the static properties of fracture, as mentioned earlier, the dynamical properties are also discussed in Chapters 6, 7, and 8. In particular, a slow drive in the external stress yields scale-free responses of breaking of solids (the so-called crackling noise). The size and waiting times of these responses, measured generally in the form of acoustic emissions, show universal characteristics (Section 6.1). To model these properties, a discrete model called fiber bundle model (FBM) proves to be very useful. The disorders are explicitly taken care of in the random failure thresholds of each element or fibers in the model. A solvable limit of the model is the long-range load transfer limit. The avalanche statistics (Eq. (6.20)) and the critical behavior in the divergence of relaxation time (Eq. (6.12)) near catastrophic failure are clearly seen in this model (discussed in detail in Section 6.2). The dynamics also gives a signature of precursor of imminent failure (Section 6.2.2). A trivial limit of the model is when all elements have equal failure thresholds. The system fails abruptly, without showing relaxation time divergence or avalanche statistics, as soon as the external stress reaches that threshold. However, it was shown that the avalanche properties are not recovered up to a finite width of the threshold distribution, up to which the failure remains abrupt (Section 6.2.4). The critical width is a tricritical point showing its own critical exponents (different from the second- order critical point).

Even if the external stress remains below the critical point, a system might fail. As can be seen from Griffith's nucleation theory (Section 4.1.1), the fracture propagation may be viewed as a barrier crossing problem. Therefore, external noise may help in crossing the barrier even for a subcritical stress value. The dynamics of strain rate — the initial Andrade creep — followed by low activities and finally acceleration toward failure are discussed in Section 7.2. Its theoretical model in terms of viscoelastic and other forms of fiber bundle model are also discussed.

As mentioned earlier, as the fracture front is slowly driven through a disordered medium, it shows intermittent dynamics. It is not clear as to which approach can provide a theoretical understanding of all the observed facts. However, in the large length scale limit, a driven elastic line makes satisfactory predictions (see Section 8.1). In particular, the range of the elastic term determines the universality class of the depinning transition associated with its dynamics. For fracture front, a long-range ($1/r^2$ type) interaction is appropriate.

In addition to the elastic line model, interface propagation is also modeled using the fiber bundles (Section 8.2). In Section 8.2.1, it was shown that by tuning the elastic properties of the bottom plate of the model, one can replicate the small and large length scale behaviors of the roughness properties of the propagating interface. A more simplified version in terms of an array of fibers with long-range load transfer and a model of self-organized interface propagation without any external dissipation scale were also discussed, respectively in Sections 8.2.2 and 8.2.3.

The presence of dislocation affects the stress–strain properties of a solid. The movements of dislocation and its conditions are discussed in Chapter 9. The necking properties in ductile fracture are also discussed.

The stress field in the solid is a tensor. A simpler analogy is found in electrical breakdown properties, where the field is a scalar. Here too, the field or current (whichever the case) concentrates around the defects and the deformation depends on the geometries of the defects. The extreme statistics also comes into play for similar reasons (Section 10.1). There are two approaches in studying this problem: the dielectric breakdown problem, where conductors are placed in a dielectric medium (nonconducting) and the random fuse model, where insulating bonds are placed in a conducting lattice. The electrical failure problem also shows avalanche dynamics and roughness properties of the broken surfaces (elaborated in Section 10.2). Particularly is of interest is the limit of high disorder, as is already pointed out, where self-averaging statistics emerges (Section 10.2.4).

The largest scale failure phenomena we commonly encounter are the earthquakes. Earthquakes, due to their devastating consequences, have been a subject of extensive studies in various disciplines, ranging from seismology to physics. In the last decade, considerable progress has been made in studying different aspects of this vast topic.

Several statistical approaches to model earthquake dynamics are discussed in Chapter 11. The Burridge–Knopoff model, a system of massive blocks connected by linear Hookean springs and slowly driven over a rough surface having velocity-dependent friction, is a simple and well-established model in this regard. The only source of nonlinearity in this model is the velocity weakening friction term. It shows the avalanche statistics reminiscent of the Gutenberg–Richter law of earthquakes (Section 11.2). A slightly different version is the train model, where the drive is applied only at one end. Interestingly, in a simplified limit where the velocity-dependent friction forces between the block and the rough surface are replaced by random pinning forces, the model becomes equivalent to the driven elastic line (in Edwards-Wilkinson class) discussed earlier for fracture front modeling (see Section 11.4). This enables one for complimentary observation of depinning transition in train models and avalanche statistics in the interface propagation model. Of particular interest is the fact that with a modification in the definition of avalanche, Omori law statistics is seen in this train model, which is not seen elsewhere in spring-block type models.

In considering the fractal nature of the tectonic plates, a simple model of an earthquake is direct overlap of two rough surfaces. A solvable case is the overlap of two Cantor sets (Section 11.2.4). It is possible to exactly calculate the magnitude statistics (Eq. (11.34)) in showing Gutenberg–Richter law. The Omori law and some other observed statistical features of earthquakes are also satisfactorily reproduced in this model. Note that the surfaces here are absolutely rigid (do not relax laterally). In the limit of finite rigidity, it reduces to the train model mentioned earlier (then it cannot be analytically solved).

A more abstract way of modeling earthquakes is the cellular automata model. In particular, the main cellular automata models (Bak-Tang-Weisenfeld, Zhang, and Manna models) and their avalanche statistics including their failure precursors are discussed in Section 11.3. For earthquakes, the dissipative model or the Olami–Feder–Christensen model is discussed in details.

Finally, toward the end, we provide some details of essential concepts used in different chapters. Particularly, the scaling theory and renormalization group analysis of percolation are provided in Appendix A. And the renormalization group analysis of rigidity percolation is briefly discussed in Appendix B. Some detailed calculations of the fiber bundle model showing its universality as well as crossover in breakdown behavior are discussed in Appendix C. The Zener breakdown and scaling theory of disordered system are discussed briefly in Appendix D. The basic definition and handling of fractal objects are discussed in Appendix E. Some detailed calculations, particularly for the discrete limit of the two-fractal overlap model of earthquakes are provided in Appendix F. The microscopic theories of friction and the effect of self-affine disorders are discussed in Appendix G.

As summarized earlier and discussed in detail in the textbook, the failure properties of disordered solids have often intriguing scaling and critical properties due to the extreme nature of the crack nucleation statistics and the manifestations of turbulent-like dynamical fluctuations captured frozen in the fractured surfaces left behind the crack fronts. Recent laboratory-scale investigations of these phenomena and the development of analytical studies of microscopic models, as in fiber bundle models, have led to considerable progress as discussed in this textbook. The recent developments in the depinning models of earthquake dynamics, as discussed in this textbook, also look very promising. We hope, the young readers will find these introductory discussions useful and inspiring, while the specialist researchers in these fields will also find this textbook helpful to take them to some of the frontier studies on these topics. It may be mentioned, almost all the contents of different chapters in this textbook are either discussed or will be discussed soon in detailed textbooks on each of them, authored by pioneers in the respective fields and published in the same series. Readers are welcome to enjoy them on their own in a direct way!

A
Percolation

Here we discuss the scaling theory and renormalization group calculations of percolation transition in some simple cases, starting with the general examples of critical exponents and framework of scaling theory.

A.1
Critical Exponent: General Examples

In general, if x is the tuning parameter of the system, changing which the system can be led to the critical point, the different quantities near critical point behaves as

$$O(x) \sim |x - x_c|^{\beta'}, \tag{A1}$$
$$\xi(x) \sim |x - x_c|^{-\nu}, \tag{A2}$$
$$\chi(x) \sim |x - x_c|^{-\gamma}, \tag{A3}$$
$$C(x) \sim |x - x_c|^{-\alpha'}, \tag{A4}$$
$$G(l, x_c) \sim l^{-(d-2+\eta)}, \tag{A5}$$

where x_c denotes the critical point. The different quantities are associated with different observables of the system: $O(x)$ denotes the order parameter, which is zero in the disordered state and nonzero in the ordered state; a phase transition is said to have occurred when the order parameter becomes nonzero. $\xi(x)$ denotes the correlation length of the system, that is, a disturbance or fluctuation in the system typically affects up to a distance ξ. The functional form suggests that this quantity diverges as the critical point is approached. $C(x)$ is called the specific heat, which can be shown to be related to fluctuations in the system. $G(x_c, l)$ is the pair correlation function, which relates some quantity (say, connection probability two points in a lattice) l distance apart, at the critical point. The exponents $\beta', \nu, \gamma, \alpha'$, and η are called critical exponents, while d denotes the spatial dimension. All these exponents are not all independent, but have some relationships among them. We discuss those relations for specific examples in appropriate places.

A.1.1
Scaling Behavior

The behaviors of the above-mentioned physical quantities can be characterized in terms of *scaling hypothesis* and finally more formally in renormalization group theory. Here, we first give a brief account of the scaling hypothesis and scaling functions, before going into the renormalization group study in detail regarding the particular case of percolation.

The scaling hypothesis states that the singular part of free energy is asymptotically a *generalized homogeneous function* near the critical point. A function f is called a homogeneous scaling function if it satisfies

$$f(\lambda^{a_1} x_1, \lambda^{a_2} x_2, \ldots, \lambda^{a_n} x_n) = \lambda^a f(x_1, x_2, \ldots, x_n) \tag{A6}$$

for any positive value of λ. In addition, its Legendre transformation (giving different free energies) and partial derivatives (giving different physical quantities) are generalized homogeneous functions. Therefore, one simply has to identify the right variables and then write down the free energies in terms of them and using the scaling hypothesis, the relationships between the critical exponents can be derived.

One of the practical uses of scaling hypothesis is the data collapse possibility. Let us describe it in terms of a two-argument function. Consider the following relation

$$f(\lambda^{a_1} x_1, \lambda^{a_2} x_2) = \lambda^a f(x_1, x_2). \tag{A7}$$

Now, this should be valid for any positive value of λ. Therefore, let us choose $\lambda = |x_1|^{-1/a_1}$. Substituting this on Eq. (A7) gives

$$|x_1|^{-a/a_1} f(x_1, x_2) = \tilde{f}(\pm 1, |x_1|^{-a_2/a_1} x_2), \tag{A8}$$

where the sign of unity is the sign of x_1. Now, the right-hand side is a function of two arguments. Hence, for different values of x_1, all curves will collapse to the function $\tilde{f}$. This idea of data collapse is used widely in critical phenomena.

A.2
Percolation Transition

Consider a nonconducting plate, on which a conducting dye is spread uniformly (see Figure A.1). If one applies a potential difference across any two opposite ends of the plate, with an ammeter in series, there will be no current initially when no dye is spread. But, if the entire area of the plate is covered by the dye, then obviously the plate conducts. Is it necessary to cover the entire plate to get a nonvanishing current? The answer is, no. Current starts to flow when there is a marginally connected path of the overlapping clusters of the dye grains across the plate. The point at which conduction first takes place is called the percolation threshold.

In order to make the discussion more quantitative and precise, let us consider the lattice percolation model. There are two versions of this model: site percolation (Figure A.2) and bond percolation (Figure A.3). In site percolation, each site of a large lattice is randomly occupied with probability p. Clusters are defined as graph of neighboring lattice sites. In bond percolation, each bond of a lattice is occupied randomly, with a probability p. A cluster is defined as a graph of overlapping bonds sharing a common site. Most of the physical properties of such random systems depend on the geometric properties of these random clusters, and in particular, on the existence of an infinite connected cluster, which spans the system. Percolation theory deals with the statistics of the clusters formed.

Let us define some quantities of interest in percolation theory. Let $n_s(p)$ denote the number of clusters (per lattice site) of size s. A detailed knowledge of $n_s(p)$ would give a lot of information about percolation statistics, as most of the quantities of interest can be extracted from various moments of the cluster size distribution n_s.

The probability that a given site (bond) is occupied and is a part of an s-size cluster is $sn_s(p)$. Let, $P(p)$ denote the probability that any occupied site (bond) belongs to the infinite (lattice spanning) cluster. Then, we have the obvious relation,

$$\sum_s sn_s + P = 1, \qquad (A9)$$

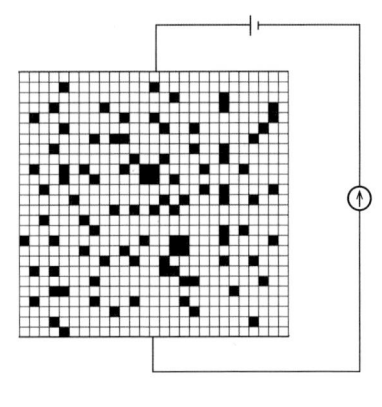

Figure A.1 Sample below and above the percolation threshold.

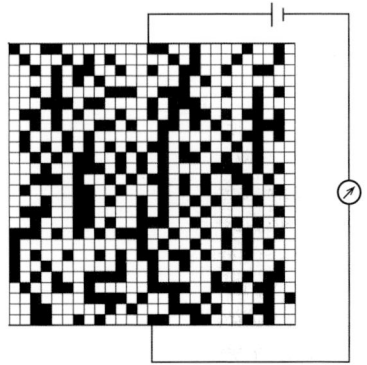

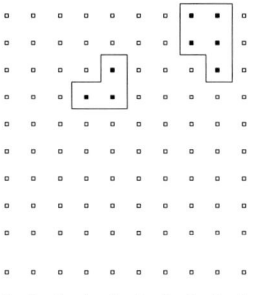

Figure A.2 Clusters in site percolation.

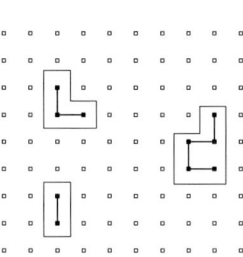

Figure A.3 Clusters in bond percolation.

where the summation extends over all finite clusters. Clearly, at $p = 1$, $P(p) = 1$ and $P(p) = 0$ for $p < p_c$ as the infinite cluster does not exist for $p < p_c$. $P(p)$ can therefore be taken as the order parameter of the percolation phase transition. Another quantity of interest is the mean size of the finite clusters, denoted by $S(p)$, which is related to $n_s(p)$ by the relation

$$S(p) = \frac{\sum_s s^2 n_s(p)}{\sum_s s n_s(p)}, \tag{A10}$$

where the summation is again over all finite clusters. One can also define a pair connectedness (or two-point correlation) function $C(p, r)$ as the probability that two occupied sites (bonds) at a distance r are members of the same cluster. The sum over the pair connectedness over all distances gives the mean cluster size: $S(p) = \sum_r C(p, r)$.

A.2.1
Critical Exponents of Percolation Transition

Most of the quantities defined above have power-law variations near p_c. For example, the variation of the total number of clusters per site $G(p) = \sum_s n_s(p)$ (sum extends over all finite clusters), the decay of the order parameter $P(p)$, and the divergence of the mean cluster size: $S(p)$, as $p \to p_c$, can be expressed by power-law variations of these quantities with the concentration interval $|p - p_c|$ as

$$G(p) \equiv \sum_s n_s(p) \sim |p - p_c|^{2-\alpha'} \tag{A11}$$

$$P(p) \sim (p - p_c)^{\beta'} \tag{A12}$$

$$S(p) \sim |p - p_c|^{-\gamma} \tag{A13}$$

$$C(p,r) \sim \frac{\exp(-r/\xi(p))}{r^{d-2+\eta}} \tag{A14}$$

where the correlation length

$$\xi(p) \sim |p - p_c|^{-\nu} \tag{A15}$$

diverges at $p = p_c$.

These powers α', β', γ, η, and ν are called the critical exponents. These exponents are observed to be universal in the sense that although p_c depends on the details of the model under study, these exponents depend only on the lattice dimensionality.

A.2.2
Scaling Theory of Percolation Transition

Scaling theory assumes that the cluster distribution function $n_s(p)$ is a homogeneous function near $p = p_c$. Thus, $n_s(p)$ is basically a function of the single scaled variable $s/S_\xi(p)$, where S_ξ denotes the typical cluster size,

$$n_s(p) \sim s^{-\beta} f\left(\frac{s}{S_\xi(p)}\right), \tag{A16}$$

with $S_\xi(p) \sim |p - p_c|^{-1/y}$. Here, β and y are two independent exponents and the scaling theory intends to relate all the above-mentioned exponents to these exponents through the scaling relations. The function, $f(x)$ is assumed to have asymptotic behavior, $f(x) \to 1$ as $x \to 0$ and $f(x) \to 0$ as $x \to \infty$. Further details of this function are unspecified in the theory. It may be noted that the above-mentioned form implies $n_s(p_c) \sim s^{-\beta}$. This has been checked by Monte Carlo simulations.

Assuming the above-mentioned scaling form for $n_s(p)$, the mth moment of $n_s(p)$ can be expressed as

$$\sum_s s^m n_s \sim \sum_s s^{m-r} f\left(\frac{s}{|p-p_c|^{1/y}}\right)$$

$$\sim |p-p_c|^{\frac{r-m-1}{y}} \int x^{m-r} f(x) dx$$

$$\sim |p-p_c|^{\frac{r-m-1}{y}}$$

assuming the integral over x $(= s/|p-p_c|^{1/y})$ converges because of the asymptotic behavior of $f(x)$. Noting that $G(p)$, $P(p)$, and $S(p)$ correspond to the zeroth, first, and second moments of n_s, respectively, we get $\alpha' = 2 - \frac{\beta-1}{y}$, $\beta' = \frac{\beta-2}{y}$, and

$\gamma = -\frac{\beta-3}{y}$. This gives the scaling relation

$$\alpha' + 2\beta' + \gamma = 2. \tag{A17}$$

This is satisfied by the observed values of the exponents. Also, since $S(p) = \sum_r C(r,p) = \int r^{d-1}C(r,p)dr$, one immediately gets

$$\begin{aligned} S(p) &\sim |p-p_c|^{-\gamma} \\ &= \int r^{d-1} \frac{e^{-r/\xi}}{r^{d-2+\eta}} dr \\ &\sim \xi^{2-\eta} \int y^{1-\eta} e^{-y} dy \\ &\sim |p-p_c|^{-\nu(2-\eta)}, \end{aligned} \tag{A18}$$

giving the scaling relation

$$\gamma = \nu(2-\eta). \tag{A19}$$

Assuming that near p_c, the typical density $sn_s \sim S_\xi^{1-\beta}$, where $S_\xi \sim |p-p_c|^{-1/y}$, one gets $sn_s \sim |p-p_c|^{(\beta-1)/y}$ for the density. Assuming further that this density scales with the inverse of the typical volume element $\xi^d \sim |p-p_c|^{-d\nu}$, we get the hyperscaling relation

$$d\nu = (\beta-1)/y = 2 - \alpha'. \tag{A20}$$

These scaling relations are satisfied by the numerically estimated values, and in special cases the exact values, of the critical exponents. The ideas of scaling and universality can be obtained formally from the renormalization group techniques. The basic idea of renormalization utilizes the fact that the correlation length of the system diverges at the critical point. The renormalization transformation in real-space changes the microscopic length scale by a factor and reduces the corresponding degrees of freedom of the system. When there is a finite correlation length in the system, this transformation changes the properties of the system as it competes with the length scale of the system, namely the correlation length. However, at critical point the correlation length scale diverges, leaving the system properties immune to scale transformation. Renormalization group thus obtains the critical point as the fixed point of the transformation. In the following, we give the general formalism of real-space renormalization and then we go over to specific applications of percolation theory.

A.3
Renormalization Group (RG) Scheme

The essential feature of the renormalization scheme is the fact that the effective number of degrees of freedom reduces as the critical point is approached. This is ensured by the presence of a finite correlation. One can define an effective interaction which gives the same partition function, that is,

$$Z(H_N) = Z(H'_{N'<N}). \tag{A21}$$

If there is 100% correlation, we could work with a single degree of freedom.

In a thermodynamic system, if the correlation length is large compared to the interatomic separation, then the experimental results for the system are obtained theoretically considering a single particle. Correlation length diverges as critical point is approached. Hence, near critical point, we need to consider, instead of Avogadro's number of particles, a single particle.

Let us define the operator $\mathcal{R}$ as

$$\mathcal{R} H_N = H'_{N'<N}. \tag{A22}$$

We keep operating this until correlation exists. Finally, we reach a point where

$$\mathcal{R} H^*_{N^*} = H^*_{N^*}. \tag{A23}$$

After this, no further reduction in the degrees of freedom is possible. This is called the fixed point, near which the relationship between the free energy densities can be written as

$$f(|\Delta T|, h) = \frac{N^*}{N} f(|\Delta T^*|, h^*). \tag{A24}$$

where $h^* = 0$ and hence $\Delta h - h$ itself.

Very close to the fixed point, $\mathcal{R}$ can always be written as a linear operator (linearization) and one can diagonalize it to write

$$\begin{pmatrix} |\Delta T'| \\ h' \end{pmatrix} = \mathcal{R} \begin{pmatrix} |\Delta T| \\ h \end{pmatrix}$$

where $|\Delta T'| = |T' - T^*|$, $|\Delta h'| = |h' - h^*| = h'$, and $|\Delta T| = |T - T^*|$, $|\Delta h| = |h - h^*| = h$. Both T and T', h and h' are very close to the critical point. In terms of the eigenvalues Λ_j of the linearized operator $\mathcal{R}$, one can write

$$h'_i = h^*_i + \sum_j \Lambda_j (h_j - h^*_j).$$

Therefore, we can write

$$f(|\Delta T|, h) = \frac{N'}{N} f(\Lambda_T |\Delta T|, \Lambda_h h).$$

Now, if the length scale is changed by a factor b, then we have $\frac{N'}{N} = b^{-d}$, where d is the dimension. Therefore,

$$f(|\Delta T|, h) = b^{-d} f(\Lambda_T |\Delta T|, \Lambda_h h).$$

Also note that $\mathcal{R}_{b_1} \mathcal{R}_{b_2} = \mathcal{R}_{b_1 b_2}$, that is, it does not matter if we change the degrees of freedom in two steps or in a single step. Therefore, we have $\Lambda_{b_1} \Lambda_{b_2} = \Lambda_{b_1 b_2}$, which implies, $\Lambda \sim b^\lambda$ and λ is independent of the choice of b. Therefore, we can write

$$f(|\Delta T|, h) = b^{-d} f(b^{\lambda_T} \Delta T, b^{\lambda_h} h).$$

Figure A.4 Schematic representation of site percolation in one dimension: nonpercolating case.

Choosing now $b^{\lambda_T}|\Delta T| = 1$, as b can be arbitrary,

$$f(|\Delta T|, h) = |\Delta T|^{d/\lambda_T} Y\left(\frac{h}{|\Delta T|^{\lambda_h/\lambda_T}}\right); \quad Y(x) \equiv f(1, x),$$

which is of the same form as that of the Widom scaling hypothesis, with the identification

$$\frac{d}{\lambda_T} = 2 - \alpha' \qquad \frac{\lambda_h}{\lambda_T} = \Delta = \frac{1}{2}(2 - \alpha' + \gamma). \tag{A25}$$

and scaling function $Y(x) = f(1, x)$. This is the same as what we obtained earlier via scaling hypothesis.

RG methods can be applied to the percolation theory to extract the critical point and critical exponents through simple calculations. For simplicity, we shall begin with the site percolation in one dimension and then go over to the treatments in two dimensions for both site and bond percolations.

A.3.1
RG for Site Percolation in One Dimension

In one dimension, site percolation is a trivial problem. It is obvious that a spanning cluster would require all the sites to be occupied (Figure A.4).

Therefore, $p_c = 1$ for this case. The connectedness or pair correlation function reads

$$C(r, p) = p^r = \exp(-r \ln(1/p))$$
$$= \exp(-r/\xi), \tag{A26}$$

with

$$\xi = \frac{1}{\ln(1/p)}$$
$$= -\frac{1}{\ln p}$$
$$= -\frac{1}{\ln(1 - (p_c - p))} \qquad p_c = 1 \quad \text{here}$$
$$= \frac{1}{(p_c - p)}.$$

Hence, we get

$$\xi = (p_c - p)^{-1},$$

giving $\nu = 1$.

A.3 Renormalization Group (RG) Scheme

Now, applying RG here, we will have the length scale scaled as $\xi' = \xi/b$. If the renormalized probability is p', we shall have

$$(p' - p_c)^{-\nu} = \frac{(p - p_c)^{-\nu}}{b}$$

$$\Rightarrow -\nu \ln\left(\frac{p' - p_c}{p - p_c}\right) = \ln\frac{1}{b} = -\ln b$$

$$\Rightarrow \nu = \frac{\ln b}{\ln \lambda},$$

with $\lambda = \frac{p' - p_c}{p - p_c} = \left(\frac{dp'}{dp}\right)_{p_c}$.

Note that p denotes the probability of occupation of a given site. In the renormalized state, a single site is formed by b number of sites in the unrenormalized lattice. Hence, we have $p' = p^b$. Therefore, for the fixed point,

$$p^* = p^{*b},$$

giving the two fixed points, $p^* = 0$ and $p^* = 1$. Now, performing the stability analysis close to the fixed point $p^* = 0$,

$$p* + \delta p' = (p^* + \delta p)^b$$

$$\Rightarrow \delta p' = (\delta p)^b.$$

In the next iteration, it becomes $\delta p'' = (\delta p)^{2b}$. Finally, the deviation will go to zero. Hence p is an irrelevant parameter for this fixed point. Therefore, we conclude that $p^* = 0$ is a stable fixed point.

Now, close to the other fixed point $p^* = 1$, we will have

$$1 - \delta p' = (1 - \delta p)^b$$
$$1 - \delta p' = 1 - b\delta p$$
$$\Rightarrow \delta p' = \delta p$$

Clearly, the deviation grows in every iteration, and the system *flows away* from this fixed point. Hence, we conclude that $p^* = 1$ is an unstable fixed point. The flow diagram is shown in Figure (A.5).

Now, we have $p' = p^b$, giving

$$\left.\frac{dp'}{dp}\right|_{p_c} = (bp^{b-1})_{p_c} = b. \tag{A27}$$

Therefore,

$$\nu = \frac{\ln b}{\ln b} = 1. \tag{A28}$$

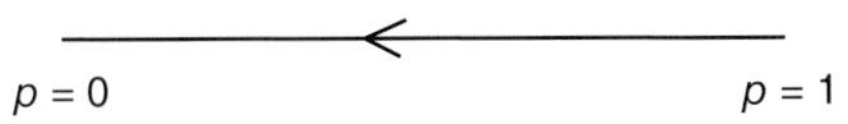

$p = 0$ ←——————————— $p = 1$

Figure A.5 Flow diagram for one-dimensional site percolation.

A Percolation

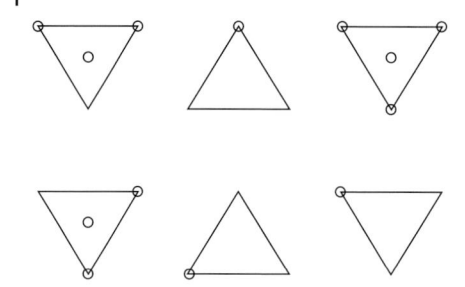

Figure A.6 Site percolation in 2-d.

A.3.2
RG for Site Percolation in Two-dimensional Triangular Lattice

Take alternate triangles in a triangular lattice and replace the three sites by a supersite (Figure (A.6)).

The supersite is occupied if two or more sites in those three sites were occupied. One can then write the renormalized probability as

$$p' = p^3 + 3p^2(1-p). \tag{A29}$$

where the first term corresponds to all sites being present, the factor 3 in the second term refers to the fact that there are three ways to keep one site empty. For the fixed point,

$$p^* = p^{*3} + 3p^{*2}(1-p^*)$$

which on solving gives $p^* = 1, 1/2, 0$. Clearly, $p^* = 1/2$ is a nontrivial fixed point (see Figure A.7 for the flow diagram).

Let us perform stability analysis around it:

$$\frac{1}{2} + \delta p' = \left(\frac{1}{2} + \delta p\right)^3 + 3\left(\frac{1}{2} + \delta p\right)^2 \left(1 - \left(\frac{1}{2} + \delta p\right)\right), \tag{A30}$$

implying,

$$\delta p' = \frac{3}{2}\delta p. \tag{A31}$$

Therefore, $\lambda = \frac{dp'}{dp} = 3/2$, giving $v = \frac{\ln b}{\ln \lambda} = \frac{\frac{1}{2}\ln 3}{\ln 3/2} \sim 1.355$, which is close to the assumed exact value 4/3. We have used $b = \sqrt{3}$ here, because $N' = N/3$ and $N' = N/b^d$. Here, the dimensionality $d = 2$, giving $b = \sqrt{3}$, where N and N' are the number of sites in the normal and super-lattices, respectively.

Figure A.7 Flow diagram for two-dimensional site percolation.

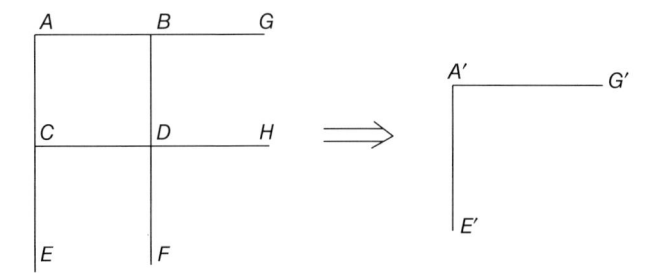

Figure A.8 Eight-bond structure: rescaling to two-bond structure.

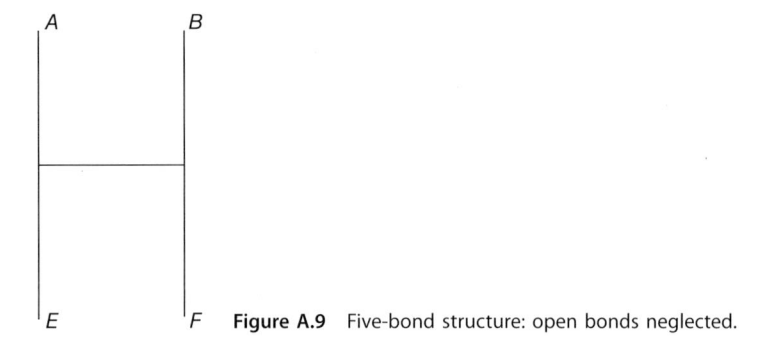

Figure A.9 Five-bond structure: open bonds neglected.

A.3.3
RG for Bond Percolation in Two-dimensional Square Lattice

Let us consider the eight-bond structure (Figure (A.8)), which can be taken as a unit cell for a square lattice.

If this eight-bond structure is replaced by the two-bond structure, it can also be taken as a unit cell. Suppose we are concerned about percolation in vertical direction. Then we can ignore the bonds BG and DH for finding the contributions to connectivity between AE and equivalently, $A'E'$ (Figure A.9).

Of course, in the large scale, those bonds could also contribute to the vertical connectivity, but here we make this approximation. Therefore, the contributions to the connection probability of AE can be written as

$$p' = p^5 + 5p^4(1-p) + 8p^3(1-p)^2 + 2p^2(1-p)^3, \tag{A32}$$

where the first term corresponds to the occupation of all the five bonds and the factor 5 corresponds to the fact that there are five ways of removing one bond and still have vertical connectivity. The other factors come from similar arguments. For the fixed point, we have

$$p^* = p^{*5} + 5p^{*4}(1-p^*) + 8p^{*3}(1-p^*)^2 + 2p^{*2}(1-p^*)^3. \tag{A33}$$

It can be seen that the physically relevant fixed points are $p = 0, 1/2, 1$ of which, $p = 1/2$ is nontrivial. Now

$$\lambda = \left.\frac{dp'}{dp}\right|_{p=1/2} = \frac{13}{8}. \tag{A34}$$

Therefore,

$$\nu = \frac{\ln 2}{\ln \frac{13}{8}} \sim 1.428, \tag{A35}$$

where we have used $N' = N/4 = N/b^d$ and $d = 2$, $b = 2$. Note that the assumed exact value is 4/3 and the estimation approximately matches.

B
Real-space RG for Rigidity Percolation

Introduced in 1984 (Feng and Sen, 1984), phase transition, related to the rigidity percolation, is experimentally very important. While the mean-field treatment works away from critical point, it is not accurately close to the critical point. A Real-space RG (RSRG) treatment of rigidity percolation is presented here, which gives exact results for the critical behavior in some models of rigidity percolation (see Barré (2009); Thorpe and Stinchcombe (2013) for details).

Consider the particular case of Berker lattice (shown in Figure B.1). This is a simple model showing the generic features of rigidity percolation, when a fraction of the bonds is removed. Since each higher generation is constructed by repeating the structure of the immediately lower generation, it is generally possible to write down the recursion relations and obtain the relevant fixed points of the model (Stinchcombe and Thorpe, 2011). To formulate the recursion relations, first consider the configurations shown in Figure B.2. In the structures, some bonds are removed. For a given number of removed bonds, a certain number of configurations is possible. For example, when one bond is removed, it can be done in eight ways and all of them keep the structure rigid. The higher number of removals also has individual weightages. Considering all the weightages, if the probability of a bond being present is p, one can connect the probability of a rigid structure in two successive generations as

$$p' = p^8 + 8p^7(1-p) + 6p^6(1-p)^2 + 2p^5(1-p)^3 = 2p^5 + 2p^7 - 3p^8. \quad (B1)$$

Hence, the recursion relation reads

$$p_{n+1} = 2p_n^5 + 2p_n^7 - 3p_n^8. \quad (B2)$$

The fixed points ($p_{n+1} = p_n = p^*$) are $p = 0, 1, 0.9446$. The third fixed point is unstable and signifies the critical point p_c. Close to this last fixed point, the above equation can be linearized as $(p_{n+1} - p_c) = \lambda_1(p_n - p_c)$ with $\lambda_1 = 10p_c^4 + 14p_c^6 - 24p_c^7 = 1.802$.

One can also find the probability $P_{n+1}(p)$ that a bond belongs to a percolating rigid cluster to follow the recursion relation

$$P_{n+1}(p) = \frac{1}{4}[5p_n^4 + 13p_n^6 - 14p_n^7]P_n(p), \quad (B3)$$

Statistical Physics of Fracture, Breakdown, and Earthquake: Effects of Disorder and Heterogeneity, First Edition.
Soumyajyoti Biswas, Purusattam Ray, and Bikas K. Chakrabarti.
© 2015 Wiley-VCH Verlag GmbH & Co. KGaA. Published 2015 by Wiley-VCH Verlag GmbH & Co. KGaA.

B Real-space RG for Rigidity Percolation

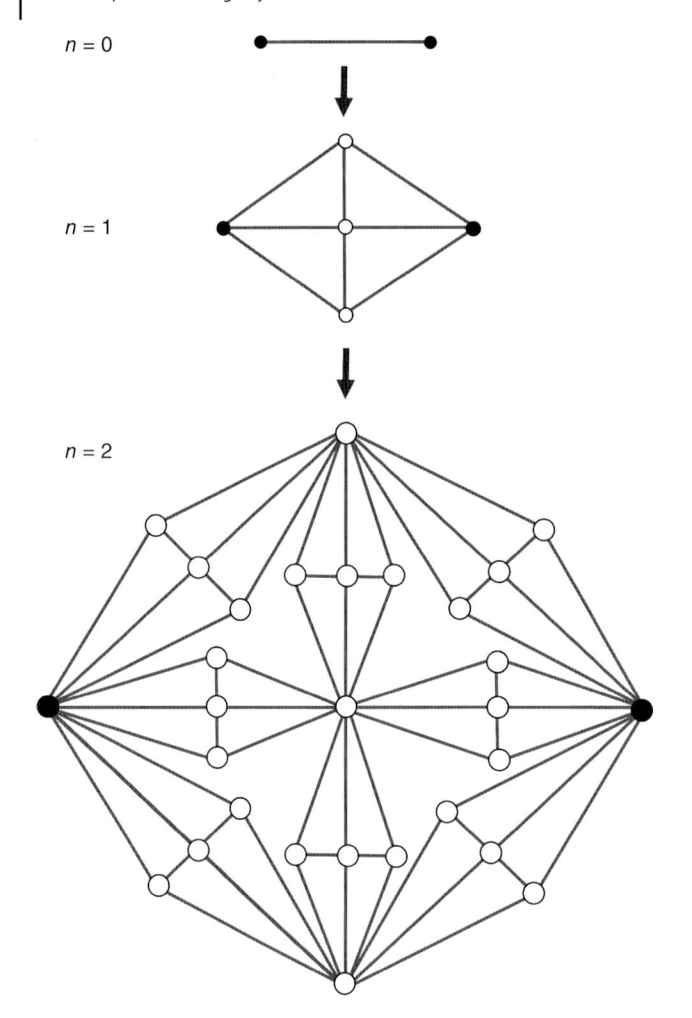

Figure B.1 Steps of forming the undiluted hierarchical Berker lattice. From Stinchcombe and Thorpe (2011).

with $P_0(p) = p$. Near the unstable fixed point, $P_{n+1}(p) = \lambda_2 P_n(p)$, where $\lambda_2 = \frac{1}{4}[5p_c^4 + 13p_c^6 - 14p_c^7] = 0.9554$. This indicates that the size of the spanning cluster goes to zero at the critical point, which is the signature of second-order phase transition. Figure B.3 shows the spanning probability for different generations. The order parameter exponent $\beta = 0.0775$ for this model.

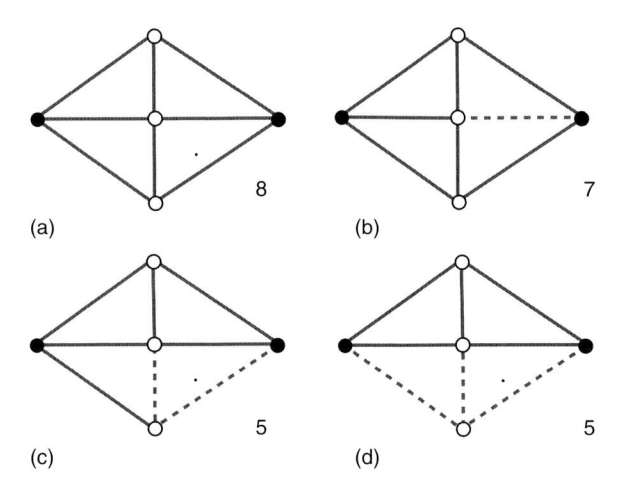

Figure B.2 The rigid configurations with some bonds removed (number of present bonds written below each configuration). The total number of ways of forming a rigid cluster gives the corresponding weightage factors in the recursion relation. From Stinchcombe and Thorpe (2011).

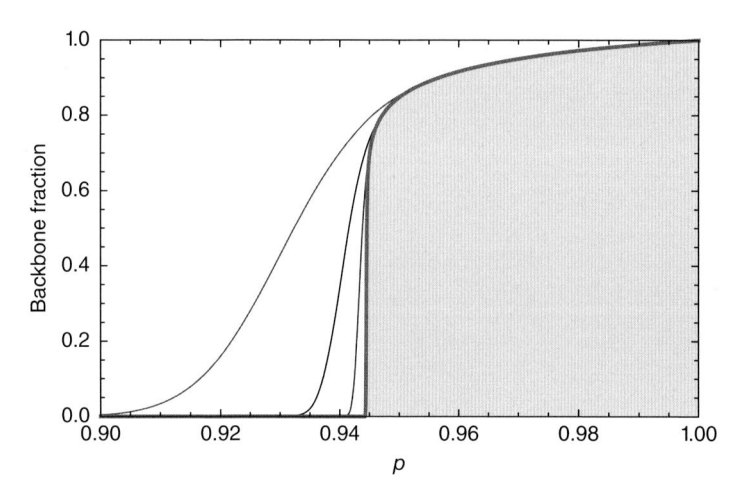

Figure B.3 Figure showing the probability of a bond belonging to the rigid spanning cluster for $n = 4, 6, 8$, and 12 terms. The singularity is at the nontrivial fixed point $p_c = 0.9446$. From Stinchcombe and Thorpe (2011).

C
Fiber Bundle Model

The fiber bundle model (FBM) of fracture is discussed in Chapter 6. Here, we elaborate the calculations for different threshold distributions. We also show how by changing the width of the threshold distribution the model shifts from ductile to quasi-brittle failure (mentioned in Section 6.2.4).

C.1
Universality Class of the Model

The behavior of the model is universal with respect to various threshold distribution functions such as (I) linearly increasing density distribution and (II) linearly decreasing density distribution within the (σ_{th}) limit 0 and 1. It can be shown that while σ_f changes with different strength distributions ($\sigma_f = \sqrt{4/27}$ for case (I) and $\sigma_f = 4/27$ for case II), the critical behavior remains unchanged: $z = 1/2 = \beta' = \gamma$, $\delta = 1$ for all these equal load sharing models (Pradhan and Chakrabarti, 2003b).

C.1.1
Linearly Increasing Density of Fiber Strength

Here, the cumulative distribution becomes (see Figure C.1)

$$P(\sigma_t) = \int_0^{\sigma_t} \rho(\sigma_{th}) d\sigma_{th} = 2 \int_0^{\sigma_t} \sigma_{th} d\sigma_{th} = \sigma_t^2. \tag{C1}$$

Therefore, U_t follows a recursion relation

$$U_{t+1} = 1 - \left(\frac{\sigma}{U_t}\right)^2. \tag{C2}$$

At the fixed point ($U_{t+1} = U_t = U^*$), the above recursion relation can be represented by a cubic equation of U^*

$$(U^*)^3 - (U^*)^2 + \sigma^2 = 0. \tag{C3}$$

Solving the aforementioned equation, we get the value of critical stress $\sigma_f = \sqrt{4/27}$, which is the strength of the bundle for the above fiber strength

Statistical Physics of Fracture, Breakdown, and Earthquake: Effects of Disorder and Heterogeneity, First Edition.
Soumyajyoti Biswas, Purusattam Ray, and Bikas K. Chakrabarti.
© 2015 Wiley-VCH Verlag GmbH & Co. KGaA. Published 2015 by Wiley-VCH Verlag GmbH & Co. KGaA.

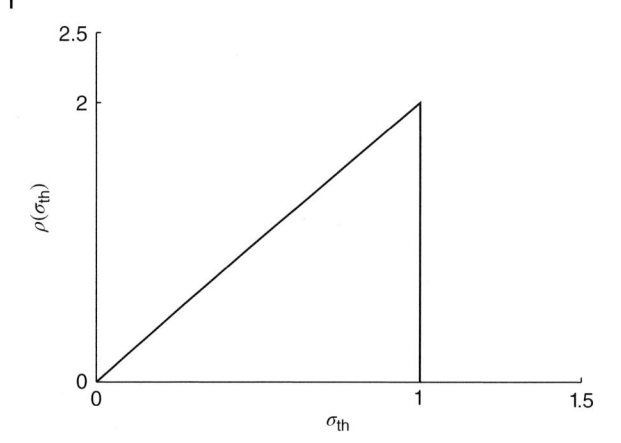

Figure C.1 The linearly increasing density $\rho(\sigma_{th})$ of the fiber strength distribution up to a cut-off strength.

distribution. Here, the order parameter can be defined as $U^*(\sigma) - U^*(\sigma_f)$ and this goes as

$$O \propto (\sigma_c - \sigma)^{\beta'}; \beta' = \frac{1}{2}. \tag{C4}$$

The susceptibility diverges as the critical point is approached from the following:

$$\chi = \left|\frac{dU^*(\sigma)}{d\sigma}\right| \propto (\sigma_f - \sigma)^{-\gamma}; \gamma = \frac{1}{2}. \tag{C5}$$

We can also show that for any $\sigma < \sigma_f$

$$U_t(\sigma) - U^*(\sigma) \approx \exp(-t/\tau), \tag{C6}$$

with

$$\tau \propto (\sigma_f - \sigma)^{-z}; z = \frac{1}{2}. \tag{C7}$$

and at $\sigma = \sigma_f$

$$U_t - U^*(\sigma_f) \propto t^{-\delta}; \delta = 1. \tag{C8}$$

C.1.2
Linearly Decreasing Density of Fiber Strength

In this case, the cumulative distribution becomes (see Figure C.2)

$$P(\sigma_t) = \int_0^{\sigma_t} \rho(\sigma_{th}) d\sigma_{th} = 2\int_0^{\sigma_t}(1-\sigma_{th})d\sigma_{th} = 2\sigma_t - \sigma_t^2 \tag{C9}$$

and U_t follows a recursion relation

$$U_{t+1} = 1 - 2\frac{\sigma}{U_t} + \left(\frac{\sigma}{U_t}\right)^2. \tag{C10}$$

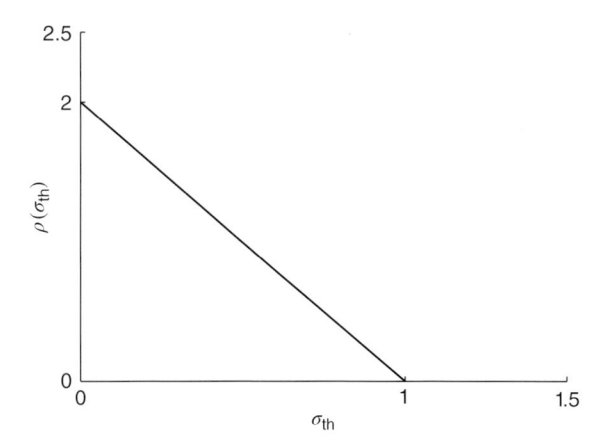

Figure C.2 The linearly decreasing density $\rho(\sigma_{th})$ of the fiber strength distribution up to a cut-off strength.

At the fixed point $(U_{t+1} = U_t = U^*)$, the above recursion relation can be represented by a cubic equation of U^* as follows:

$$(U^*)^3 - (U^*)^2 + 2\sigma U^* - \sigma^2 = 0. \tag{C11}$$

Solution of the aformentioned equation suggests the value of critical stress $\sigma_f = 4/27$, which is the strength of the bundle for the above-mentioned fiber strength distribution. Also, the order parameter goes as

$$O \equiv [U^*(\sigma) - U^*(\sigma_f)] \propto (\sigma_c - \sigma)^{\beta'}; \beta' = \frac{1}{2} \tag{C12}$$

and the susceptibility diverges with the similar power-laws as before when the critical point is approached from the following:

$$\chi = \left|\frac{dU^*(\sigma)}{d\sigma}\right| \propto (\sigma_f - \sigma)^{-\gamma}; \gamma = \frac{1}{2}. \tag{C13}$$

Here also for any $\sigma < \sigma_f$

$$U_t(\sigma) - U^*(\sigma) \approx \exp(-t/\tau), \tag{C14}$$

where

$$\tau \propto (\sigma_c - \sigma)^{-z}; z = \frac{1}{2}. \tag{C15}$$

and at $\sigma = \sigma_f$

$$U_t - U^*(\sigma_f) \propto t^{-\delta}; \delta = 1. \tag{C16}$$

Thus, the democratic fiber bundles (for different fiber strength distributions) show phase transition with a well-defined order parameter that shows similar power-law variation on the way the critical point is approached. The susceptibility and relaxation time also diverge with same power exponent for all the cases. Therefore, failure of democratic fiber bundles belongs to a universality class characterized by the universal values of the associated exponents (z, β', γ, and δ).

C.1.3
Nonlinear Stress–Strain Relationship

The stress–strain relationship is generally nonlinear for these models and it depends on the form of the threshold distribution. For a realistic form of the stress–strain curve, we consider a uniform density distribution of fiber strength, having a lower cut-off. Until failure of any of the fibers (due to this lower cut-off), the bundle shows linear elastic behavior. As soon as the fibers start failing, the stress–strain relationship becomes nonlinear. The dynamic critical behavior remains essentially the same and the static (fixed point) behavior shows elastic/plastic-like deformation before rupture of the bundle.

Here, the fibers are elastic in nature having identical force constant κ and the random fiber strengths distributed uniformly in the interval $[\sigma_L, 1]$ with $\sigma_L > 0$; the normalized distribution of the threshold stress of the fibers thus has the form (see Figure C.3):

$$\rho(\sigma_{\text{th}}) = \begin{cases} 0, & 0 \leq \sigma_{\text{th}} \leq \sigma_L \\ \frac{1}{1-\sigma_L}, & \sigma_L < \sigma_{\text{th}} \leq 1 \end{cases}. \tag{C17}$$

For an applied stress $\sigma \leq \sigma_L$, none of the fibers break, though they are elongated by an amount $\delta l = \sigma/\kappa$. The dynamics of breaking starts when applied stress σ is greater than σ_L. Now, for $\sigma > \sigma_L$, the fraction of unbroken fibers follows a recursion relation (for $\rho(\sigma_{\text{th}})$ as in Figure C.3):

$$U_{t+1} = 1 - \left[\frac{F}{NU_t} - \sigma_L\right]\frac{1}{1-\sigma_L} = \frac{1}{1-\sigma_L}\left[1 - \frac{\sigma}{U_t}\right], \tag{C18}$$

which has stable fixed points:

$$U^*(\sigma) = \frac{1}{2(1-\sigma_L)}\left[1 + \left(1 - \frac{\sigma}{\sigma_f}\right)^{1/2}\right]; \sigma_c = \frac{1}{4(1-\sigma_L)}. \tag{C19}$$

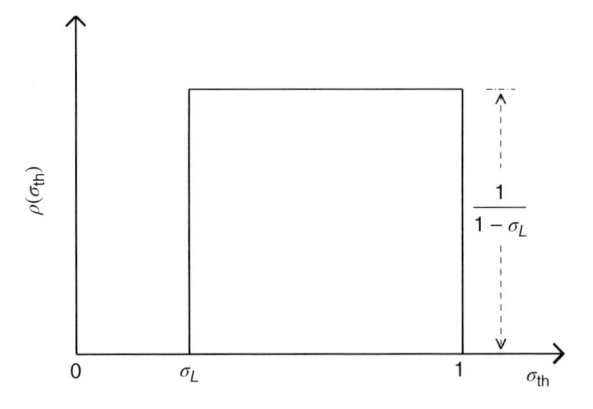

Figure C.3 The fiber breaking strength distribution $\rho(\sigma_{\text{th}})$ considered for studying elastic/plastic-type nonlinear deformation behavior of the ELS model.

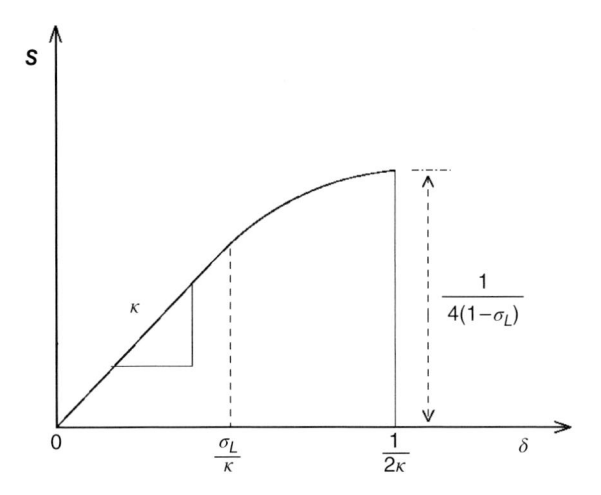

Figure C.4 Schematic stress (S)–strain (δ) curve of the bundle (shown by the solid line), following Eq. (C22), with the fiber strength distribution following Eq. (C17) (as shown in Figure C.3).

The model now has a critical point $\sigma_f = 1/[4(1-\sigma_L)]$ beyond which total failure of the bundle takes place. The aformentioned equation also requires that $\sigma_L \leq 1/2$ (to keep the fraction $U^* \leq 1$). As one can easily see, the dynamics of U_t for $\sigma < \sigma_f$ and also at $\sigma = \sigma_f$ remains the same as discussed in the earlier section. At each fixed point, there will be an equilibrium elongation $\delta l(\sigma)$, and a corresponding stress $S = U^* \kappa \delta l(\sigma)$ develops in the system (bundle). This $\delta l(\sigma)$ can be easily expressed in terms of $U^*(\sigma)$. This requires the evaluation of σ^*, the internal stress per fiber developed at the fixed point, corresponding to the initial (external) stress $\sigma (= F/N)$ per fiber applied on the bundle when all the fibers were intact. From the first part of Eq. (C19), one then gets (for $\sigma > \sigma_L$)

$$U^*(\sigma) = 1 - \frac{\sigma^* - \sigma_L}{(1-\sigma_L)} = \frac{1-\sigma^*}{1-\sigma_L}. \tag{C20}$$

Consequently,

$$\kappa \delta l(\sigma) = \sigma^* = 1 - (1-\sigma_L)U^*(\sigma). \tag{C21}$$

It may be noted that the internal stress $\sigma_f^* (= \sigma_f/U^*(\sigma_f))$ is universally equal to 1/2 at the failure point $\sigma = \sigma_f$ of the bundle. This finally gives the stress–strain relation for the ELS model as follows:

$$S = \begin{cases} \kappa \delta l, & 0 \leq \sigma \leq \sigma_L \\ \kappa \delta l(1-\kappa\delta l)/(1-\sigma_L), & \sigma_L \leq \sigma \leq \sigma_f \\ 0, & \sigma > \sigma_f \end{cases} \tag{C22}$$

This stress–strain relationship is shown in Figure C.4. The initial linear region (Hookean) has a slope κ (the spring constant for every fiber). This region for stress S continues up to the strain value $\delta l = \sigma_L/\kappa$, until which there is no failure event

($U^*(\sigma) = 1$). After this, nonlinearity appears due to the failure of a few fibers and the consequent decrease of $U^*(\sigma)$ (from unity). It finally drops to zero discontinuously by an amount $\sigma_f^* U^*(\sigma_f) = 1/[4(1 - \sigma_L)] = \sigma_f$ at the global failure point $\sigma = \sigma_f$ or $\delta l = \sigma_f^*/\kappa = 1/2\kappa$ for the bundle. This indicates that the stress drop at the final failure point of the bundle is related to the extent (σ_L) of the linear region of the stress–strain curve of the same bundle.

C.2
Brittle to Quasi-brittle Transition and Tricritical Point

As discussed in Section 6.2.4, in brittle fracture, failure occurs abruptly without any precursor, as soon as the global failure threshold is reached. The stress–strain curve is linear before the failure point and the initial state of the material is restored when the external load (below its critical failure value) is withdrawn. In the quasi-brittle region, however, the stress–strain curve becomes nonlinear beyond a certain point. Before the final failure, there are precursors in terms of avalanches. But the initial point of the stress–strain curve is reached when load is withdrawn. Here, we discuss these features in the fiber bundle model, addressing particularly the role of disorder in bringing the system from brittle to quasi-brittle region through a tricritical point.

Consider a rectangular threshold distribution having a lower cut-off at σ_L and upper cut-off at σ_R. Let it can be represented as

$$\begin{aligned}\rho(\sigma_{\text{th}}) &= 0 \quad \text{for} \quad 0 \le \sigma_{\text{th}} < \sigma_L \\ &= \frac{1}{\sigma_R - \sigma_L} \quad \text{for} \quad \sigma_L \le \sigma_{\text{th}} \le \sigma_R \\ &= 1 \quad \text{for} \quad \sigma_R < \sigma_{\text{th}}.\end{aligned} \quad \text{(C23)}$$

The recursion relation (see Eq. (6.4)) for a fraction of surviving fibers when an initial stress $\sigma_L \le \sigma \le \sigma_R$ is applied will then read (Bhattacharyya et al., 2003)

$$U_{t+1} = \frac{1}{\sigma_R - \sigma_L}\left(\sigma_R - \frac{\sigma}{U_t}\right). \quad \text{(C24)}$$

The physically relevant fixed point will be

$$U^* = \frac{\sigma_R}{2(\sigma_R - \sigma_L)} + \frac{1}{(\sigma_R - \sigma_L)^{1/2}}\left[\frac{\sigma_R^2}{4(\sigma_R - \sigma_L)} - \sigma\right]^{1/2}. \quad \text{(C25)}$$

The usual driving parameter is σ and as can be easily seen, its critical value is $\sigma_f = \frac{\sigma_R^2}{4(\sigma_R - \sigma_L)}$ and the order parameter $O = U^*(\sigma) - U^*(\sigma_f)$ goes as $(\sigma_f - \sigma_0)^{1/2}$.

However, one can also look at the transition in terms of strength of disorder. Particularly, one can increase σ_R (effectively making the disorder stronger by drawing it from a broader distribution) to see the transition. Of course, in real situation, it may not be possible to tune directly the width of disorder. However, it is of interest to see at what point the failure behavior qualitatively changes. One can write the fixed-point solution as

$$U^*(\sigma, \sigma_L, \sigma_R) = \frac{\sigma_R}{2(\sigma_R - \sigma_L)} + \frac{1}{(\sigma_R - \sigma_L)^{1/2}}$$
$$\left[\left\{\sigma_R - (2\sigma + 2\sqrt{\sigma^2 - \sigma_L\sigma})\right\}\left\{\sigma_R - (2\sigma - 2\sqrt{\sigma^2 - \sigma_L\sigma})\right\}\right]$$
(C26)

The critical value of $\sigma_{R,f}$ is $\left(2\sigma + 2\sqrt{\sigma^2 - \sigma_L\sigma}\right)$. Therefore, the order parameter

$$O = U^*(\sigma_R) - U^*(\sigma_{R,f}) \sim (\sigma_R - \sigma_{R,f})^{1/2}, \tag{C27}$$

which is also the usual critical exponent obtained for load-induced failure.

As for the dynamics, one can write the recurrence relation Eq. (C24) in the form of a differential equation as follows

$$\frac{dU(t)}{dt} = -\frac{(\sigma_R - \sigma_L)U^2(t) - \sigma_R U(t) + \sigma}{(\sigma_R - \sigma_L)U(t)}. \tag{C28}$$

Close to the fixed point, one can write $U(t) = U^* + \epsilon(t)$, where $\epsilon(t) \to 0$. Then, the aforementioned equation becomes

$$\frac{d\epsilon(t)}{dt} = -\epsilon(t)\frac{2U^*(\sigma_R - \sigma_L) - \sigma_R}{(\sigma_R - \sigma_L)U^*}. \tag{C29}$$

Using the fixed-point solution from Eq. (C25), one can write the aforementioned equation as

$$\frac{d\epsilon(t)}{dt} = -\epsilon(t)\frac{2\left[\frac{\sigma_R^2}{4(\sigma_R-\sigma_L)} - \sigma\right]^{1/2}}{(\sigma_R - \sigma_L)^{1/2}\left[\frac{\sigma_R}{(\sigma_R-\sigma_L)} + \frac{1}{(\sigma_R-\sigma_L)^{1/2}}\left\{\frac{\sigma_R^2}{4(\sigma_R-\sigma_L)} - \sigma\right\}^{1/2}\right]}$$
$$= -\epsilon(t)A. \tag{C30}$$

Now, close to the critical point, $\frac{1}{(\sigma_R-\sigma_L)^{1/2}}\left\{\frac{\sigma_R^2}{4(\sigma_R-\sigma_L)} - \sigma\right\}^{1/2} \to 0$. Therefore,

$$A \approx \frac{2(\sigma_R - \sigma_L)^{1/2}}{\sigma_R}\left[\frac{\sigma_R^2}{4(\sigma_R - \sigma_L)} - \sigma\right]^{1/2}$$
$$= \frac{2(\sigma_R - \sigma_L)^{1/2}}{\sigma_R}\left[\left(\sigma_R - 2\sigma - 2\sqrt{\sigma^2 - \sigma_L\sigma}\right)\left(\sigma_R - 2\sigma + 2\sqrt{\sigma^2 - \sigma_L\sigma}\right)\right]^{1/2}. \tag{C31}$$

Now, the solution of Eq. (C30) is of the form

$$\epsilon(t) = C \exp(-At), \tag{C32}$$

implying a timescale $\tau \sim 1/A$. As can be easily seen from Eq. (C31), the timescale goes as

$$\tau \sim (\sigma_R - \sigma_{R,f})^{-1/2}, \tag{C33}$$

where $\sigma_{R,f} = 2\sigma + 2\sqrt{\sigma^2 - \sigma_L\sigma}$.

C.2.1
Abrupt Failure and Tricritical Point

It is clear that there exists a trivial limit $\sigma_L = \sigma_R$ (disorder distribution is a delta function) when the failure will be abrupt. On the other hand, when the distribution is sufficiently broad, one can see by plotting the fixed-point solution Eq. (C25) that the system may remain stable under partially damaged condition, implying in such cases the failure is not abrupt. It is interesting to note that this transition from nonabrupt to abrupt failure occurs at a finite width of the threshold distribution. Since there are partially damaged stable configurations in the nonabrupt failure region, the stress–strain curve will become nonlinear for $\sigma > \sigma_L$, hence showing a quasi-brittle behavior. On the other hand, for the abrupt failure, the stress–strain curve is linear up to the failure point, signifying a brittle rupture. This transition can therefore be also looked at as brittle to quasi-brittle transition. The failure transition is continuous in the quasi-brittle region (for a suitably defined order parameter) and is first order in the brittle region (all fibers breaking suddenly for a given load). The point separating these two regimes is a tricritical point having its own set of critical exponents, as we discuss now.

The fact that there will be a tricritical point separating the abrupt and nonabrupt failure was noted by Andersen et al. (1997). Roy and Ray (2014) further quantified this point by noting the critical exponents and related extreme statistics. In particular, it is easy to see that the transition becomes abrupt when $U^*(\sigma_{R,f}^{TCP}) = 1$, that is, the entire system collapses at one step (no partially damaged stable state exists). For that, one should have $\frac{\sigma_{R,f}^{TCP}}{2(\sigma_{R,f}^{TCP} - \sigma_L)} = 1$, giving $\sigma_{R,f}^{TCP} = 2\sigma_L$ and also for the external force equation $\sigma = \sigma_L$. These two conditions together classify the tricritical point at which the system just fails abruptly. Substituting these conditions in Eq. (C26), one gets

$$O \sim \left(\sigma_R - \sigma_{R,f}^{TCP}\right). \tag{C34}$$

Also, regarding the timescale divergence, substituting the tricritical point conditions in Eq. (C31), one gets

$$\tau \sim \left(\sigma_R - \sigma_{R,f}^{TCP}\right)^{-1}. \tag{C35}$$

The numerical observation of this divergence was noted by Roy and Ray (2014), Roy (2012). The set of exponents for order parameter and timescale divergence is different from the set when the failure is nonabrupt.

D
Quantum Breakdown

As discussed in Chapter 10, electrical conduction through a composite of dielectric and conductors, a percolating path (see Appendix A) must appear that connects the two electrodes, when the electrons are considered as classical particles. If p is the concentration of conductors in the sample, then the conductivity goes to zero as p approaches p_c (the percolation threshold) (Stauffer and Aharony, 1994) from above, following $|p - p_c|^{t_c}$. For $p < p_c$, the sample is nonconducting to begin with, however it becomes conducting after local failures of the dielectrics if an external field is applied. The global breakdown voltage decreases to zero following $|p - p_c|^{\phi_b}$ if p approaches p_c from below.

However, Anderson localization (see Lee and Ramakrishnan, 1985 for example) asserts that an electron will be localized even in a percolating lattice for backscattering in random clusters with dimension less than 3. Hence the electrons, being quantum particles, cannot diffuse through disorder (although percolating) lattices.

In three dimensional system with disorder, there is a well defined mobility edge ϵ_c in the the conduction band, below which all states are localized and all states above it are extended. The nonconducting and conducting phases appear when the Fermi energy ϵ_f is below or above this mobility edge respectively. The Anderson transition from the insulating to the conducting (metallic) phase has already been studied (Lee and Ramakrishnan, 1985). It was found that the conductivity in the metallic phase grows as $|\epsilon_f - \epsilon_c|^{t_q}; t_q \neq t_c$.

In the classical case, there exists a breakdown voltage that scales as $E_b \sim |p_c - p|^\nu$. Analogously in quantum case, one can think of an applied field that causes breakdown and consequently conduction in an otherwise insulator phase (i.e. $\epsilon_f < \epsilon_c$).

For usual insulators, the states of the electrons are localized exponentially within the atomic distances (a). However, if a field E is applied and the energy gained by the electron in traveling the distance a is larger or equal to the band gap $\triangle \epsilon_b$ (or when $eEa \geq \triangle \epsilon_b$), there is a possibility of tunneling of the electron states across this band gap. In the case of Zener breakdown (Zener, 1932, 1934), the band is tilted towards the applied electric field. The effective band gap reduces from $\triangle \epsilon_b$ to $\triangle \epsilon_{br}$. If the effective width of the reduced band gap ($w = \triangle \epsilon_{br}/eE$) is of the order of $\xi_q a$ (with ξ_q denoting the correlation length and a denotes the

Statistical Physics of Fracture, Breakdown, and Earthquake: Effects of Disorder and Heterogeneity, First Edition.
Soumyajyoti Biswas, Purusattam Ray, and Bikas K. Chakrabarti.
© 2015 Wiley-VCH Verlag GmbH & Co. KGaA. Published 2015 by Wiley-VCH Verlag GmbH & Co. KGaA.

D Quantum Breakdown

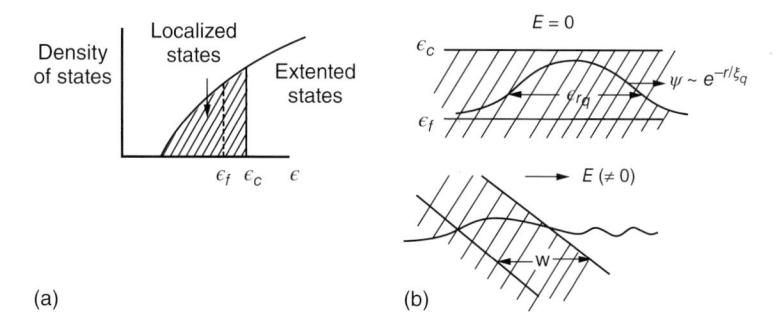

Figure D.1 Schematic density of states for Anderson insulators ($d > 2$) shown in (a), where the Fermi label ϵ_f is below the mobility edge ϵ_c (metallic phase for $\epsilon_f > \epsilon_c$). For strong electric field E (b), the band of (localized) states get tilted and tunnelling occurs when the effective width $w (= |\Delta \epsilon_m|/eE)$ of the mobility gap $\Delta \epsilon_m$, is less than or equal to the localization length ξ_q.

lattice separation), interband tunnleing can take place, breaking the insulation. Therefore, the scaling of the Zener breakdown field is quadratic with the band gap: $E_b = \Delta \epsilon_b / e \xi_q$ a $\sim |\Delta \epsilon_b|^2$, as $\xi_q \sim |\Delta \epsilon_b|^{-1}$.

Similar breakdown can occur in the case of Anderson insulator for more than two dimensions under the application of an external field. In this case, the states are localized within a length scale ξ_q, which scales as $\xi_q \sim |\Delta \epsilon_m|^{-v_q}$. Here, $\Delta \epsilon_m \equiv \epsilon_c - \epsilon_f$ is the mobility gap. In presence of an electric field, the gap gets tilted along the field direction (see Figure D.1). The effective width w can be estimated assuming that the energy Eew gained by the electron while travelling through the width is comparable with the gap $\Delta \epsilon_m$. This gives $w = \Delta \epsilon_m / Ee$. Now, the quantum mechanical tunneling rate through an energy barrier of width w is given by $\exp(-c\kappa w)$, where c is a dimensionless constant, κ is the 'imaginary momentum'. Now, if the only length scale near the breakdown point is given by ξ_q, then the width will be scaled by ξ_q and the momentum will be inverse of this length scale. Hence, the dimensionless tunneling rate expression takes the form $g(E) \sim \exp(-c(a/\xi_q)(w/\xi_q))$ (Chakrabarti and Samanta, 2010), where a is the lattice constant, the only other length scale (microscopic). This on simplification, gives $g(E) \sim \exp(-(ca/Ee)\Delta \epsilon_m^{T_q})$, with $T_q = 1 + 2v_q$. For $v_q \sim 0.9$ (Lee and Ramakrishnan, 1985; Meir et al., 1986), $T_q \sim 2.8$ in three dimensions.

The cumulative distribution of the failure probability is Gumbel, as in the case of fracture discussed in Section 4.2, for the sysetm of size L and is given by Chakrabarti (1994)

$$F_L(E) \sim 1 - \exp[-L^d g(E)]$$

$$\sim 1 - \exp\left[-L^d \exp\left(-\frac{|\Delta \epsilon_m|^{T_q}}{E}\right)\right]. \qquad (D1)$$

This implies

$$E_b \sim \frac{|\Delta \epsilon_m|^{T_q}}{\ln L}, \qquad (D2)$$

as the size dependence of the typical value of the breakdown field on and above of which $F_L(E)$ becomes significant.

E
Fractals

In the familiar Euclidean dimensions of space, the number of independent variables to specify the dynamics of the system gives the dimension. But, one can also define dimensions in a different way. Suppose we decrease the linear size of a system by a factor $1/b$. Then it is obvious that a quantity $Q(L_1)$, measured with the initial length scale L_1, is related to the same quantity $Q(L_2)$ measured with the changed length scale $L_2 (= L_1/b)$, by the following relation:

$$\frac{Q(L_1)}{Q(L_2)} = b^{-D} \tag{E1}$$

where D is the dimension of the system. Defined in this way, D need not be an integer. We consider a few examples when this is so. These fractional dimensional objects are called *fractals*.

- **Example 1:** Consider a square divided into four equal parts. Now, drop the fourth quadrant and do the same for the remaining three. If this is continued up to infinite steps, one can generate an object which is a fractal (Figure E.1). This statement is justified from the fact that if we reduce the length to half of the original one, then the mass of the system reduces to one-third of the original value. Hence,

$$\left(\frac{1}{2}\right)^D = \frac{1}{3}.$$

Giving,

$$D = \frac{\ln 3}{\ln 2} = 1.5849 \ldots \tag{E2}$$

Note that the fractal dimension of the object is less than the embedding Euclidean dimension.

- **Example 2:** Consider an equilateral triangle. Mark the middle points of each side and join them. We now have three triangles where each side is halved. One can continue the same process with these three triangles. Eventually, one would again reach a fractal object (Figure E.2). In this object too, if the length is reduced to half of its original value, the mass of the object reduces to one-third of the original value. Therefore, as before,

$$D = \frac{\ln 3}{\ln 2}, \tag{E3}$$

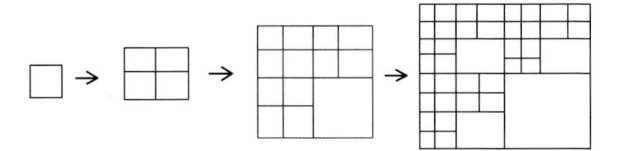

Figure E.1 Construction of a fractal object starting from a square. (shown up to third generation) The fractal (self-similar at all length scales) is formed as the generation number goes to infinity.

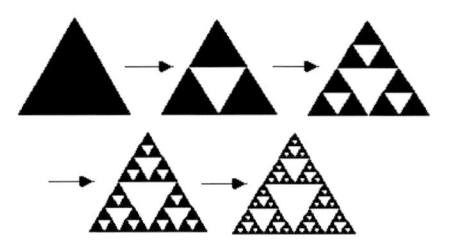

Figure E.2 The steps of formation of the Sierpinski triangle; the fractal is formed at infinite step or generation number.

Figure E.3 Steps of forming the Cantor set; the fractal set is formed for $n \to \infty$.

implying that we again have the fractal dimension less than the Euclidean dimension. This particular structure, which is an example of fractal, is called a *Sierpinski triangle*.

- **Example 3:** Consider a line segment of unit length. Cut it into three pieces and remove the middle piece. One can continue to do this indefinitely (Figure E.3). In this case, the length is scaled by a factor $1/3$ and the number of line segment increases by a factor 2. Hence, we again have,

$$\left(\frac{1}{2}\right)^D = \left(\frac{1/3}{1}\right),$$

implying

$$D = \frac{\ln 2}{\ln 3}, \tag{E4}$$

which is also not an integer. This example of fractal object is called a *Cantor set*.

F
Two-fractal Overlap Model

The seismic faults are known to posses self-affine roughness properties (see e.g., Santucci et al., 2007 and references therein). Hence, the effect of roughness on earthquake phenomenon is an interesting question. Chakrabarti and Stinchcombe (1999) considered two rough surfaces, modeled by Cantor sets, being dragged one over the other and calculated the overlap function and related it with the energy released in the motion. Hence, its statistics is relevant for earthquakes.

Cantor set (regular) is one of the simplest examples of fractals. It is constructed in the following way: a line segment in $[0:1]$ is taken and is divided into three equal parts. Then the middle part is removed. The same operation is performed on the two remaining parts and so on. This process is required to be carried out ad infinitum until the remaining set becomes a true fractal. However, in reality, there is always a lower (as well as upper) bound for self-similarity arising out of other physical constraints. Hence, we work with a finite generation Cantor set only.

In this model, both the surfaces involved in earthquake, namely the tectonic plate and earth's crust, are considered as Cantor sets in the simplified limit (Figure F.1). The energy released in a slip event was related to the amount of overlaps between the two surfaces. The two main statistical laws for earthquakes, namely the Gutenberg-Richter law and Omori law were found to be valid here.

F.1
Renormalization Group Study: Continuum Limit

Let us consider the generator for the nth generation within $[0:1]$ $G_0 = [0,1]$, $G_1 \equiv RG_0 = [0,a] \cup [b,1], \ldots, G_{n+1} = RG_n, \ldots$. Denoting the mass density of the set by $D_n(r)$ one has $D_n(r) = 1$ only if r is in the part which is not removed from the set up to nth generation. The overlap magnitude between two such sets is of the form $s_n(r) = \int dr' D_n(r') D_n(r - r')$ (for symmetric fractals).

The overlap integral s_1 in the first generation can be expressed by the projection of the shaded region along the vertical diagonals in Figure F.2(a). This gives the form in Figure F.2(b). For $a = b \leq 1/3$, the nonvanishing $s_1(r)$ regions are symmetric and do not overlap. The slope of the middle curve is double of that in the sides. It can be checked that the distribution $\rho_1(s)$ of the overlaps s at this generation

Statistical Physics of Fracture, Breakdown, and Earthquake: Effects of Disorder and Heterogeneity, First Edition.
Soumyajyoti Biswas, Purusattam Ray, and Bikas K. Chakrabarti.
© 2015 Wiley-VCH Verlag GmbH & Co. KGaA. Published 2015 by Wiley-VCH Verlag GmbH & Co. KGaA.

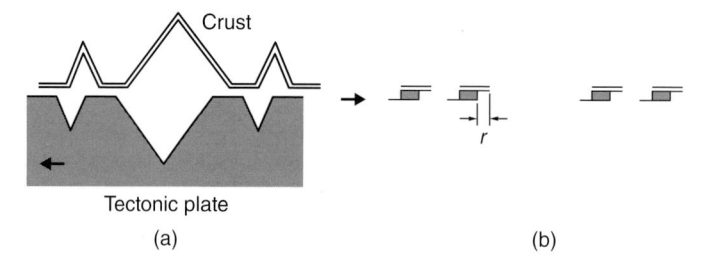

Figure F.1 (a) Schematic representation of the rough earth surface and the tectonic plate. (b) The one-dimensional projection of the surfaces form overlapping Cantor sets.

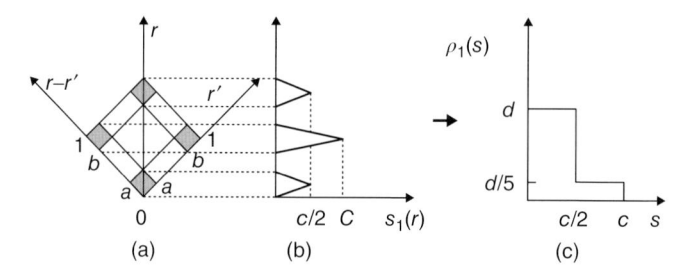

Figure F.2 (a) Two Cantor sets along the axes r and $r - r'$. (b) The overlap $s_1(r)$ along the diagonal. (c) The corresponding density $\rho_1(s)$.

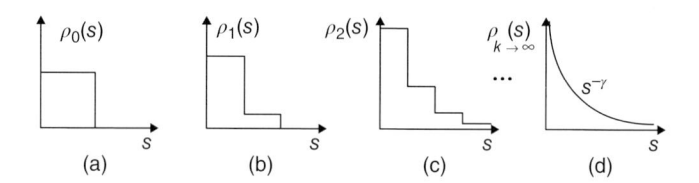

Figure F.3 The overlap densities (probabilities) $\rho(s)$ at various generations: (a) zeroth, (b) first, (c) second, and (d) infinite generation.

is shown in Figure F.2. Here, c and d are both greater than unity and for normalization $cd = 5/3$. For successive generations, the overlaps can be represented by Figure F.3, where

$$\rho_{n+1}(s) = \tilde{R}\rho_n(s) \equiv \frac{d}{5}\rho_n\left(\frac{s}{c}\right) + \frac{4d}{5}\rho_n\left(\frac{2s}{c}\right). \tag{F1}$$

Assuming $\rho^*(s)$ as the fixed-point distribution of the renormalization group (RG) equation and considering the form $\rho^*(s) \sim s^{-\beta}\tilde{\rho}(s)$, one gets $(d/5)c^\beta + (4d/5)(c/2)^\beta = 1$. Here, $\tilde{\rho}(s)$ represents an arbitrary modular function that includes a logarithmic correction for large s. The normalization condition $cd = 5/3$ is satisfied for $\beta = 1$ implying

$$\rho^*(s) \equiv \rho(s) \sim s^{-\beta}; \qquad \beta = 1 \tag{F2}$$

F.2
Discrete Limit

Apart from the continuous motion of the plates, the discrete case can be exactly solved (Bhattacharyya, 2005). Let $s_n(t)$ be the overlap between the two sets at time t. At $t = 0$, the two sets are placed on top of each other and the overlap is maximum (2^n). Also, the step size is chosen to be $1/3^n$ such that two elements can either overlap fully (if they are on top of each other) or does not overlap at all (no partial overlap). Because of the regular nature of the sets, the total overlap can have discrete values such as $s_n = 2^{n-k}$ and $k = 0, \ldots, n$.

If $Nr(s_n)$ is the number of times the overlap s_n has taken place during a complete interval (i.e., 3^n time steps), then

$$Nr(2^{n-k}) = {}^nC_k 2^k, \quad k = 0, \ldots, n \tag{F3}$$

Therefore, if $\text{Prob}(s_n)$ is the probability that the overlap is s_n after time t, then for the general case $s_n = 2^{n-k}, k = 0, \ldots, n$, it has the following form:

$$\text{Prob}(2^{n-k}) = \frac{Nr(2^{n-k})}{\sum_{k=0}^{n} Nr(2^{n-k})}$$

$$= \frac{2^k}{3^n} {}^nC_k$$

$$= {}^nC_{n-k} \left(\frac{1}{3}\right)^{n-k} \left(\frac{2}{3}\right)^k \tag{F4}$$

F.2.1
Gutenberg-Richter Law

As the allowed values for the overlaps are $s_n = 2^{n-k}, k = 0, \ldots, n$, one can write $\log_2 s_n = n - k$. Then, the above equation takes the form

$$\text{Prob}(s_n) = {}^nC_{\log_2 s_n} \left(\frac{1}{3}\right)^{\log_2 s_n} \left(\frac{2}{3}\right)^{n-\log_2 s_n}$$

$$\equiv F(\log_2 s_n). \tag{F5}$$

Now, near the maxima, it has the form

$$F(M) = \frac{3}{2\sqrt{n\pi}} \exp\left[-\frac{9}{4}\frac{(M - n/3)^2}{n}\right], \tag{F6}$$

with $M = \log_2 s_n$. In order to get the GR law, this distribution is to be integrated between M and ∞

$$F_{\text{cum}}(M) = \int_M^\infty F(M') dM'$$

$$= \int_M^\infty \frac{3}{2\sqrt{n\pi}} \exp\left(-\frac{9}{4}\frac{(M' - n/3)^2}{n}\right) dM'. \tag{F7}$$

Substituting $p = \frac{3}{2\sqrt{n}}(M' - n/3)$, one gets

$$F_{\text{cum}}(M) = \frac{1}{\sqrt{\pi}} \int_{\frac{3}{2\sqrt{n}}(M-n/3)}^{\infty} \exp(-p^2) dp. \tag{F8}$$

This can be simplified as

$$F_{\text{cum}}(M) = \frac{1}{3}\sqrt{\frac{n}{\pi}} \exp\left[-\frac{9}{4}\frac{(M-n/3)^2}{n}\right](M-n/3)^{-1}. \tag{F9}$$

$F_{\text{cum}}(M)$ indicates that the "average" quakes are of magnitude $n/3$, while

$$F_{\text{cum}}(M) \sim \exp[-(9/4)(M-n/3)^2/n] \tag{F10}$$

can be further simplified for large M. Using $e^{-a^2} = (1/\sqrt{2\pi})\int_{-\infty}^{+\infty} e^{-x^2/2+\sqrt{2}ax} dx$ and $\int_{-\infty}^{+\infty} e^{-f(x)} dx \sim e^{-f(x_0)}$, where x_0 refers to the extremal point with $\partial f/\partial x|_{x=x_0} = 0$, one finds $F_{\text{cum}}(M) \sim e^{-(9/4)[M(m_0/n)-2M/3]} \sim e^{-3M/4}$ with $x_0 = \left(\frac{3}{\sqrt{2n}}\right)m_0$; $m_0 = n$. Using it, one gets Bhattacharya et al. (2009)

$$\log F_{\text{cum}}(M) = A - \frac{3}{4}M, \tag{F11}$$

where A is a constant that depends on n. This is the Gutenberg-Richter law, which holds for the large M values for the model, as expected. Also, the magnitude M can be related to the energy released E by noting $M = \log_2 s$ here. The overlap is related to the energy giving $M \sim \log E$, implying $F_{\text{cum}} \sim E^{-3/4}$.

The Gutenberg–Richter law obtained here is not limited to the simple construction of the regular Cantor set, but is generally valid for many other similar situations such as random Cantor sets, Sierpinski gasket, and carpet sliding over their respective replica (Pradhan et al., 2003), and a fractional Brownian profile overlapping on another (De Rubeis et al., 1996). This general validity suggests that the GR law can be related to the fractal geometry of the fault surfaces. Note that identifying the aftershocks as these adjusted overlaps, with average size given by Eq. (F9), an average magnitude ($n/3$) can be defined, which is dependent on the fractal geometry generator fraction ($= 1/3$ in the present case) and the generation number (n). This agrees with the observed data quite satisfactorily (see Bhattacharya et al., 2011).

F.2.2
Omori Law

If $N^{(M_c)}(t)$ is the cumulative number of aftershocks having magnitude $M \geq M_c$, where M_c is some threshold, the Omori law gives

$$\frac{dN^{(M_c)}(t)}{dt} = \frac{1}{t^p}. \tag{F12}$$

The value of p is close to unity, but has a spread. In the present case, $p = 0$ if $M_c = 1$, since aftershock occurs at every instance of time. But, if M_c is fixed at the second highest value, that is, $n-1$, then for $t = 2.3^{r_1}$ (where $r_1 = 0, \ldots, n-1$), there is an aftershock having magnitude $n-1$. Hence, the aftershocks in geometric progression with common ratio 3 lead to the general rule $N(3t) = N(t) + 1$, giving

$$N(t) = \log_3(t). \tag{F13}$$

Omori law, on integration, gives $N(t) = t^{1-p}$. Hence, from this model, we get $p = 1$ for the particular threshold. By varying the threshold, the values between 0 and 1 can be obtained.

G
Microscopic Theories of Friction

While friction is a complex phenomenon relevant for different length scales, its microscopic origin remains a point of interest both because of its technological importance in nano devices and also it is often the base of understanding this phenomenon in higher scales. Questions regarding the atomistic origin of friction can now be asked even experimentally (due to advanced Atomic Force Microscopy (AFM) studies). Furthermore, molecular dynamic simulations in large scale can test model hypothesis formulated in theory (Bhushan et al., 1994; Braun and Naumovets, 2006; Hölscher et al., 2008). Such simulations consider two atomically smooth surfaces being dragged one over the other. The effects of inhomogeneity in terms of lubrication and/or disorders (such as vacancies) are also addressed.

One of the simplest and earliest models for atomistic friction was proposed by Tomlinson (1929). Two linear arrays of contacts were considered. The lower array was modeled as periodically placed atoms or for that matter simply a sinusoidal potential and the upper array were thought of as a series of mutually independent beads that are connected elastically to the bulk above. The oversimplified picture did not consider the interaction between these arrays of atoms.

G.1
Frenkel-Kontorova Model

This model (Frenkel and Kontorova, 1938) addresses some of the simplifications made in the earlier model. Here, the sliding array of atoms is connected by linear springs with the nearest neighbors. Hence, there is a possibility of interesting cooperative phenomenon occurring in certain conditions of the model. The bottom array is still taken as a sinusoidal potential. The Hamiltonian of the system reads

$$H = \sum_{i=1}^{N} \left(\frac{1}{2}K(x_{i+1} - x_i - a)^2 + V(x_i)\right), \tag{G1}$$

where x_i is the position of the ith atom and a is the equilibrium spacing between the atoms. The substrate potential, as mentioned earlier, is of the form $V(x) = -V_0 \cos(\frac{2\pi x}{b})$. Therefore, there are two competing scales in the system: the equilibrium spacing of the upper chain and the period of the substrate potential. While the harmonic springs will try to keep the separation of the atoms unchanged, the substrate potential minima will attract the atoms. Simultaneous satisfaction of these two competing forces is possible only if the ratio a/b is commensurate. In that case, the atoms will arrange periodically (with the common period of the two potentials) and a finite force will be required to move the chain over the potential. On the other hand, if the ratio is incommensurate, for a given value of the spring constant, there is a critical value of the amplitude of the substrate potential up to which the upper chain is "fee." This means up to that amplitude value, the chain can move under application of any infinitesimal applied force. Up to this point, the hull function remains analytic (Peyrard and Aubry, 1983). Beyond that amplitude, the analyticity of the hull function is broken and the upper chain becomes pinned in the sense that a finite force is now required to move it. This is called the breaking of analyticity transition or Aubry transition (Peyrard and Aubry, 1983).

G.2
Two-chain Model

The Frenkel-Kontorova model mentioned earlier can be extended in many ways, namely studies in higher dimensions, effect of impurity, and so on. But, one shortcoming that remained is the rigidity of the lower surface. The substrate potential was always taken as a fixed sinusoidal function, which rules out any interaction and consequent readjustments in the positions of the lower atoms. This question was addressed in the two-chain model (Matsukawa and Fukuyama, 1994). Here, both the surfaces were taken in the same footing in the sense that both were modeled as harmonically coupled chains. The equations of motion for the chains read

$$m_a \gamma_a (\dot{x}_i - \langle \dot{x}_i \rangle) = K_a (x_{i+1} + u_{i-1} - 2x_i)$$
$$+ \sum_{j \in b}^{N_b} F_I(x_i - y_j) + F_{ex}, \quad \text{(G2)}$$

$$m_b \gamma_b (\dot{y}_i - \langle \dot{y}_i \rangle) = K_b (y_{i+1} + y_{i-1} - 2y_i)$$
$$+ \sum_{j \in a}^{N_a} F_I(y_i - x_j) - K_s(y_i - ic_b) \quad \text{(G3)}$$

where the suffixes a and b denote the upper and lower chains, respectively and other symbols have usual meanings. The interatomic force F_I is derived from the potential

$$U_I = -\frac{K_I}{2} \exp\left(-4\left(\frac{x}{c_b}\right)\right), \quad \text{(G4)}$$

where K_I is the interaction strength.

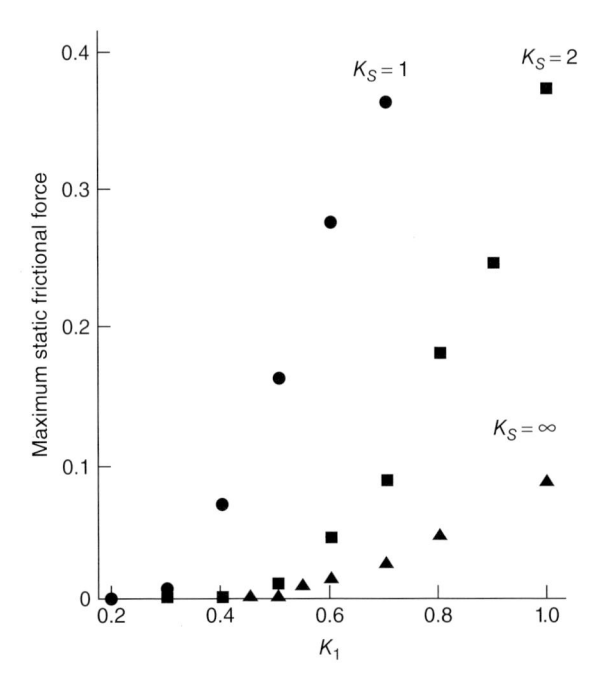

Figure G.1 The variations of the maximum static friction with the amplitude (K_I) of the interchain potential for different values of the lower-chain stiffness (K_s). The limit $K_s \to \infty$ corresponds to Frenkel-Kontorova model. But it is clearly seen that even for finite K_s (i.e., when the lower chain can relax) there is a finite value of interchain potential amplitude up to which the static friction is practically zero and it increases afterward, signifying Aubry transition (Matsukawa and Fukuyama, 1994), with kind permission from American Physical Society. Figure courtesy H. Matsukawa. is still

The friction force was argued to be of the form

$$-\sum_i \sum_j \langle F_I(x_i - y_j)\rangle_t = N_a \langle F_{ex}\rangle_t. \tag{G5}$$

It was numerically shown that the velocity dependence of the kinetic friction vanishes as the static friction was increased (one of the Amonton-Coulomb law).

The breaking-of-analyticity transition is still observed in this model for finite rigidity of the lower chain (see Figure G.1)

G.2.1
Effect of Fractal Disorder

The effects of disorder are significant even in the microscale. Eriksen et al. (2010) considered the effect of self-affine roughness. The surface, in its simplest form, was taken as a random Cantor set. It was constructed in the following way: a line segment [0 : 1] was taken and was divided into s equal segments. Then $s - r$ of those segments were randomly removed. In the next generation, the same operation was

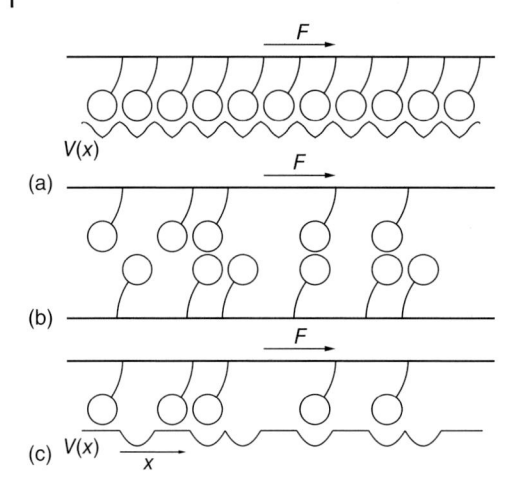

Figure G.2 Schematic representation of the two-chain version of the Tomlinson model with (a) no disorder, (b) Cantor set disorder, and (c) the effective substrate potential. From Eriksen et al. (2010), with kind permission from American Physical Society.

performed on all remaining segments and so on. This gives rise to a self-similar structure in the statistical sense.

The roughness was introduced on both chains of a Tomlinson model in this way. Then the interaction between the atoms in the upper and lower chains was taken as very short ranged (see Figure G.2) such that there is an attractive interaction when an atom comes right on top of another and apart from that there is no other interaction. It was found that the static frictional force had a distribution, which

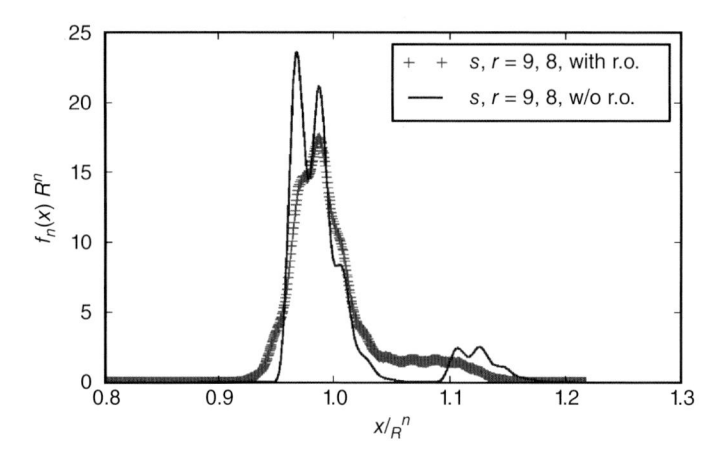

Figure G.3 The overlap distribution for $s = 9, r = 8$ is shown. The dotted curve shows the average distribution with random offset, and the continuous curve is that without random offset. The distribution is qualitatively different from the Gaussian distribution expected for random disorder. From Eriksen et al. (2010), with kind permission from American Physical Society.

is qualitatively different from what one would expect for a random disorder. The distribution looks like (Eriksen et al., 2010)

$$f^{s,r}(x/R)/R = \sum_{j=1}^{r} \tilde{c}^{s,r}(j)(f^{s,r} \ldots j-1 \text{ terms} \ldots f^{s,r})(x), \quad (G6)$$

where $R = r^2/s$ and $\tilde{c}^{s,r}(x) = \frac{{}^rC_x^{s-r} C_{r-x}}{{}^sC_r}$. The function is shown in Figure G.3 for a given combination of s,r (9, 8). Clearly, the distribution is very different from the Gaussian distribution expected for random defects.

References

Acharyya, M. and Chakrabarti, B.K. (1996a) Cluster statistics in dielectric breakdown. *Physica A*, **224**, 287

Acharyya, M. and Chakrabarti, B.K. (1996b) Growth of breakdown susceptibility in random composites and the stick-slip model of earthquakes: prediction of dielectric breakdown and other catastrophes. *Phys. Rev. E*, **53**, 140.

Acharyya, M. and Chakrabarti, B.K. (1996c) Response of random dielectric composites and earthquake models to pulses: prediction possibilities. *Physica A*, **224**, 254.

Alava, M., Nukala, P.K.V.V., and Zapperi, S. (2009a) Size effects in statistical fracture. *J. Phys. D*, **42**, 214012.

Alava, M.J., Nukala, P.K.V.V., and Zapperi, S. (2009b) Size effects in statistical fracture. *J. Phys. D*, **42**, 214012.

Alava, M.J. and Zapperi, S. (2004) Comment on 'roughness of interfacial crack front: stress-weighted percolation in the damage zone'. *Phys. Rev. Lett.*, **92**, 049601.

Allen, M.P. and Tildesley, D.J. (1987) *Computer Simulation of Liquids*, Oxford University Press.

Ambegaokar, V., Halperin, B.I., Nelson, D.R., and Siggia, E.D. (1980) Dynamics of superfluid films. *Phys. Rev. B*, **21**, 1806.

Amitrano, D., Grasso, J.R., and Senfaute, G. (2005) Seismic precursory patterns before a cliff collapse and critical point phenomena. *Geophys. Res. Lett.*, **32**, L08314.

Amitrano, D. and Helmstetter, A. (2006) Brittle creep, damage, and time to failure in rocks. *J. Geophys. Res.*, **111**, B11201.

Andersen, J.V., Sornette, D., and Leung, K. (1997) Tricritical behavior in rupture induced by disorder. *Phys. Rev. Lett.*, **78**, 2140.

Anderson, T.L. (1995) *Fracture Mechanics: Fundamentals and Applications*, CRC Press.

Andrade, E.Nd.C. (1910) The viscous flow in metals and allied phenomena. *Proc. R. Soc. London, Ser. A*, **84**, 1.

Ansari-Rad, M., Allaei, S.M.V., and Sahimi, M. (2012) Nonuniversality of roughness exponent of quasistatic fracture surfaces. *Phys. Rev. E*, **85**, 021121.

de Arcangelis, L., Redner, S., and Coniglio, A. (1985a) Anomalous voltage distribution of random resistor networks and a new model for the backbone at the percolation threshold. *Phys. Rev. B*, **31**, 4725.

de Arcangelis, L., Redner, S., and Herrmann, H.J. (1985b) A random fuse model for breaking processes. *J. Phys. Lett.*, **46**, 585.

Bak, P., Christensen, K., Danon, L., and Scanlon, T. (2002) Unified scaling law for earthquakes. *Phys. Rev. Lett.*, **88**, 178501.

Bak, P., and Tang, C. (1989) Earthquakes as a self-organized critical phenomenon. *J. Geo. Res.*, **94** (B11), 15,635–15,637.

Bak, P., Tang, C., and Wiesenfeld, K. (1987) Self-organized criticality: an explanation of the 1/f noise. *Phys. Rev. Lett.*, **59**, 381.

Baldassarri, A., Colaiori, F., and Castellano, C. (2003) Average shape of a fluctuation: universality in excursions of stochastic processes. *Phys. Rev. Lett.*, **90**, 060601.

Barai, P., Nukala, P.K.V.V., Alava, M.J., and Zapperi, S. (2013) Role of the sample thickness in planar crack propagation. *Phys. Rev. E*, **88**, 0422411.

Baró, J., Corral, A., Illa, X., Planes, A., Salje, E.K.H., Schranz, W., Soto-Parra, D.E., and Vives, E. (2013) Statistical similarity between the compression of a porous material and earthquakes. *Phys. Rev. Lett.*, **110**, 088 702.

Barré, J. (2009) Hierarchical models of rigidity percolation. *Phys. Rev. E*, **80**, 061 108.

Basquin, O.H. (1910) The exponential law of endurance tests. *Porc. Am. Soc. Test. Mater.*, **10**, 625.

Batrouni, G.G., Hansen, A., and Schmittbuhl, J. (2002) Heterogeneous interfacial failure between two elastic blocks. *Phys. Rev. E*, **65**, 036 126.

Baudet, C., Charaix, E., Clement, E., Guyon, E., Hulin, F.P., and Leroy, C. (1985) *Scaling Phenomena in Disordered Systems*, Plenum Publishing Corporation, New York, p. 399.

Bazant, Z.P. and Planas, J. (1997) *Fracture and Size Effect in Concrete and Other Quasibrittle Materials*, CRC Press LLC.

Beale, P.D. and Duxbury, P.M. (1988) Theory of dielectric breakdown in metal-loaded dielectrics. *Phys. Rev. B*, **37**, 2785.

Benguigui, L. (1986) Lattice and continuum percolation transport exponents: experiments in two-dimensions. *Phys. Rev. B*, **34**, 8176.

Benguigui, L., Ron, P., and Bergman, D.J. (1987) Strain and stress at the fracture of percolative media. *J. Phys. France*, **48**, 1547.

Bergman, D.J. and Stroud, D. (1992) *Solid State Physics: Applied in Research and Applications*, Vol. 46, Academic Press, New York, p. 147.

Bhattacharyya, P. (2005) Of overlapping cantor sets and earthquakes: analysis of the discrete chakrabarti-stinchcombe model. *Physica A*, **348**, 199.

Bhattacharya, P., Chakrabarti, B.K., and Kamal (2011) A fractal model of earthquake occurrence: theory, simulations and comparisons with the aftershock data. *J. Phys. Conf. Ser.*, **319**, 012 004.

Bhattacharyya, P., Chakrabarti, B.K., Kamal, and Samanta, D. (2009) *Reviews of Nonlinear Dynamics and Complexity*, Wiley-VCH Verlag GmbH, Berlin.

Bhattacharya, K. and Manna, S.S. (2007) Self-organized critical models of earthquakes. *Physica A*, **384**, 15.

Bhattacharyya, P., Pradhan, S., and Chakrabarti, B.K. (2003) Phase transition in fiber bundle models with recursive dynamics. *Phys. Rev. E*, **67**, 046 122.

Bhushan, B., Israelachvili, J.N., and Landman, U. (1994) Nanotribology: friction, wear and lubrication at the atomic scale. *Nature*, **374**, 607.

Binder, K. and Heermann, D. (1988) *Monte Carlo Simulations in Statistical Physics*, Springer, Berlin.

Biswas, S. and Chakrabarti, B.K. (2011) Depinning transitions in elastic strings. arXiv, 1108.1707.

Biswas, S. and Chakrabarti, B.K. (2013a) Crossover behaviors in one and two dimensional heterogeneous load sharing fiber bundle models. *Eur. Phys. J. B*, **86**, 160.

Biswas, S. and Chakrabarti, B.K. (2013b) Self-organized dynamics in local load sharing fiber bundle models. *Phys. Rev. E*, **88**, 042 112.

Biswas, S., Ray, P., and Chakrabarti, B.K. (2013) Equivalence of the train model of earthquake and boundary driven edwards-wilkinson interface. *Eur. Phys. J. B*, **86**, 388.

Biswas, S., Roy, S., and Ray, P. (2014) Nucleation versus percolation: scaling criterion for failure in disordered solids. arXiv:1411.7827.

Boffa, J.M., Allain, C., and Hulin, J.P. (2000) Experimental analysis of self-affine fractured rock surfaces through shadow length measurements. *Physica A*, **278**, 65.

Bonamy, D. (2009) Intermittency and roughening in the failure of brittle heterogeneous materials. *J. Phys. D: Appl. Phys.*, **42**, 214 014.

Bonamy, D. and Bouchaud, E. (2011) Failure of heterogeneous materials: a dynamic phase transition? *Phys. Rep.*, **498**, 1.

Bonamy, D., Ponson, L., Prades, S., Bouchaud, E., and Guillot, C. (2006) Scaling exponents for fracture surfaces in homogeneous glass and glassy ceramics. *Phys. Rev. Lett.*, **97**, 135 504.

Bouchaud, E. (1997) Scaling properties of cracks. *J. Phys. Condens. Matter*, **9**, 4319.

Bouchaud, J.P., Bouchaud, E., Lapasset, G., and Planes, J. (1993) Models of fractal cracks. *Phys. Rev. Lett.*, **71**, 2240.

Bouchaud, E., Lapasset, G., and Planè, J. (1990) Fractal dimension of fractured surfaces: a universal values? *Eur. Phys. Lett.*, **13**, 73.

Bouchaud, E. and Paun, F. (1999) Fracture and damage at a microstructural scale. *Comput. Sci. Eng.*, **1**, 32.

Bouchbinder, E., Mathiesen, J., and Procaccia, I. (2004) Roughening of fracture surfaces: the role of plastic deformation. *Phys. Rev. Lett.*, **92**, 245 505.

Bouchbinder, E., Procaccia, I., Santucci, S., and Vanel, L. (2006) Fracture surfaces as multiscaling graphs. *Phys. Rev. Lett.*, **96**, 055 509.

Bowman, D.R. (1989) Model for dielectric breakdown in metal-insulator composites. *Phys. Rev. B*, **40**, 4641.

Bradley, R.M. and Wu, K. (1994) Dynamic fuse model for electromigration failure of polycrystalline metal films. *Phys. Rev. E*, **50**, 631(R).

Braun, O.M. and Naumovets, A.G. (2006) Nanotribology: microscopic mechanisms of friction. *Surf. Sci. Rep.*, **60**, 79.

Brown, S. and Scholz, C.H. (1985) Broad bandwidth study of the topography of natural rock surfaces. *J. Geophys. Res.*, **90**, 12 575.

Burridge, R. and Knopoff, L. (1967) Model and theoretical seismicity. *Bull. Seismol. Soc. Am.*, **57**, 341.

Carlson, J.M. and Langer, J.S. (1989a) Mechanical model of an earthquake fault. *Phys. Rev. A*, **40**, 6470.

Carlson, J.M. and Langer, J.S. (1989b) Properties of earthquakes generated by fault dynamics. *Phys. Rev. Lett.*, **62**, 2632.

Carlson, J.M., Langer, J.S., and Shaw, B.E. (1994) Dynamics of earthquake faults. *Rev. Mod. Phys.*, **66**, 657.

Chakrabarti, B.K. (1994) in *Non-Linearity and Breakdown in Soft Condensed Matter* (eds K.K. Bardhan, B.K. Chakrabarti, and A. Hansen), Springer-Verlag, Heidelberg, p. 171.

Chakrabarti, B.K. and Benguigui, L.G. (1997) *Statistical Physics of Fracture and Breakdown in Disordered Systems*, Oxford University Press.

Chakrabarti, B.K., Chowdhury, D., and Stauffer, D. (1986) Molecular dynamics study of fracture in 2d disordered elastic lennard-jones solids. *Z. Phys. B: Condens. Matter*, **62**, 343.

Chakrabarti, B.K., Roy, A.K., and Manna, S.S. (1988) Breakdown exponents in lattice and continuum percolation. *J. Phys. C*, **21**, L65.

Chakrabarti, B.K. and Samanta, D. (2010) Scaling theory of quantum breakdown in solids. *Phys. Rev. B*, **81**, 052301.

Chakrabarti, B.K. and Stinchcombe, R.B. (1999) Stick-slip statistics for two fractal surfaces: a model for earthquakes. *Physica A*, **270**, 27.

Chauve, P., Le Doussal, P., and Wiese, K.J. (2001) Renormalization of pinned elastic systems: how does it work beyond one loop? *Phys. Rev. Lett.*, **86**, 1785.

Chayes, J.T., Chayes, L., and Durrett, R. (1986) Critical behavior of the two-dimensional first passage time. *J. Stat. Phys.*, **45**, 933.

Chen, Y.C., Lu, Z., Nomura, K., Wang, W., Kalia, R.K., Nakano, A., and Vashishta, P. (2007) Interaction of voids and nanoductility in silica glass. *Phys. Rev. Lett.*, **99**, 155 506.

Chen, Y.C., Nomura, K., Kalia, R.K., Nakano, A., and Vashishta, P. (2009) Void deformation and breakup in shearing silica glass. *Phys. Rev. Lett.*, **103**, 035 501.

Chessa, A., Vespignani, A., and Zapperi, S. (1999) Critical exponents in stochastic sandpile models. *Comput. Phys. Commun.*, **121**, 299.

Chianca, C.V., Martins, J.S.S., and de Oliveira, P.M.C. (2009) Mapping the train model of earthquakes onto the stochastic sandpile model. *Eur. Phys. J. B*, **68**, 549.

Christensen, K., Danon, L., Scanlon, T., and Bak, P. (2002) Unified scaling law for earthquakes. *Proc. Natl. Acad. Sci. U.S.A.*, **99**, 2509.

Christophorou, L.G. (ed.) (1982) *Gaseous Dielectrics III*, Pergamon Press, New York.

Cieplak, M., Maritan, A., and Banavar, J.R. (1996) Invasion percolation and eden growth: geometry and universality. *Phys. Rev. Lett.*, **76**, 3754.

Ciliberto, S., Guarino, A., and Scorretti, R. (2001) The effect of disorder on the fracture nucleation process. *Physica D*, **158**, 83.

Cohen, D., Lehmann, P., and Or, D. (2009) Fiber bundle model for multiscale modeling of hydromechanical triggering of shallow landslides. *Water Resour. Res.*, **45**, W10 436.

Coniglio, A. (1982) Cluster structure near the percolation threshold. *J. Phys. A: Math. Gen.*, **15**, 3829.

Corral, A. (2003) Local distributions and rate fluctuations in a unified scaling law for earthquakes. *Phys. Rev. E*, **68**, 035 102(R).

Corral, A. (2004) Long-term clustering, scaling and universality in the temporal occurrence of earthquakes. *Phys. Rev. Lett.*, **92**, 108 501.

Corral, A. (2006a) *Modelling Critical and Catastrophic Phenomena in Geoscience*, Springer-Verlag, p. 191.

Corral, A. (2006b) Universal earthquake-occurrence jumps, correlations with time, and anomalous diffusion. *Phys. Rev. Lett.*, **97**, 178 501.

Cottrell, A.H. (1953) *Dislocation and Plastic Flow in Crystal*, Clarendon Press, Oxford.

Cramer, T., Wanner, A., and Gumbsch, P. (2000) Energy dissipation and path instabilities in dynamic fracture of silicon single crystals. *Phys. Rev. Lett.*, **85**, 788.

Cule, D. and Hwa, T. (1996) Tribology of sliding elastic media. *Phys. Rev. Lett.*, **77**, 278.

Daguier, P., Bouchaud, E., and Lapasset, G. (1995) Roughness of a crack front pinned by microstructural obstacles. *Europhys. Lett.*, **31**, 367.

Daguier, P., Nghiem, B., Bouchaud, E., and Creuzel, F. (1997) Pinning and depinning of crack fronts in heterogeneous materials. *Phys. Rev. Lett.*, **78**, 1062.

Dalmas, D., Lelarge, A., and Vandembroucq, D. (2008) Crack propagation through phase-separated glasses: effect of the characteristic size of disorder. *Phys. Rev. Lett.*, **101**, 255 501.

Daniels, H.E. (1945) The statistical theory of the strength of bundles of threads, I. *Proc. R. Soc. London, Ser. A*, **183**, 404.

Danku, Z. and Kun, F. (2013a) Creep rupture as a non-homogeneous poissonian process. *Sci. Rep.*, **3**, 2688.

Danku, Z. and Kun, F. (2013b) Temporal and spatial evolution of bursts in creep rupture. *Phys. Rev. Lett.*, **111**, 084 302.

Dauchot, O., Karmakar, S., Procaccia, I., and Zylberg, J. (2011) Athermal brittle-to-ductile transition in amorphous solids. *Phys. Rev. E*, **84**, 046 105.

De Rubeis, V., Hallgass, R., Loreto, V., Paladin, G., Pietronero, L., and Tosi, P. (1996) Self-affine asperity model for earthquakes. *Phys. Rev. Lett.*, **76**, 2599.

Delaplace, A., Schimittbuhl, J., and Måløy, K.J. (1999) High resolution description of a crack front in a heterogeneous plexiglas block. *Phys. Rev. E*, **60**, 1337.

Dhar, D. (1990) Self-organized critical state of sandpile automaton models. *Phys. Rev. Lett.*, **64**, 1613.

Dhar, D. (1999) The abelian sandpile and related models. *Physica A*, **263**, 4.

Dhar, D. and Ramaswamy, R. (1989) Exactly solved model of self-organized critical phenomena. *Phys. Rev. Lett.*, **63**, 1659.

Diaz-Guilera, A. (1992) Noise and dynamics of self-organized critical phenomena. *Phys. Rev. A*, **45**, 8551.

Diaz-Guilera, A. (1994) Dynamic renormalization group approach to self-organized critical phenomena. *Europhys. Lett.*, **26**, 177.

Dieter, G.E. (1928) *Mechanical Metallurgy*, McGraw-Hill Book Company.

Duemmer, O. and Krauth, W. (2007) Depinning exponents of the driven long-range elastic string. *J. Stat. Mech.: Theory Exp.*, **2007**, P01 019.

Durin, G. and Zapperi, S. (2000) Scaling exponents for barkhausen avalanches in polycrystalline and amorphous ferromagnets. *Phys. Rev. Lett.*, **84**, 4705.

Duxbury, P.M., Beale, P.D., and Leath, P.L. (1986) Size effects of electrical background in quenched random media. *Phys. Rev. Lett.*, **57**, 1052.

Duxbury, P.M. and Leath, P.L. (1987) The failure distribution in percolation models of breakdown. *J. Phys. A: Math. Gen.*, **20**, L411.

Duxbury, P.M., Leath, P.L., and Beale, P.D. (1987) Breakdown properties of quenched random systems: the random-fuse network. *Phys. Rev. B*, **36**, 367.

Edwards, S.F. and Wilkinson, D.R. (1982) The surface statistics of a granular aggregate. *Proc. R. Soc. London, Ser. A*, **381**, 17.

Eriksen, J.A., Biswas, S., and Chakrabarti, B.K. (2010) Effect of fractal disorder on static friction in the tomlinson model. *Phys. Rev. E*, **82**, 041 124.

Ertas, D. and Kardar, M. (1994) Critical dynamics of contact line depinning. *Phys. Rev. E*, **49**, R2532.

Essam, J.W. (1980) Percolation theory. *Rep. Prog. Phys.*, **43**, 833.

Falk, M.L. (1999) Molecular-dynamics study of ductile and brittle fracture in model noncrystalline solids. *Phys. Rev. B*, **60**, 7062.

Family, F. and Vicsek, T. (1985) Scaling of the active zone in the eden process on percolation networks and the ballistic deposition model. *J. Phys. A: Math. Gen.*, **18**, L75.

Family, F., Zhang, Y.C., and Vicsek, T. (1986) Invasion percolation in an external field: dielectric breakdown in random media. *J. Phys. A: Math. Gen.*, **19**, L733.

Feng, S. and Sen, P.N. (1984) Percolation on elastic networks: new exponent and threshold. *Phys. Rev. Lett.*, **52**, 216.

Fisher, D.S. (1985) Sliding charge-density waves as a dynamic critical phenomenon. *Phys. Rev. B*, **31**, 1396.

Flekkøy, E.G. and Malthe-Sørenssen, A. (2002) Modelling hydrofracture. *J. Geophys. Res.*, **107**, 2151.

Frenkel, Y. and Kontorova, T. (1938) *Zh. Eksp. Teor. Fiz.*, **8**, 1340.

Galindo Torres, S.A. and Mu noz Casta no, J.D. (2007) Simulation of the hydraulic fracture process in two dimensions using a discrete element method. *Phys. Rev. E*, **75**, 066 109.

Gao, H. (1996) A theory of local limiting speed in dynamic fracture. *J. Mech. Phys. Solids*, **44**, 1453.

Gao, H. and Rice, J.R. (1989) A first order perturbation analysis on crack trapping by arrays of obstacles. *J. Appl. Mech.*, **56**, 828.

Garcimartin, A., Guarino, A., Belton, L., and Ciliberto, S. (1997) Statistical properties of fracture precursors. *Phys. Rev. Lett.*, **79**, 3202.

de Gennes, P.G. (1976) On a relation between percolation theory and the elasticity of gels. *J. Phys. Lett.*, **37**, 1.

Ghosh, M., Chakrabarti, B.K., Majumdar, K.K., and Chakrabarti, R.N. (1989) Stress relaxation in anelastic percolating solids. *Solid State Commun.*, **70**, 229.

Gjerden, K.S., Stormo, A., and Hansen, A. (2012) A model for stable interfacial crack growth. arXiv:1205.6661.

Gjerden, K.S., Stormo, A., and Hansen, A. (2013) Universality classes in constrained crack growth. *Phys. Rev. Lett.*, **111**, 135 502.

Gómez, J.B., I niguez, D., and Pacheco, A.F. (1993) Solvable fracture model with local load transfer. *Phys. Rev. Lett.*, **71**, 380.

Grassberger, P. (1994) Efficient large-scale simulations of a uniformly driven system. *Phys. Rev. E*, **49**, 2436.

Griffith, A.A. (1921) The phenomena of rupture and flow in solids. *Philos. Trans. R. Soc. London, Ser. A*, **221**, 163.

Guarino, A., Garcimartin, A., and Ciliberto, S. (1998) An experimental test of the critical behaviour of fracture precursors. *Eur. Phys. J. B*, **6**, 13.

Guarino, A., Ciliberto, S., and Garcimartin, A. (1999a) Failure time and microcrack nucleation. *Europhys. Lett.*, **47**, 456.

Guarino, A., Scorretti, R., and Ciliberto, S. (1999b) Material failure time and the fiber bundle model with thermal noise. arXiv:cond-mat/9908329.

Gumbel, E.J. (1958) *Statistics of Extremes*, Columbia University Press, New York.

Gumbsch, P., Zhou, S.J., and Hollan, B.L. (1997) Molecular dynamics investigation of dynamic crack stability. *Phys. Rev. B*, **55**, 3445.

Gutenberg, B. and Richter, C.F. (1944) Frequency of earthquakes in california. *Bull. Seismol. Soc. Am.*, **34**, 185.

Gutenberg, B. and Richter, C.F. (1956) Earthquake magnitude, intensity, energy and acceleration (second paper). *Bull. Seismol. Soc. Am.*, **46**, 105.

Halász, Z., Danku, Z., and Kun, F. (2012) Competition of strength and stress disorder in creep rupture. *Phys. Rev. E*, **85**, 016 116.

Hallgass, R., Loreto, V., Mazzella, O., Paladin, G., and Pietronero, L. (1997) Earthquake statistics and fractal faults. *Phys. Rev. E*, **56**, 1346.

Halperin, B.I., Shechao, F., and Sen, P.N. (1985) Differences between lattice and continuum percolation transport exponents. *Phys. Rev. Lett.*, **54**, 2391.

Halpin-Healy, T. (1989) Diverse manifolds in random media. *Phys. Rev. Lett.*, **62**, 442.

Hansen, A. and Hemmer, P.C. (1994) Burst avalanches in bundles of fibers: local load sharing. *Phys. Lett. A*, **184**, 394.

Hansen, A., Hinrichsen, E.L., and Roux, S. (1991) Roughness of crack interfaces. *Phys. Rev. Lett.*, **66**, 2476.

Hansen, A., Plouraboué, F., and Stéphane, R. (1995) Shadows in a self-affine landscape. *Fractals*, **3**, 91.

Hansen, A. and Schmittbuhl, J. (2003) Origin of the universal roughness exponent of brittle fracture surfaces: stress-weighted percolation in the damage zone. *Phys. Rev. Lett.*, **90**, 045 504.

Hao, S.W., Zhang, B.J., and Tian, J.F. (2012) Relaxation creep rupture of heterogeneous material under constant strain. *Phys. Rev. E*, **85**, 012 501.

Hemmer, P.C. and Hansen, A. (1992) The distribution of simultaneous fiber failures in fiber bundles. *J. Appl. Mech.*, **59**, 909.

Hergarten, S. and Neugebauer, H.J. (2002) Foreshocks and aftershocks in the olami-feder-christensen model. *Phys. Rev. Lett.*, **88**, 238 501.

Herrmann, H.J. and Roux, S. (eds) (1990) *Statistical Models for the Fracture of Disordered Media*, Elsevier Amsterdam.

Hidalgo, R.C., Kovács, K., Pagonabarraga, I., and Kun, F. (2008a) Universality class of fiber bundles with strong heterogeneities. *Eur. Phys. Lett.*, **81**, 54 005.

Hidalgo, R.C., Zapperi, S., and Herrmann, H.J. (2008b) Discrete fracture model with anisotropic load sharing. *J. Stat. Mech.: Theory Exp.*, **2008**, P01 004.

Hidalgo, R.C., Kun, F., and Herrmann, H.J. (2002a) Creep rupture of viscoelastic fiber bundles. *Phys. Rev. E*, **65**, 032 502.

Hidalgo, R.C., Moreno, Y., and Herrmann, H.J. (2002b) Fracture model with variable range interaction. *Phys. Rev. E*, **65**, 046 148.

Hirsch, P.B. and Roberts, S.G. (1996) Comment on the brittle-to-ductile transition: a cooperative dislocation generation instability; dislocation dynamics and the strain-rate dependence of the transition temperature. *Acta Mater.*, **44**, 2361.

Hölscher, H., Schirmeisen, A., and Schwarz, U.D. (2008) Principles of atomic friction: from sticking atoms to superlubric sliding. *Philos. Trans. R. Soc. London, Ser. A*, **366**, 1383.

Ivashkevich, E.V. (1996) Critical behavior of the sandpile model as a self-organized branching process. *Phys. Rev. Lett.*, **76**, 3368.

Jagla, E.A. (2011) Creep rupture of materials: insights from a fiber bundle model with relaxation. *Phys. Rev. E*, **83**, 046 119.

Janssen, H.K. (1998) Influence of long-range interactions on the critical behavior of systems with negative fisher exponent. *Phys. Rev. E*, **58**, 2673(R).

Joós, B. and Duesbery, M.S. (1997) The peierls stress of dislocations: an analytic formula. *Phys. Rev. Lett.*, **78**, 266.

Kahng, B., Batrouni, G.G., de Arcangelis, L., and Herrmann, H.J. (1988) Electrical breakdown in a fuse network with random, continuously distributed breaking strengths. *Phys. Rev. B*, **37**, 7625.

Kaiser, J. (1953) Erkenntnisse und folgerungen aus der messung von geräuschen bei zugbeanspruchung von metallischen werkstoffen. *Arch. Eisenhüttenw.*, **24**, 43.

Kalia, R.K., Nakano, A., Omeltchenko, A., Tsuruta, K., and Vashishta, P. (1997) Role of ultrafine microstructures in dynamic fracture in nanophase silicon nitride. *Phys. Rev. Lett.*, **78**, 2144.

Kardar, M. (1998) Nonequilibrium dynamics of interfaces and lines. *Phys. Rep.*, **301**, 85.

Karimi, M., Roarty, T., and Kaplan, T. (2006) Molecular dynamics simulations of crack propagation in ni with defects. *Modell. Simul. Mater. Sci. Eng.*, **14**, 1409.

Karpas, E. and Kun, F. (2011) Disorder-induced brittle-to-quasi-brittle transition in fiber bundles. *Eur. Phys. Lett.*, **95**, 16 004.

Katzav, E. and Adda-Bedia, M. (2013) Stability and roughness of tensile cracks in disordered materials. *Phys. Rev. E*, **88**, 052 402.

Kawamura, H. (2006) in *Modeling Critical and Catastrophic Phenomena in Geoscience* (eds P. Bhattacharyya and B.K. Chakrabarti), Springer-Verlag, Heidelberg, pp. 223–257.

Kawamura, H., Hatano, T., Kato, N., Biswas, S., and Chakrabarti, B.K. (2012) Statistical physics of fracture friction and earthquakes. *Rev. Mod. Phys.*, **84**, 839.

Kawamura, H., Yamamoto, T., Kotani, T., and Yoshino, H. (2010) Asperity characteristics of the olami-feder-christensen model of earthquakes. *Phys. Rev. E*, **81**, 031 119.

Khantha, M., Pope, D.P., and Vitek, V. (1995) Dislocation generation instability and the brittle-to-ductile transition. *Mater. Sci. Eng., A*, **192-193**, 435.

Khantha, M., Pope, D.P., and Vitek, V. (1997) Cooperative generation of dislocation loops and the brittle-to-ductile transition. *Mater. Sci. Eng., A*, **234-236**, 629.

Kierfeld, J. and Vinokur, V.M. (2006) Slow crack propagation in heterogeneous materials. *Phys. Rev. Lett.*, **96**, 175 502.

Kikuchi, H., Kalia, R.K., Nakano, A., Vashishta, P., Branicio, P.S., and Shimojo, F. (2005) Brittle dynamic fracture of crystalline cubic silicon carbide (3c-sic) via molecular dynamics simulation. *J. Appl. Phys.*, **98**, 103 524.

Knackstedt, M.A., Sheppard, A.P., and Pinczewski, W.V. (1998) Simulation of mercury porosimetry on correlated grids: evidence for extended correlated heterogeneity at the pore scale in rocks. *Phys. Rev. E*, **58**, 6923(R).

Knackstedt, M.A., Sheppard, A.P., and Sahimi, M. (2001) Pore network modelling of two-phase flow in porous rock: the effect of correlated heterogeneity. *Adv. Water Resour.*, **24**, 257.

Kolton, A.B., Rosso, A., Giamarchi, T., and Krauth, W. (2006) Dynamics below the depinning threshold in disordered elastic systems. *Phys. Rev. Lett.*, **97**, 057 001.

Kun, F., Carmona, H.A., Andrade, J.S. Jr., and Herrmann, H.J. (2008) Universality behind basquin's law of fatigue. *Phys. Rev. Lett.*, **100**, 094 301.

Kun, F., Costa, M.H., Costa Filho, R.N., Andrade, J.S. Jr., Soares, J.B., Zapperi, S., and Herrmann, H.J. (2007) Fatigue failure of disordered materials. *J. Stat. Mech.: Theory Exp.*, **2007**, P02 003.

Kun, F., Hidalgo, R.C., Herrmann, H.J., and Pál, K.F. (2003a) Scaling laws of creep rupture of fiber bundles. *Phys. Rev. E*, **67**, 061 802.

Kun, F., Moreno, Y., Hidalgo, R.C., and Herrmann, H.J. (2003b) Creep rupture has two universality classes. *Eur. Phys. Lett.*, **63**, 347.

Kunz, K. and Souilard, B. (1978) Essential singularity in the percolation model. *Phys. Rev. Lett.*, **40**, 133.

Kurita, K. and Fujii, N. (1979) Stress memory of crystalline rocks in acoustic emission. *Geophys. Res. Lett.*, **6**, 9.

Langer, J.S. and Lobkovsky, A.E. (1999) Rate- and-state theory of plastic deformation near a circular hole. *Phys. Rev. E*, **60**, 6978.

Laurson, L., Illa, X., and Alava, M.J. (2009) The effect of thresholding on temporal avalanche statistics. *J. Stat. Phys.*, **2009**, P01 019.

Laurson, L., Illa, X., Santucci, S., Tallakstad, K.T., Måløy, K.J., and Alava, M.J. (2013) Evolution of the average avalanche shape with the universality class. *Nat. Commun.*, **4**, 2927.

Laurson, L., Santucci, S., and Zapperi, S. (2010) Avalanches and clusters in planar crack front propagation. *Phys. Rev. E*, **81**, 046 116.

Laurson, L. and Zapperi, S. (2010) Roughness and multiscaling of planar crack fronts. *J. Stat. Mech.: Theory Exp.*, **2010**, P11 014.

Lavrov, A. (2003) The kaiser effect in rocks: principles and stress estimation techniques. *Int. J. Rock Mech. Min. Sci.*, **40**, 151.

Le Doussal, P., Wiese, K.J., Moulinet, S., and Rolley, E. (2009) Height fluctuations of a contact line: a direct measurement of the renormalized disorder correlator. *Eur. Phys. Lett.*, **87**, 56 001.

Lechenault, F., Pallares, G., George, M., Rountree, C., Bouchaud, E., and Ciccotti, M. (2010) Effect of finite probe size on self-affine roughness measurements. *Phys. Rev. Lett.*, **104**, 025 502.

Lee, P.A. and Ramakrishnan, T.V. (1985) Disordered electronic systems. *Rev. Mod. Phys.*, **57**, 287.

Lemerle, S., Ferré, J., Chappert, C., Mathet, V., Giamarchi, T., and Le Doussal, P. (1998) Domain wall creep in an ising ultrathin magnetic film. *Phys. Rev. Lett.*, **80**, 849.

Leocmach, M., Perge, C., Divoux, T., and Manneville, S. (2014) Creep and brittle failure of a protein gel under stress. arXiv, 1401.8234.

Li, Y.S. (1987) Size and location of the largest current in a random resistor network. *Phys. Rev. B*, **36**, 5411.

Lima, A.R., Moukarzei, C.F., Grosse, I., and Penna, T.J.P. (2000) Sliding blocks with random friction and absorbing random walks. *Phys. Rev. E*, **61**, 2267.

Ling, M. (1999) Cooperative dislocation generation under applied loads via kosterlitz-thouless mechanism: a monte carlo study. Dissertations available from ProQuest, p. AAI9926164.

Lobb, C.J., Hui, P.M., and Stroud, D. (1987) Nonuniversal breakdown behavior in superconducting and dielectric composites. *Phys. Rev. B*, **36**, 1956.

Lockner, D.A. (1998) A generalized law for brittle deformation of westerly granite. *J. Geophys. Res.*, **103**, 5107.

Lockner, D.A., Byerlee, J.D., Kuksenco, V., Ponomarev, A., and Sidorin, A. (1991) Quasi-static fault growth and shear fracture energy in granite. *Nature*, **350**, 39.

Lübeck, S. (1997) Large-scale simulations of the zhang sandpile model. *Phys. Rev. E*, **56**, 1590.

Lübeck, S. (2004) Universal scaling behavior of non-equilibrium phase transitions. *Int. J. Mod. Phys. B*, **18**, 3977.

Lübeck, S. and Usadel, K.D. (1997) Numerical determination of the avalanche exponents of the Bak-Tang-Wiesenfeld model. *Phys. Rev. E*, **55**, 4095.

Måløy, K.J., Hansen, A., Hinrichsen, E.L., and Roux, S. (1992) Experimental measurements of the roughness of brittle cracks. *Phys. Rev. Lett.*, **68**, 213.

Måløy, K.J., Santucci, S., Schmittbuhl, J., and Toussaint, R. (2006) Local waiting time fluctuations along a randomly pinned crack front. *Phys. Rev. Lett.*, **96**, 045 501.

Måløy, K.J. and Schmittbuhl, J. (2001) Dynamical event during slow crack propagation. *Phys. Rev. Lett.*, **87**, 105 502.

Maes, C., Van Moffaert, A., Frederix, H., and Strauven, H. (1998) Criticality in creep experiments on cellular glass. *Phys. Rev. B*, **57**, 4987.

Main, I.G. (2000) A damage mechanics model for power-law creep and earthquake aftershock and foreshock sequences. *Geophys. J. Int.*, **142**, 151.

Majumdar, S.N. and Dhar, D. (1991) Height correlations in the abelian sandpile model. *J. Phys. A: Math. Gen.*, **24**, L357.

Malthe-Sørenssen, A. (1999) Tilted sandpiles, interface depinning, and earthquake models. *Phys. Rev. E*, **59**, 4169.

Mandelbrot, B.B., Passoja, D.E., and Paullay, A.J. (1984) Fractal character of fracture surfaces of metals. *Nature*, **308**, 721.

Manna, S.S. (1991) Two-state model of self-organized criticality. *J. Phys. A: Math. Gen.*, **24**, L363.

Manna, S.S. and Bhattacharya, K. (2006) Self-organized critical earthquake model with moving boundary. *Eur. Phys. J. B*, **54**, 493.

Manna, S.S. and Chakrabarti, B.K. (1987) Dielectric breakdown in the presence of random conductors. *Phys. Rev. B*, **36**, 4078.

Matsukawa, H. and Fukuyama, H. (1994) Theoretical study of friction: one-dimensional clean surfaces. *Phys. Rev. B*, **49**, 17 286.

Meakin, P. (1985) The structure of two-dimensional witten-sander aggregates. *J. Phys. A: Math. Gen.*, **18**, L661.

Meakin, P. (1998) *Fractal Scaling and Growth Far from Equilibrium*, Cambridge University Press.

Meir, Y., Aharony, A., and Harris, A.B. (1986) Quantum percolation in magnetic fields. *Phys. Rev. Lett.*, **56**, 976.

Mendelson, K.S. (1975) A theorem on the effective conductivity of a two-dimensional heterogeneous medium. *J. Appl. Phys.*, **46**, 4740.

Michlmayr, G., Cohen, D., and Or, D. (2012) Sources and characteristics of acoustic emissions from mechanically stressed geologic granular media – a review. *Earth Sci. Rev.*, **112**, 97.

Minnhagen, P., Westman, O., Jonsson, A., and Olsson, P. (1995) New exponent for the nonlinear iv characteristics of a two dimensional superconductor. *Phys. Rev. Lett.*, **74**, 3672.

Moreira, A.A., Oliveira, C.L.N., Hansen, A., Araújo, N.A.M., Herrmann, H.J., and Andrade, J.S. Jr. (2012) Fracturing highly disordered materials. *Phys. Rev. Lett.*, **109**, 255 701.

Moreno, Y., Gómez, J.B., and Pacheco, A.F. (1999) Self-organized criticality in a fiber-bundle-type model. *Physica A*, **274**, 400.

Mori, T. and Kawamura, H. (2006) Simulation study of the one-dimensional burridge-knopoff model of earthquakes. *J. Geophys. Res.*, **111**, B07 302.

Mori, T. and Kawamura, H. (2008a) Simulation study of earthquakes based on the two-dimensional burridge-knopoff model with long-range interactions. *Phys. Rev. E*, **77**, 051 123.

Mori, T. and Kawamura, H. (2008b) Simulation study of the two-dimensional burridge-knopoff model of earthquakes. *J. Geophys. Res.*, **113**, B06 301.

Mott, N.F. (1948) Fracture of metals: some theoretical considerations. *Engineering*, **165**, 16-18.

Nakano, A., Kalia, R.K., and Vashishta, P. (1995) Dynamics and morphology of brittle cracks: a molecular dynamics study of silicon nitride. *Phys. Rev. Lett.*, **75**, 3138.

Narayan, O. and Fisher, D.S. (1992) Critical behavior of sliding charge-density waves in 4-e dimensions. *Phys. Rev. B*, **46**, 11 520.

Narayan, O. and Fisher, D.S. (1993) Threshold critical dynamics of driven interfaces in random media. *Phys. Rev. B*, **48**, 7030.

Nattermann, T., Stepanow, S., Tang, L.H., and Leschorn, H. (1992) Dynamics of interface depinning in a disordered medium. *J. Phys. II France*, **2**, 1483.

Nechad, H., Helmstetter, A., Guerjouma, R.E., and Sornette, D. (2005a) Andrade and critical time-to-failure laws in fiber-matrix composites: experiments and model. *J. Mech. Phys. Solids*, **53**, 1099.

Nechad, H., Helmstetter, A., Guerjouma, R.E., and Sornette, D. (2005b) Creep ruptures in heterogeneous materials. *Phys. Rev. Lett.*, **94**, 045 501.

Niemeyer, L., Pietronero, L., and Wiesmann, H.J. (1984) Fractal dimension of dielectric breakdown. *Phys. Rev. Lett.*, **52**, 1033.

Nukala, P.K.V.V. and Šimunović, S. (2004) An efficient block-current predictor for simulating fracture using large fuse networks. *J. Phys. A: Math. Gen.*, **37**, 2093.

Nukala, P.K.V.V. and Simunovic, S. (2004) Scaling of fracture strength in disordered quasi-brittle materials. *Eur. Phys. J. B*, **37**, 91.

Nukala, P.K.V.V., Zapperi, S., Alava, M.K., and Simunovic, S. (2008) Anomalous roughness of fracture surfaces in 2d fuse models. *Int. J. Fract.*, **154**, 119.

Nukala, P.K.V.V., Zapperi, S., and Šimunović, S. (2006) Crack surface roughness in three-dimensional random fuse networks. *Phys. Rev. E*, **74**, 026 105.

Olami, Z., Feder, H.J.S., and Christensen, K. (1992) Self-organized criticality in a continuous, nonconservative cellular automaton modeling earthquakes. *Phys. Rev. Lett.*, **68**, 1244.

Omori, F. (1895) On the aftershocks of earthquakes. *J. Coll. Sci. Imp. Univ. Tokyo*, 7, 111.

Paczuski, M. and Boettcher, S. (1996) Universality in sandpiles, interface depinning, and earthquake models. *Phys. Rev. Lett.*, **77**, 111.

Papanikolaou, S., Bohn, F., Sommer, R.L., Durin, G., Zapperi, S., and Sethna, J.P. (2011) Universality beyond power laws and the average avalanche shape. *Nat. Phys.*, **7**, 316.

Parsons, W.B. (1939) *Engineers and Engineering in the Renaissance*, MIT Press, Cambridge, MA, p. 661.

Pauchard, L. and Meunier, H. (1993) Instantaneous and time-lag breaking of a two-dimensional solid rod under a bending stress. *Phys. Rev. Lett.*, **70**, 3565.

Peirce, F.T. (1926) Theorems on the strength of long and of composite specimens. *J. Text. Ind.*, **17**, 355.

Petri, A., Paparo, G., Vespignani, A., Alippi, A., and Costantini, M. (1994) Experimental evidence for critical dynamics in microfracturing processes. *Phys. Rev. Lett.*, **73**, 3423.

Peyrard, M. and Aubry, S. (1983) Critical behavior at the transition by breaking of analyticity in the discrete frenkel-kontorova model. *J. Phys. C*, **16**, 1593.

Picallo, C.B., López, J.M., Zapperi, S., and Alava, M.J. (2010) From brittle to ductile fracture in disordered materials. *Phys. Rev. Lett.*, **105**, 155 502.

Pietronero, L., Vespignani, A., and Zapperi, S. (1994) Renormalization scheme for self-organized criticality in sandpile models. *Phys. Rev. Lett.*, **72**, 1690.

Politi, A., Ciliberto, S., and Scorretti, R. (2002) Failure time in the fiber-bundle model with thermal noise and disorder. *Phys. Rev. E*, **66**, 026 107.

Pomeau, Y. (1992) Brisure spontanée des cristaux bidimensionnel courbés. *C.R. Acad. Sci. Paris II*, **314**, 553.

Ponson, L. (2009) Depinning transition in the failure of inhomogeneous brittle materials. *Phys. Rev. Lett.*, **103**, 055 501.

Ponson, L., Auradou, H., Pessel, M., Lazarus, V., and Hulin, J.P. (2007) Failure mechanisms and surface roughness statistics of fractured fontainebleau sandstone. *Phys. Rev. Lett.*, **76**, 036 108.

Ponson, L., Auradou, H., Vié, P., and Hulin, J.P. (2006a) Low self-affine exponents of fractured glass ceramics surfaces. *Phys. Rev. Lett.*, **97**, 125 501.

Ponson, L., Bonamy, D., Auradou, H., Mourot, G., Morel, S., Bouchaud, E., Guillot, C., and Hulin, J.P. (2006b)

Anisotropic self-affine properties of experimental fracture surfaces. *Int. J. Fract.*, **140**, 27.

Ponson, L., Bonamy, D., and Barbier, L. (2006c) Cleaved surface of i-AlPdMn quasicrystals: influence of the local temperature elevation at the crack tip on the fracture surface roughness. *Phys. Rev. B*, **74**, 184 205.

Ponson, L., Bonamy, D., and Bouchaud, E. (2006d) Two-dimensional scaling properties of experimental fracture surfaces. *Phys. Rev. Lett.*, **96**, 035 506.

Pontuale, G., Colalori, F., and Petri, A. (2013) Slow crack propagation through a disordered medium: Critical transition and dissipation. *Eur. Phys. Lett.*, **101**, 16 005.

Power, W.L., Tullis, T.E., Brown, S.R., Boitnott, G.N., and Scholz, C.H. (1987) Roughness of natural fault surfaces. *Geophys. Res. Lett.*, **14**, 29.

Pradhan, S., Bhattacharyya, P., and Chakrabarti, B.K. (2002) Dynamic critical behavior of failure and plastic deformation in the random fiber bundle model. *Phys. Rev. E*, **66**, 016 116.

Pradhan, S. and Chakrabarti, B.K. (2001) Precursors of catastrophe in the Bak-Tang-Wiesenfeld, manna, and random-fiber-bundle models of failure. *Phys. Rev. E*, **65**, 016 113.

Pradhan, S. and Chakrabarti, B.K. (2003a) Failure due to fatigue in fiber bundles and solids. *Phys. Rev. E*, **67**, 046 124.

Pradhan, S. and Chakrabarti, B.K. (2003b) Failure properties of fiber bundle models. *Int. J. Mod. Phys. B*, **17**, 5565.

Pradhan, S., Chakrabarti, B.K., and Hansen, A. (2005a) Crossover behavior in a mixed-mode fiber bundle model. *Phys. Rev. E*, **71**, 036 149.

Pradhan, S., Hansen, A., and Hemmer, P.C. (2005b) Crossover behavior in burst avalanches: signature of imminent failure. *Phys. Rev. Lett.*, **95**, 125 501.

Pradhan, S., Chakrabarti, B.K., Ray, P., and Dey, M.K. (2003) Magnitude distribution of earthquakes: two fractal contact area distribution. *Phys. Scr. T*, **106**, 77.

Pradhan, S., Chandra, A.K., and Chakrabarti, B.K. (2013) Noise-induced rupture process: phase boundary and scaling of waiting time distribution. *Phys. Rev. E*, **88**, 012 123.

Pradhan, S., Hansen, A., and Chakrabarti, B.K. (2010) Failure processes in elastic fiber bundles. *Rev. Mod. Phys.*, **82**, 499.

Pradhan, S., Hansen, A., and Hemmer, P.C. (2006) Crossover behavior in failure avalanches. *Phys. Rev. E*, **74**, 016 122.

Pradhan, S. and Hemmer, P.C. (2007) Relaxation dynamics in strained fiber bundles. *Phys. Rev. E*, **75**, 056 112.

Pradhan, S. and Hemmer, P.C. (2008) Energy bursts in fiber bundle models of composite materials. *Phys. Rev. E*, **77**, 031 138.

Pradhan, S. and Hemmer, P.C. (2009) Breaking-rate minimum predicts the collapse point of overloaded materials. *Phys. Rev. E*, **79**, 041 148.

Pradhan, S. and Hemmer, P.C. (2011) Prediction of the collapse point of overloaded materials by monitoring energy emissions. *Phys. Rev. E*, **83**, 041 116.

Priezzhev, V.B. (1994) Structure of two-dimensional sandpile. I. Height probabilities. *J. Stat. Phys.*, **74**, 955.

Procaccia, I. and Zylberg, J. (2013) Propagation mechanism of brittle cracks. *Phys. Rev. E*, **87**, 012 801.

Rabinovitch, A., Friedman, M., and Bahat, D. (2004) Failure time in heterogeneous materials- non-homogeneous nucleation. *Europhys. Lett.*, **67**, 969.

Räisänen, V.I., Alava, M.J., and Nieminen, R.M. (1998) Fracture of three-dimensional fuse networks with quenched disorder. *Phys. Rev. B*, **58**, 14 288.

Raischel, F., Kun, F., and Herrmann, H.J. (2006) Local load sharing fiber bundles with a lower cutoff of strength disorder. *Phys. Rev. E*, **74**, 035 104(R).

Ramanathan, S. and Fisher, D.S. (1998) Onset of propagation of planar cracks in heterogeneous media. *Phys. Rev. B*, **58**, 6026.

Ramos, O., Cortet, P.P., Ciliberto, S., and Vanel, L. (2013) Experimental study of the effect of disorder on subcritical crack growth dynamics. *Phys. Rev. Lett.*, **110**, 165 506.

Ray, P. (2006) Breakdown of heterogeneous materials. *Comput. Mater. Sci.*, **37**, 141.

Ray, P. and Chakrabarti, B.K. (1985a) The critical behaviour of fracture properties of dilute brittle solids near the percolation threshold. *J. Phys. C*, **18**, L185.

Ray, P. and Chakrabarti, B.K. (1985b) A microscopic approach to the statistical fracture analysis of disordered brittle solids. *Solid State Commun.*, **53**, 477.

Ray, P. and Chakrabarti, B.K. (1988) Strength of disordered solids. *Phys. Rev. B*, **38**, 715.

Reiweger, I., Schweizer, J., Dual, J., and Herrmann, H.J. (2009) Modelling snow failure with a fiber bundle model. *J. Glasiol.*, **55**, 997.

Richter, C.F. (1935) An instrumental earthquake magnitude scale. *Bull. Seismol. Soc. Am.*, **25** (1-2), 1-32.

Rolley, E., Guthmann, C., Gombrowicz, R., and Repain, V. (1998) Roughness of the contact line on a disordered substrate. *Phys. Rev. Lett.*, **80**, 2865.

Rosso, A., Hartmann, A.K., and Krauth, W. (2003) Depinning of elastic manifolds. *Phys. Rev. E*, **67**, 021 602.

Rosso, A. and Krauth, W. (2002) Roughness at the depinning threshold for a long-range elastic string. *Phys. Rev. E*, **65**, 025 101(R).

Rosti, J., Illa, X., Koivisto, J., and Alava, M.J. (2009) Crackling noise and its dynamics in fracture of disordered media. *J. Phys. D*, **42**, 214 013.

Rosti, J., Koivisto, J., Laurson, L., and Alava, M.J. (2010) Fluctuations and scaling in creep deformation. *Phys. Rev. Lett.*, **105**, 100 601.

Rountree, C.L., Kalia, R.K., Lidorikis, E., Nakano, A., Van Brutzel, L., and Vashishta, P. (2002) Atomistic aspects of crack propagation in brittle materials: multimillion atom molecular dynamics simulations. *Annu. Rev. Mater. Res.*, **32**, 377.

Rountree, C.L., Prades, S., Bonamy, D., Bouchaud, E., Kalia, R., and Guillot, C. (2007) A unified study of crack propagation in amorphous silica: using experiments and simulations. *J. Alloys Compd.*, **434-435**, 60.

Roux, S. (2000) Thermally activated breakdown in the fiber-bundle model. *Phys. Rev. E*, **62**, 6164.

Roux, S., Hansen, A., Herrmann, H.J., and Guyon, E. (1988) Rupture of heterogeneous media in the limit of infinite disorder. *J. Stat. Phys.*, **52**, 237.

Roy, C., Kundu, S., and Manna, S.S. (2013) Scaling forms for relaxation times of the fiber bundle model. *Phys. Rev. E*, **87**, 062 137.

Roy, S. (2012) Ductile-brittle transition: a statistical mechanical approach. MSc Thesis. Institute of Mathematical Sciences, Chennai.

Roy, S. and Ray, P. (2014) Criticality in fiber bundle model. arXiv:1412.1211.

Rundle, J.B., Klein, W., Gross, S., and Turcotte, D.L. (1995) Boltzmann fluctuations in numerical simulations of nonequilibrium lattice threshold systems. *Phys. Rev. Lett.*, **75**, 1658.

Sahimi, M. (1994) *Application of Percolation Theory*, Taylor and Francis, London.

Sahimi, M. (2002) *Heterogeneous Materials II: Nonlinear and Breakdown Properties and Atomistic Melting*, Springer.

Sahimi, M. and Tajer, S.E. (2005) Self-affine fractal distributions of the bulk density, elastic moduli, and seismic wave velocities of rock. *Phys. Rev. E*, **71**, 046 301.

Saichev, A. and Sornette, D. (2005) Andrade, omori, and time-to-failure laws from thermal noise in material rupture. *Phys. Rev. E*, **71**, 016 608.

Samanta, D., Chakrabarti, B.K., and Ray, P. (2009) Classical and quantum breakdown in disordered materials, in *Quantum and Semi-Classical Percolation and Breakdown in Disordered Solids* (eds A.K. Sen, K.K. Bardhan, and B.K. Chakrabarti), Springer.

Santucci, S., Grob, M., Toussaint, R., Schmittbuhl, J., Hansen, A., and Måløy, K.J. (2010) Fracture roughness scaling: a case study on planar cracks. *Eur. Phys. Lett.*, **92**, 44 001.

Santucci, S., Måløy, K.J., Delaplace, A., Mathiesen, J., Hansen, A., Haavig Bakke, J.O., Schmittbuhl, J., Vanel, L., and Ray, P. (2007) Statistics of fracture surfaces. *Phys. Rev. E*, **75**, 016 104.

Sawada, Y., Ohta, S., Yamazaki, M., and Honjo, H. (1982) Self-similarity and phase-transition-like behavior of a randomly growing structure governed by a nonequilibrium parameter. *Phys. Rev. A*, **26**, 3557.

Schittbuhl, J., Hansen, A., and Batrouni, G.G. (2003) Roughness of interfacial crack front: stress-weighted percolation in the damage zone. *Phys. Rev. Lett.*, **90**, 045 505.

Schmittbuhl, J., Hansen, A., and Batrouni, G.G. (2004) Reply to comment on 'roughness of interfacial crack front:

stress-weighted percolation in the damage zone'. *Phys. Rev. Lett.*, **92**, 049 602.

Schmittbuhl, J. and Måløy, K.J. (1997) Direct observation of self-affine crack propagation. *Phys. Rev. Lett.*, **78**, 3888.

Schmittbuhl, J., Roux, S., Vilotte, J.P., and Måløy, K.J. (1995a) Interfacial crack pinning: effect of nonlocal interactions. *Phys. Rev. Lett.*, **74**, 1787.

Schmittbuhl, J., Schmitt, F., and Scholz, C. (1995b) Scaling invariance of crack surfaces. *J. Geophys. Res.*, **100**, 5953.

Schmittbuhl, J., Vilotte, J.P., and Roux, S. (1995c) Reliability of self-affine measurements. *Phys. Rev. E*, **51**, 131.

Schweizer, J. (1997) Laboratory experiments on shear failure of snow. *Ann. Glaciol.*, **26**, 97.

Scorretti, R., Ciliberto, S., and Guarino, A. (2001) Disorder enhances the effects of thermal noise in the fiber bundle model. *Europhys. Lett.*, **55**, 626.

Seppälä, E.T., Räisänen, V.I., and Alava, M.J. (2000) Scaling of interfaces in brittle fracture and perfect plasticity. *Phys. Rev. E*, **61**, 6312.

Sethna, J.P., Dahmen, K.A., and Myers, C.R. (2001) Crackling noise. *Nature*, **410**, 242.

Sezer, A. (2012) Employing cantor sets for earthquake time series analysis in two zones of western turkey. *Sci. Iran.*, **19**, 1456.

Shcherbakov, R. and Turcotte, D.L. (2003) Damage and self-similarity in fracture. *Theor. Appl. Fract. Mech.*, **39**, 245.

Shekhawat, A., Zapperi, S., and Sethna, J.P. (2013) From damage percolation to crack nucleation through finite size criticality. *Phys. Rev. Lett.*, **110**, 185 505.

Shimada, M. (2000) *Mechanical Behavior of Rocks Under High Pressure Conditions*, Taylor and Francis.

Skal, A. and Shklovskii (1975) *Sov. Phys. Semicond.*, **8**, 1029.

Sornette, D. (1988) Weibull-like distribution induced by fluctuations in percolation. *J. Phys.*, **49**, 889.

Sornette, D. (1989) Elasticity and failure of a set of elements loaded in parallel. *J. Phys. A*, **22**, L243.

de Sousa Vieira, M. (1992) Self-organized criticality in a deterministic mechanical model. *Phys. Rev. A*, **46**, 6288.

Stanley, H.E. (1977) Cluster shapes at the percolation threshold: and effective cluster dimensionality and its connection with critical-point exponents. *J. Phys. A*, **10**, L211.

Stauffer, D. (1979) Scaling theory of percolation clusters. *Phys. Rep.*, **54**, 1.

Stauffer, D. and Aharony, A. (1994) *Introduction to Percolation Theory*, 2nd edn, Taylor and Francis, London.

Stinchcombe, R.B., Duxbury, P.M., and Shukla, P.K. (1986) The minimum gap on diluted cayley trees. *J. Phys. A*, **19**, 3903.

Stinchcombe, R.B. and Thorpe, M.F. (2011) Renormalization group treatment of rigidity percolation. arXiv, 1107.4982.

Stormo, A., Gjerden, K.S., and Hansen, A. (2012) Onset of localization in heterogeneous interfacial failure. *Phys. Rev. E*, **86**, 025 101(R).

Swadener, J.G., Baskes, M.I., and Nastasi, M. (2002) Molecular dynamics simulation of brittle fracture in silicon. *Phys. Rev. Lett.*, **89**, 085 503.

Takayasu, H. (1985) Simulation of electric breakdown and resulting variant of percolation fractals. *Phys. Rev. Lett.*, **54**, 1099.

Tallakstad, K.T., Toussaint, R., Santucci, S., and Måløy, K.J. (2013) Non-gaussian nature of fracture and the survival of fat-tail exponents. *Phys. Rev. Lett.*, **110**, 145 501.

Tallakstad, K.T., Toussaint, R., Santucci, S., Schmittbuhl, J., and Måløy, K.J. (2011) Local dynamics of a randomly pinned crack front during creep and forced propagation: an experimental study. *Phys. Rev. E*, **83**, 046 108.

Tanguy, A., Gounelle, M., and Roux, S. (1998) From individual to collective pinning: effect of long-range elastic interactions. *Phys. Rev. E*, **58**, 1577.

Tantot, A., Santucci, S., Ramos, O., Deschanel, S., Verdier, M.A., Mony, E., Wei, Y., Ciliberto, S., Vanel, L., and Di Stefano, P.C.F. (2013) Sound and light from fractures in scintillators. *Phys. Rev. Lett.*, **111**, 154 301.

Thorpe, M.F. and Stinchcombe, R.B. (2013) Two exactly soluble models of rigidity percolation. *Philos. Trans. R. Soc. London, Ser. A*, **372**, 20120 038.

Tomlinson, G.A. (1929) *Philos. Mag.*, **7**, 905.

Tu, K.N. (2003) Recent advances on electromigration in very-large-scale-integration of interconnects. *J. Appl. Phys.*, **94**, 5451.

Tzschichholz, F. and Herrmann, H.J. (1995) Simulations of pressure fluctuations and acoustic emission in hydraulic fracturing. *Phys. Rev. E*, **51**, 1961.

Vanel, L., Ciliberto, S., Cortet, P., and Santucci, S. (2009) Time-dependent rupture and slow crack growth: elastic and viscoplastic dynamics. *J. Phys. D*, **42**, 214 007.

Vashishta, P., Kalia, R.K., Rino, J.P., and Ebbsjö, I. (1990) Interaction potential for sio2: a molecular-dynamics study of structural correlations. *Phys. Rev. B*, **41**, 12 197.

Wangen, M. (2013) Finite element modeling of hydraulic fracturing in 3d. *Comput. Geosci.*, **17**, 647.

Weibull, W. (1961) *Fatigue Testing and Analysis of Results*, Pergamon, New York.

Wiederhorn, S.M. (1984) Brittle fracture and toughening mechanisms in ceramics. *Annu. Rev. Mater. Sci.*, **14**, 373.

Witten, T.A. and Sander, L.M. (1981) Diffusion-limited aggregation a kinetic critical phenomenon. *Phys. Rev. Lett.*, **47**, 1400.

Wu, K. and Bradley, R.M. (1994) Theory of electromigration failure in polycrystalline metal films. *Phys. Rev. B*, **50**, 12 468.

Xia, J., Gould, H., Klein, W., and Rundle, J.B. (2005) Simulation of the burridge-knopoff model of earthquakes with variable range stress transfer. *Phys. Rev. Lett.*, **95**, 248 501.

Xia, J., Gould, H., Klein, W., and Rundle, J.B. (2008) Near-mean-field behavior in the generalized burridge-knopoff earthquake model with variable-range stress transfer. *Phys. Rev. E*, **77**, 031 132.

Yoshioka, N., Kun, F., and Ito, N. (2012) Time evolution of damage in thermally induced creep rupture. *Eur. Phys. Lett.*, **97**, 26 006.

Zapperi, S. and Nukala, P.K.V.V. (2006) Fracture statistics in the three-dimensional random fuse model. *Int. J. Fract.*, **140**, 99.

Zapperi, S., Nukala, P.K.V.V., and Simunovic, S. (2005a) Crack avalanches in the three-dimensional random fuse model. *Physica A*, **357**, 129.

Zapperi, S., Nukala, P.K.V.V., and Simunovic, S. (2005b) Crack roughness and avalanche precursors in the random fuse model. *Phys. Rev. E*, **71**, 026 106.

Zapperi, S., Ray, P., Stanley, H.E., and Vespignani, A. (1997a) First-order transition in the breakdown of disordered media. *Phys. Rev. Lett.*, **78**, 1408.

Zapperi, S., Vespignani, A., and Stanley, H.E. (1997b) Plasticity and avalanche behaviour in microfracturing phenomena. *Nature*, **388**, 658.

Zapperi, S., Ray, P., Stanley, H.E., and Vespignani, A. (1999a) Analysis of damage clusters in fracture processes. *Physica A*, **270**, 57.

Zapperi, S., Ray, P., Stanley, H.E., and Vespignani, A. (1999b) Avalanches in breakdown and fracture processes. *Phys. Rev. E*, **59**, 5049.

Zener, C. (1932) Non-adiabatic crossing of energy levels. *Proc. R. Soc. London, Ser. A*, **137**, 696.

Zener, C. (1934) A theory of the electrical breakdown of solid dielectrics. *Proc. R. Soc. London, Ser. A*, **145**, 523.

Zhang, Y.C. (1989) Scaling theory of self-organized criticality. *Phys. Rev. Lett.*, **63**, 470.

Index

a
Abelian 229
abrupt failure 292
abrupt rupture 85–87
acoustic emission 4, 70, 71, 118, 161, 188, 266
aftershock 226, 242, 248, 263
amorphous defect 265
Anderson localization 293, 294
Anderson transition 13
Andrade creep 129–134, 266
anisotropic scaling 54
anisotropy 65
atomistic friction 303
avalanche 4, 69, 71, 74, 81, 85, 88, 90, 92, 94, 97, 101, 104, 106, 107, 134, 149, 150, 152, 154, 190, 210, 218, 221–223, 229, 230, 235, 237, 242, 250–252, 254, 259, 261–263, 266, 267
– map 139
– size distribution 119, 139–141, 156, 160, 188, 197
average breaking stress 42
averaged wavelet coefficient 148

b
backbone mass 195
Bak Tang Wiesenfeld (BTW) model 228–230, 232, 268
Basquin law 125, 128, 134
Berker lattice 281
bond percolation 22, 279, 280
breakdown susceptibility 80, 204
breakdown voltage 178, 198, 200, 203, 204, 293
breaking rate 84
brittle-ductile transition
– confining pressure 173
– dislocation density 173
– statistical mechanics 174, 175
– temperature 173
– vibration 173
brittle failure 179
brittle fracture 1, 12, 40
brittle material 1, 9, 11, 12, 14, 17, 30, 32, 33, 131, 171, 172
Burgers vector 19, 20, 168, 169
Burridge-Knopoff model 215, 216, 218, 219, 246, 248, 261, 262, 267
burst size distribution 82, 84

c
Cantor set 297, 305
cellular automata model 228
chemical length 181, 184
cluster(s) 271
cluster statistics 37, 204
conductivity 179, 184, 198, 293
continuum limit 297, 298
continuum model 183, 184
correlation length 23, 24, 38, 43, 65, 66, 72, 180, 181, 183, 185, 186, 199, 203, 238–241, 265, 269, 275
crack front propagation model 5
crack front roughness 57
crack propagation 45
creep 4, 60, 111, 122, 129, 131, 266
creep and stress rupture tests 7
creep rupture 4
creep test 113, 117
critical exponent 24, 25, 38, 266, 269, 273, 276, 292
critical point 183
crossover 4, 35–38, 43, 50, 60, 62, 66, 68, 100, 131, 138, 187, 191, 196, 266
crossover behavior 84

crystal   12, 17, 18, 20, 40, 165, 167, 169–171, 174
cumulative distribution   113, 120, 123, 125, 132, 183, 186, 200, 204, 225, 285
cup and cone fracture   12

d
data collapse   270
defects   108
depinning   136, 140, 142, 144–147, 163
dielectric breakdown   178, 197, 202, 267
– continuum model   200, 201
– critical point   199
– cumulative distribution   200
– dilute limit   198
– field   203
– vs. fuse failure   177
– sample size   199
– shortest path   201
– stochastic model   201, 202
diffusion limited aggregation model   202
dilute limit   179, 180
discrete limit   299
dislocation   27
dislocation dynamics
– applied stress   172
– brittle-ductile transition, See brittle-ductile transition
– grain boundaries   171
– non-linear stress-strain response   165
– plastic strain   169, 170
– slip, See slip
– temperature   172
– width   170
dislocation mobility   170
disorder   1, 177–179, 187, 190, 193, 195, 196, 205, 249, 251, 263, 265–267, 290
disordered chain   250
dissipation   40, 139, 149, 150, 154, 229, 240, 242, 262
divergence   23, 24, 152, 175, 186, 198, 239–241, 266, 292
driven fluctuating line   135
ductile fracture   1, 12, 14, 15
ductile material   9, 11, 40, 165

e
earthquake   5, 6
– cellular automata model, See cellular automata model
– frequency statistics   85
– spring-block model, See spring-block model
– train model, See train model
edge dislocation   18–21

Edwards-Wilkinson model   248, 261
elastic deformation   8
elastic modulus   32, 42, 87, 88, 112, 148
elastic region   1, 8, 9
electromigration   184, 185
equal load sharing (ELS) model   75, 76, 84, 108, 285, 289
exponent
– dielectric breakdown field in discrete model   200
– fuse failure current in continuum model   182
– fuse failure current in discrete model   182
extreme fluctuations   265
extreme statistics   27, 33–35, 37, 42, 43, 182, 199, 205, 265, 292
Eyring rheology   132

f
failure current   178
failure probability   33
failure threshold   35, 112, 114, 115, 120, 125, 127, 149, 152, 154, 158, 190, 266
failure time   112
fatigue   111, 119, 179
fatigue tests   7
fiber bundle model   2, 34, 36, 43, 71, 74, 84, 113, 125, 131, 188, 240, 266, 268, 285
– equal load sharing process   75
– heterogeneous load sharing   92, 93
– local load sharing   96
– mixed load sharing   101
– mixed mode load sharing   90–92
– one dimensions case   94–96
– power-law load sharing   89, 90
– two dimensions case   96, 99, 100
finite clusters   272
foreshock   242
fractal   295, 296
fracture front   56, 69, 135
fracture process zone   68
Frenkel Kontorova model   303, 304
friction
– Frenkel Kontorova model   303, 304
– two-chain model   304, 305
fuse current   178
fuse failure   177

g
Gaussian distribution function   50
global breakdown voltage   293
global load sharing model   36, 37, 118, 123, 188
global roughness exponent   193

grain boundaries   17, 20, 21, 171
Griffith's theory   1, 15, 27, 30–33, 40, 42, 43,
   111, 113, 119, 134, 266
Gumbel distribution   112, 183
Gumbell statistics   32, 33
Gutenberg-Richter law   5, 208, 209, 213, 225,
   248, 261, 262, 267, 268, 299, 300

h
heterogeneous load sharing   92, 93
homogeneous scaling function   270
Hooke's law   8–10, 20, 166
hull function   304
hydraulic fracture   161–163

i
impact tests   7
impurities   18
in-plane roughness   57, 58
interface depinning   253
interface propagation   267
intergranular fracture   12, 48, 52
interpolations   88
interstitial   17, 18

k
Kaiser effect   118
Kostrerlitz-Thouless limit   174

l
Laplace equation   28, 201, 202
lattice defect   1, 22, 265
Lennard-Jones (LJ) potential   34, 39, 42
Lifhitz scale   265
linear elastic fracture mechanics   1
linear elastic region   9
line defect   18
local load sharing model   35, 96, 109,
   126

m
Manna model   203, 234–236, 240, 268
mechanical fracture   5
microcracks   17, 21, 27, 105, 111, 125
minimum gap   203
mixed load sharing   101
mixed-mode load sharing   90–92, 108
modified Omori law   209
molecular dynamics   39, 41, 43, 66, 67, 102
molecular dynamic simulation   303
mono-affinity   49
Monte Carlo simulations   89, 174
multi affinity   54

n
necking   15
node-link-blob model   180
non-abrupt failure   292
non-linear stres-strain response   165

o
Olami Feder Christensen (OFC) model   240,
   242
Omori law   5, 208, 211, 224, 226, 242, 248,
   254, 261–263, 300
order parameter   23, 24, 77, 87, 146, 269, 286,
   287
Oslo rice pile SOC model   221, 223
Oslo sand pile model   222
out of plane fracture roughness   47
overhangs   57, 147, 148, 258

p
Peierls-Nabarro theory   170
percolating lattice model   37–39
percolation   22
– threshold   22, 34, 37, 38, 42, 177, 180,
   185–187, 205, 270
– transition   271
phase boundary   120, 127, 128
planar defects   20
plane roughness   56
plastic deformation   9
plastic region   1, 8, 10
plastic strain   9, 169, 170
point defect   17
power-law decay   210
power-law load sharing   89, 90, 108
power spectrum   48, 58, 137, 191, 192
precursor   4, 132, 237, 268
– BTW model   238, 239
– fiber bundle model   240
– Manna model   240
probability density function   158, 159
probe size   62

q
quantum breakdown   178, 293, 294
quasi-brittle behavior   290, 292
quasi-brittle materials   11

r
random fuse model   196
random fuse network   185, 187–189, 191
random threshold spring model   109
– breakdown events   102, 105
– vs. fiber bundle model   102
– molecular dynamics   102, 103
– ruptured bonds   104, 105

relaxation dynamics   79
renormalization group (RG)   136, 225, 268, 270, 274, 298
– bond percolation   279, 280
– rigidity percolation   281, 282
– site percolation   276–278
rheological fiber bundle model   122
Richter magnitude scale   209
rigidity percolation   22, 268, 281, 282
roughness exponent   3–5, 48
rupture stress   8

s

sandpile model   230
scale divergence   87
scaling behavior   270
scaling function   210
scaling relation   24, 25, 181, 213, 231, 236, 250
scaling theory   273
screw dislocation   17, 18, 20
self-affine   3, 45, 47, 49, 54, 55, 58, 63, 64, 148, 191, 225, 250, 268, 305
self-affine objects   47
self-averaging statistics   35, 36, 43, 265, 267
self-similar   46, 180
Shannon entropy   72
shearing stress   166
Sierpinski triangle   296
site percolation   22, 276–278
slip
– atom movement   167–169
– planes   165
– shear stress   166, 167
solid   130
spatial correlation   65
spatial non-uniformity   108
spring-block model   214, 215, 246, 247, 262
stochastic model   201, 202, 234, 247
strain hardening coefficient   10
strain rate   131, 266
stress concentration   5, 16, 25, 27, 28, 30, 35, 36, 126, 135, 178, 189, 196
stress intensity factor   32, 135, 138, 139
stress-strain curve   8, 9, 11

stress-strain relation   288, 289
subcritical behavior   216, 218
subcritical failure   4, 134
super-critical behavior   216, 218

t

threshold percolation   22
tilt boundaries   17, 21
Tomlinson model   306
toughness factor   136
train model   219–222, 246, 248, 250, 254, 255, 261–263, 267
transgranular fracture   12, 48, 52
tricritical point   86, 292
true stress   11
two-chain model   304, 305
two fractal overlap model   223
– Cantor set   297
– continuum limit   297, 298
– discrete limit   299

u

universality   4

v

vacancies   17, 38
Verlet's algorithm   103
voids   27

w

waiting time distribution   60, 71, 117, 120, 122, 210
waiting time matrix   60, 139
wavelet coefficient   58, 148
Weibull distribution   183
Weibull statistics   32, 33
Weibull threshold distribution   83, 121

y

Young's modulus   9, 42, 87, 148

z

Zener breakdown   268, 294
Zhang model   232, 233, 242, 244